6

Topics in Medicinal Chemistry

Neurodegenerative Diseases

Volume Editor: Celia Dominguez

With contributions by

R. W. Bürli · K. R. Bales · V. Beaumont · S. Courtney · A. Duplantier · M. E. Gurney · A. S. Kiselyov · I. Muñoz-Sanjuan · F. S. Menniti · N. Plath · E. Schaeffer · A. Scheel · N. Svenstrup · E. Thomas

Editor
Dr. Celia Dominguez
CHDI Management Inc.
Advisors to
CHDI Foundation, Inc.
Center Drive 6080
90045 Los Angeles California
Suite 100
USA
Celia.Dominguez@chdifoundation.org

ISSN 1862-2461 e-ISSN 1862-247X
ISBN 978-3-642-16757-7 e-ISBN 978-3-642-16758-4
DOI 10.1007/978-3-642-16758-4
Springer Heidelberg Dordrecht London New York

Cover design: KünkelLopka GmbH, Heidelberg, Germany
Typesetting and Production: SPi Publisher Services

Springer is part of Springer Science+Business Media (www.springer.com)

Volume Editor

Dr. Celia Dominguez

CHDI Management Inc.
Advisors to
CHDI Foundation, Inc.
Center Drive 6080
90045 Los Angeles California
Suite 100
USA
Celia.Dominguez@chdifoundation.org

Editorial Board

Topics in Medicinal Chemistry Also Available Electronically

Topics in Medicinal Chemistry is included in Springer's eBook package *Chemistry and Materials Science.* If a library does not opt for the whole package the book series may be bought on a subscription basis. Also, all back volumes are available electronically.

For all customers who have a standing order to the print version of *Topics in Medicinal Chemistry*, we offer the electronic version via SpringerLink free of charge.

If you do not have access, you can still view the table of contents of each volume and the abstract of each article by going to the SpringerLink homepage, clicking on "Browse by Online Libraries", then "Chemical Sciences," and finally by selecting *Topics in Medicinal Chemistry.*

You will find information about the

- Editorial Board
- Aims and Scope
- Instructions for Authors
- Sample Contribution

at springer.com using the search function by typing in *Topics in Medicinal Chemistry.*

Color figures are published in full color in the electronic version on SpringerLink.

Aims and Scope

Drug research requires interdisciplinary team-work at the interface between chemistry, biology and medicine. Therefore, the new topic-related series *Topics in Medicinal Chemistry* will cover all relevant aspects of drug research, e.g. pathobiochemistry of diseases, identification and validation of (emerging) drug targets, structural biology, drugability of targets, drug design approaches, chemogenomics, synthetic chemistry including combinatorial methods, bioorganic chemistry, natural compounds, high-throughput screening, pharmacological in vitro and in vivo investigations, drug-receptor interactions on the molecular level, structure-activity relationships, drug absorption, distribution, metabolism, elimination, toxicology and pharmacogenomics.

In general, special volumes are edited by well known guest editors.

In references *Topics in Medicinal Chemistry* is abbreviated *Top Med Chem* and is cited as a journal.

Preface to the Series

Medicinal chemistry is both science and art. The science of medicinal chemistry offers mankind one of its best hopes for improving the quality of life. The art of medicinal chemistry continues to challenge its practitioners with the need for both intuition and experience to discover new drugs. Hence sharing the experience of drug discovery is uniquely beneficial to the field of medicinal chemistry.

The series Topics in Medicinal Chemistry is designed to help both novice and experienced medicinal chemists share insights from the drug discovery process. For the novice, the introductory chapter to each volume provides background and valuable perspective on a field of medicinal chemistry not available elsewhere. Succeeding chapters then provide examples of successful drug discovery efforts that describe the most up-to-date work from this field.

The editors have chosen topics from both important therapeutic areas and from work that advances the discipline of medicinal chemistry. For example, cancer, metabolic syndrome and Alzheimer's disease are fields in which academia and industry are heavily invested to discover new drugs because of their considerable unmet medical need. The editors have therefore prioritized covering new developments in medicinal chemistry in these fields. In addition, important advances in the discipline, such as fragment-based drug design and other aspects of new lead-seeking approaches, are also planned for early volumes in this series. Each volume thus offers a unique opportunity to capture the most up-to-date perspective in an area of medicinal chemistry.

Dr. Peter R. Bernstein
Prof. Dr. Armin Buschauer
Prof. Dr. Gunda J. Georg
Dr. John Lowe
Dr. Hans Ulrich Stilz

Foreword to Volume 6

The attrition rate in the clinical development of new chemical entities (NCEs) has increased over time, in spite of escalating funds allocated to research and development. Less than 10% of all NCEs succeed in effectively treating some clinical symptom. NCEs develop to treat central nervous system (CNS) disorders, together with oncology, exhibit the greatest attrition of all. One can argue that the major explanation for this lack of success is the complexity of the biological mechanisms underlying CNS indications and our lack of understanding of the aetiology of these disorders. Therefore, it is essential that an early assessment of compound efficacy in Phase II is sought. Most marketed CNS drugs modulate G-protein-coupled receptors (GPCRs) or protein transporters associated with the major neurotransmitters in the brain. It is only recently that molecular research led to the identification of novel targets implicated in neurodegenerative conditions. Many of these targets are intracellular and require specific approaches to assess drug efficacy, since many of these mechanisms cannot be directly monitored through non-invasive means in human subjects. The CNS drug development process is most likely to be successful, the more similar the human and animal models are to one another. For this reason, focusing on specific genetic conditions (for which the aetiology is known) as a way to assess drug efficacy and safety might lead to a more rapid evaluation of specific mechanisms.

The *major challenge* for CNS drug discovery is the lack of understanding of the human biological mechanisms underlying the various stages of these diseases. Considerable more effort needs to be placed on investigating whether the human biology supports specific target hypotheses underlying novel drug discovery campaigns. Non for-profit Foundation and Government researchers are much more likely to work with early symptomatic patient populations, and it is essential that specific hypotheses are investigated before clinical development. In addition, to expedite the development of clinical lead molecules for early human studies, a few critical domains are worth highlighting as having a major impact in the evaluation of novel chemical series. These are: advances in ADMET (absorption, distribution, metabolism, excretion, and toxicity), and specifically in understanding interactions with the major drug efflux mechanisms present at the blood–brain barrier, computer modelling of drug–target interactions and crystal structure aided drug design and the development of on-target occupancy endpoints to understand the pharmacokinetic/pharmacodynamic (PK/PD) relationship for a given molecule: the necessary occupancy to determine a suitable therapeutic window between efficacy in a given biological mechanism and potential side effects.

It is likely that drugs developed to treat one condition might be beneficial for other indications. This evokes a sense of hope that the efforts in developing new treatments will have broad impact from a medical and societal standpoint. This book highlights different approaches to mitigate the cellular processes thought to be affected regardless of which animal model (typically disease-specific) is first used to identify lead clinical candidates. In essence, it is my belief that disease animal models should be selected for any given target based on the molecular similarities to the human condition. In the context of lead evaluation, argue that an animal disease-specific model might not be necessary. Rather, a *mechanism* model, based on a proximal and specific readout for a specific target, allows researchers to quickly screen through potential leads and establish a PK/PD correlation as rapidly as possible. Once a suitable lead is identified, costly and lengthy experiments can be conducted with the disease model. In some instances, the most faithful disease models in terms of pathology or molecular changes require many months of drug administration. Therefore, the utilization of mechanism models to identify clinical leads is essential.

Neurodegenerative disorders have been typically grouped based on pathological findings and neurological symptoms. They are generally characterized by a slow symptom progression, adult or juvenile onset, specific neuronal vulnerability, and the presence of aggregated proteins identified as inclusion bodies after histological analysis. They all share age as the major risk factor, and in most cases (a notable exception being Huntington's disease, HD) have a mixed aetiology from a molecular perspective. While in all cases, familial forms of these disorders have been documented, and these account for a small percentage of all reported cases. For example, in spite of the broad incidence of Alzheimer's disease (AD) and Parkinson's disease (PD), only between 1% and 5% of all cases display a Mendelian inheritance pattern.

The existence of familial cases prompted a strong emphasis in the research community to identify the molecular basis for these disorders. These efforts have largely been successful in identifying the genes causative for the various conditions. For instance, mutations in amyloid precursor protein (APP) were identified in the case of AD; Huntingtin in the case of HD; α-synuclein, DJ-1, LRRK2, and Pink1, among others, for PD; TDP-43 and SOD1, among others, for amyotropic lateral sclerosis (ALS); and SMN-1 for spinal muscular atrophy (SMA). Because of this, clinical development strategies have shifted from the application of drugs developed to treat the symptoms of the disease (traditionally investigated through a re-purposing of existing psychotropic molecules developed for psychiatric conditions), to strategies aimed at modulating the main biochemical mechanisms thought to be affected by these proteins. This, thus far, has also proven unsuccessful from a clinical standpoint, as no effective new treatments have yet been identified. However, many are in clinical development.

In spite of their broad prevalence in the general population, neurodegenerative diseases are very inefficiently treated. Some of the challenges in drug discovery can be ascribed to the fact that most causative genes for neurodegenerative disorders cannot be targeted through traditional pharmacological means. In spite of more recent efforts to develop molecular therapies to eliminate expression of the mutated proteins, the majority of current clinical development strategies are aimed at "restoring" normal neuronal or glial function based on the cellular mechanisms now thought to underlie the toxic effects that arise from the mutant proteins.

There are some important commonalities in the molecular mechanisms thought to underlie these disorders. Many of the genes identified through positional cloning as causative of these set of disorders encode proteins that were shown to aggregate in in vitro and in vivo models and form the pathological inclusions traditionally used to diagnose these diseases. This was unexpected and argued that perhaps the propensity of these proteins to aggregate or form multimeric species had direct relevance to their toxic properties. In addition, many of the cellular mechanisms identified as being affected in rodent models for these diseases (generated through genetic means by introducing a mutant gene) show perplexing similarities. Among the mechanisms identified, mitochondrial disturbances, deficits in axonal transport and synaptogenic mechanisms, autophagy, protein folding, and transcriptional dysregulation are affected in all these diseases. Therefore, there is typically a convergence of strategies being developed to treat these disorders. A caveat in all of these approaches is the fact that, with the sole exception of HD and SMA, the aetiology of these disorders is mixed, and the majority of cases originate without much evidence for mutations or deregulation of the pathways linked to the mutant proteins which cause the familial (typically of earlier onset and faster progression) cases.

The converging field of synaptic dysregulation and transcriptional adaptations to changes in neurotransmitter tone is exemplified by the alterations known to exist in terms of key transcriptional effector molecules, such as histone acetylation and cAMP response element-binding protein (CREB) signalling. In this regard, two chapters are dedicated to the development of isotype selective modulation of phospodiesterase (PDE) inhibitors to modulate neurotransmission (through effects on cAMP signalling) and transcriptional modulation (Chap. 2), and to the identification of subtype-specific inhibition of histone deacetylases (HDACs; Chap. 1). The large repertoire of various enzyme subtypes found in mammalian neurons is a key factor in the development of effective treatments, without significant side effects already identified for non-subtype selective small molecule modulators of these classes of key enzymes needed for normal brain function. The main challenge here is twofold: to identify which enzymes need to be specifically targeted for each indication (as signalling in neurons is localized to specific domains coupled to selective signal transduction mechanisms); and to develop early measures of target engagement and efficacy in clinical trials through imaging studies or other measures to monitor changes in brain activity in response to a drug effect, such as quantitative electro-encephalogram (EEG). The initial findings for potential efficacy in neurodegeneration for the HDAC inhibitors originated from studies using non-selective molecules, which had significant toxicities associated with them after prolonged administration in animals. The effects of these molecules in various cognitive rodent models and in pathological analysis encouraged neuroscientists to try to identify class-selective molecules with good brain exposure, of increased potency and with fewer peripheral side effects. Similarly, the early clinical data surrounding rolipram (PDE4 inhibitor) in treating depression and displaying a pro-cognitive effect, together with many converging aspects of cAMP cascade deficits in various disease models, prompted the development of very specific active site inhibitors for this broad family of signalling molecules. Some of the recent advances in the development of selective brain penetrant PDE inhibitors are highlighted in Chap. 2, with an emphasis on cognitive enhancement for AD. However, many of these molecules are likely to exhibit activities beneficial to other neurodegenerative conditions, and are currently being tested for other indications in animal models.

In terms of synaptic biology, a key finding is the vulnerability of specific neuronal populations that die selectively in the various disorders. Presumably, this vulnerability will eventually be found to originate from the specific role of each mutated protein within these cells, or to the properties of the neuronal and glial cells found in the circuitry affected in each disorder. In essence, the spectrum of clinical symptoms used to define these diseases can be largely explained by the pathological findings of neuronal death and gliosis affecting the relevant circuitry. A key approach for treating these disorders is therefore based on improving the function of the circuits affected in each disease, and specific neurotransmitter modulators are being developed based on the cells most affected by each condition.

Within synaptic modulation, one of the predominant hypothesis common to all neurodegenerative conditions is the specific vulnerability of neurons to deregulated calcium signalling, and specifically calcium signalling mediated by glutamate receptors. This theory, termed *excitotoxicity*, is far from proven, but appears common to many of these disorders. Excessive or extrasynaptic calcium entry has become one of the major strategies for the treatment of neurodegenerative diseases. In the case of ALS, riluzole, the only approved drug for this disease, is a sodium channel blocker thought to modulate excessive calcium entry. Chapter 3 specifically reviews the theory of excitotoxicity and abnormal glutamate signalling mostly in the context of Alzheimer's disease, although similar principles (and drugs) are relevant to other indications. Indeed, many of these are currently in clinical development for various CNS disorders. For instance, both memantine (an NMDA receptor antagonist) and mGluR5 antagonists are in clinical development for PD and HD. However, the essential role for glutamate signalling in brain function makes this a difficult mechanism to modulate with an acceptable therapeutic window. The complexities in glutamate signalling in various circuits affected in these disorders require a deeper investigation of the changes during disease progression in human subjects, to better predict whether a specific drug might lead to clinical improvement.

Other strategies aimed at the indirect modulation of glutamate, acetylcholine and other major neurotransmitter systems that are being prosecuted, which might be associated with more tolerable adverse effects. For instance, the role for metabolites of the kynurenine pathway (a product of tryptophan degradation), shown to be neuroactive and specifically altered in human subjects and animal models for some diseases, is described in Chap. 4. This novel approach to modulate synaptic transmission through a specific modulation in key metabolic enzymes (such as kynurenine mono-oxygenase or KMO; kynurenine amino transferase or KAT) highlights some of the new avenues taken by industry experts to uncover novel methods for treating these difficult diseases. Chapter 4 focuses on the development of KMO inhibitors specifically for HD. However, changes in kynurenine metabolites have been reported in many neurodegenerative indications. In addition to the role of kynurenine pathway metabolites in synaptic transmission, this pathway has been implicated in the modulation of the immune response, a biological area of active investigation to decrease neuronal cell death.

Finally, Chap. 5 highlights recent developments in the treatment of SMA. The genetic cause of this disease is well understood (lack of expression of the gene SMN1 due to a missense mutation), and current efforts are aimed at enhancing expression of a gene, SMN2, which can act to compensate for the loss of the SMN1 gene. This chapter illustrates the strength of focused efforts on overcoming the cause of the disease, through various means. All share in common the utilization of cellular and animal models focused on understanding the effects of novel molecules in increasing expression of SMN1. Similarly, in HD and familial models of AD and PD, a specific emphasis on demonstrated genetic contributions to the disease is key to develop novel drugs with a well-validated biological principle. In the case of other, genetically heterogeneous disorders (AD, ALS and PD), the applicability of such strategies rests perilously on the assumption that similar biological principles will apply in idiopathic cases where the cause of the pathology is unknown. Overall, the various approaches highlighted here serve to illustrate new directions in CNS drug discovery, and leads to an emphasis in a deeper understanding of the molecular causes for these disorders as a more efficient way to overcome the inherent difficulties in treating, or preventing, neurodegeneration.

August 2010 Ignacio Muñoz-Sanjuan

Contents

Top Med Chem 6: 1–56
DOI: 10.1007/7355_2010_10

Published online: 10 September 2010

The Role of Histone Deacetylases in Neurodegenerative Diseases and Small-Molecule Inhibitors as a Potential Therapeutic Approach

Roland W. Bürli, Elizabeth Thomas, and Vahri Beaumont

Abstract Neurodegenerative disorders are devastating for patients and their social environment. Their etiology is poorly understood and complex. As a result, there is clearly an urgent need for therapeutic agents that slow down disease progress and alleviate symptoms. In this respect, interference with expression and function of multiple gene products at the epigenetic level has offered much promise, and histone deacetylases play a crucial role in these processes. This review presents an overview of the biological pathways in which these enzymes are involved and illustrates the complex network of proteins that governs their activity. An overview of small molecules that interfere with histone deacetylase function is provided.

Keywords Epigenetics, Metallo enzyme, Polyglutamine, Sirtuin, Transcriptional regulation

Contents

R.W. Bürli, E. Thomas
BioFocus, Chesterford Research Park, Saffron Walden, Essex CB10 1XL, UK

V. Beaumont (✉)
CHDI Foundation, 6080 Center Drive, Los Angeles CA 90045, USA
e-mail: Vahri.Beaumont@CHDIFoundation.org

Abbreviations

AD Alzheimer's disease
ADME Absorption, distribution, metabolism, excretion
ALS Amyotrophic lateral sclerosis
BDNF Brain-derived neurotrophic factor
CBP CREB-binding protein
CNS Central nervous system
DNA Deoxyribonucleic acid
DRPLA Dentatorubral pallidoluysian atrophy
FDA Food and Drug Administration
FRDA Friedreich's ataxia
FXN Frataxin
GDNF Glial cell-derived neurotrophic factor
HAT Histone acetylase
HD Huntington's disease
HDAC Histone deacetylase
Htt Huntingtin
MBG Metal-binding group
MEF2 Myocyte-enhancing factor 2
PD Parkinson's disease
PolyQ Polyglutamine
RTS Rubenstein–Taybi syndrome
SBMA Spinal bulbar muscular atrophy
SCA Spinocerebellar ataxia
SUMO Small ubiquitin-like modifier
TBP TATA-binding protein
UPS Ubiquitin–proteosome system

1 Neurodegeneration

"Neurodegeneration" is the umbrella term for the progressive loss of structure or function of neurons, leading to their eventual death and brain atrophy. Many neurodegenerative diseases arise as a result of genetic mutations in seemingly unrelated genes. For instance, a collection of neurodegenerative diseases fall into the category of "polyglutamine (polyQ) diseases," where expansions in polyQ-encoding CAG tracts in discrete genes have been identified as the primary insult leading to neurodegeneration. The lengths of CAG repeats in genes susceptible to expansion are polymorphic in the unaffected population. At some point though, expansions of CAG repeats cross a pathogenic threshold, acquiring a propensity for further expansion, and encoding proteins with neurotoxic activity and a proclivity for self aggregation. These include Huntington's disease (HD; expansion >37 CAG in the Htt gene encoding Huntingtin), dentatorubral pallidoluysian atrophy (DRPLA; expansion in the atrophin-1 protein), spinal bulbar muscular atrophy (SBMA or "Kennedy's disease"; CAG expansion in the androgen receptor) and a subset of the spinocerebellar ataxias (SCA); where SCA1-3 and 7 are caused by expansion in ataxin proteins, SCA-6 is caused by expansion in the calcium channel subunit CACNA1A, SCA-12 is caused by expansion in the PP2R2B protein, and SCA-17 is caused by expansion in the TATA-binding protein (TBP). Other neurodegenerative diseases are not caused by CAG tract expansion, but nevertheless, abnormal accumulation of protein is implicated. In Parkinson's disease (PD), accumulation of the protein α-synuclein into Lewy bodies is a hallmark of the disease. In Alzheimer's disease (AD), the abnormal accumulation of β-amyloid into plaques and hyper-phosphorylated tau proteins into tangles appears to be the causal insult.

While the central nervous system (CNS) region-specific expression of the aberrant protein or the differential susceptibility of neuronal subsets to the primary insult varies, key themes start to emerge for these seemingly disparate neurodegenerative diseases, which are collected under the catch-all title of "CNS proteopathies." Among these themes, irregular protein folding, interference of intracellular vesicle trafficking, disruption of proteosomal degradation pathways, altered subcellular localization (especially neuronal inclusions), and abnormal protein interactions have been repeatedly implicated. This, coupled with the potential gain or loss of function of mutated protein compared to "wild-type" protein, appears to contribute to the neurodegenerative pathological cascade.

Ultimately, there is surprising congruence in the downstream pathophysiological cascades, which precede neuronal apoptosis and brain atrophy. Synaptic dysfunction, mitochondrial energy metabolism defects, and transcriptional dysregulation are common to many of these diseases. The resultant symptoms are manifested as disturbances in motor, psychiatric, and cognitive impairment, which progress in severity as the diseases progress. Manifestation of any of these symptoms is a consequence of the neuronal subsets primarily affected. For instance, if neurodegeneration is prevalent in basal ganglia, midbrain, or cerebellar structures, key regions involved in the planning, execution, and control of fine movements, motor control is clearly impacted (in SCA diseases, HD and PD). In the case of amyotrophic lateral sclerosis (ALS) and SBMA, it is the lower motor neurons that are targeted and degenerate. Cognitive impairment and dementia are also major components of many of the neurodegenerative diseases, when cortical and subcortical nuclei are compromised (particularly so in AD, PD, and HD). Comorbidity with psychiatric disturbances such as anxiety and depression are also common in these latter diseases, but these symptoms are often overlooked or undertreated.

The ultimate outcome of all of the neurodegenerative diseases described above is enormous pressure on the social environment of patients and premature death. Our pharmacological arsenal for treatment remains poor, despite years of intensive research in these areas. The disease burden of neurodegenerative disorders to the worldwide healthcare industry is very high [3.4% of all deaths in high-income countries were attributed to AD and other dementias (WHO Global Burden of Disease 2004)]. As the average survival age increases, this statistic is likely to rise.

While a "magic bullet" for the treatment of neurodegenerative disorders is highly unlikely, histone deacetylase inhibition is a promising avenue of investigation for multiple neurodegenerative disorders. This review focuses on the potential amelioration of symptoms common to many of these diseases, by assessing the role that HDAC inhibition may have in the treatment of key aspects of disease pathophysiology common to the CNS proteopathies, with specific examples taken from each disease. Current and future directions in the development of brain-penetrant subtype selective HDAC inhibition are discussed.

2 Overview of Histone Deacetylases

2.1 Classification

HDACs are a complex superfamily of proteins, sharing a common deacetylase activity of acetylated ε-amino groups of lysine side chains. Originally, histones were found to be their substrates, but more recently it has been shown that they act on a large set of diverse proteins. To date, 3,600 lysine acetylation sites across 1,750 proteins are known. This collection has been dubbed the "lysine acetylome," with particular prevalence in large macromolecular complexes [1]. Hence, from a functional point of view, it would be more appropriate to refer to HDACs as lysine deacetylases (or KDACs) [1]. Moreover, it is clear that the role of HDACs extends significantly beyond their catalytic activity and forms part of an intricate regulation process, much of which is not fully understood. While HDACs do not display classical DNA binding domains, they are incorporated into large multicomponent complexes that govern transcriptional processes. As a result, the activity of HDACs is determined by their environment and the variety, quantity, and identity of partner proteins present in any given location. These complex interactions are involved in a diverse array of pathways from activation of cell death through caspase-mediated cleavage [2] to muscle differentiation in concert with the transcription factor MEF2 (myocyte-enhancing factor 2) [3]. Subcellular distribution also plays a key role in HDAC function and is varied temporally and spatially both by cell type and intracellular targeting. Although HDACs are classically viewed as repressors of transcription, their inhibition is reported to give rise to as much gene upregulation as downregulation [4]. This is due to their complex network of interactions, and as a result HDAC transcriptional effects should be viewed as modulation rather than solely repression or inhibition.

The HDAC superfamily consists of 18 members originating from two different evolutionary starting points which exhibit a common lysine deacetylase activity. The classical HDAC family is characterized by a well-conserved Zn^{2+} catalytic domain (Table 1 classes I, IIa, IIb, and IV). The sirtuins (class III HDACs) comprise a distinct subfamily of HDACs, which use NAD^{+} as cofactor.

Class I HDACs (1, 2, 3, and 8) are defined by their similarity to the yeast RPD3 transcriptional factor. They are expressed ubiquitously across cell types and have an average length of 443 amino acids [5, 6]. HDACs 1, 3, and 8 possess a nuclear localization signal motif. While HDAC1 and 2 proteins lack a nuclear export signal and thus are exclusively nuclear in location, HDAC3 has both nuclear import and export signals and can shuttle between the nucleus and the cytoplasm. Class I deacetylases contain an N-terminal catalytic domain which constitutes most of their length. They demonstrate high enzymatic activity to histone substrates. HDACs 1, 2, and 3 form part of large multi-protein regulatory complexes with partner proteins such as Sin3, NURD, CoREST, and PRC2 [7, 8]. No protein complexes for HDAC8 have yet been identified.

Class IIa HDACs (4, 5, 7, and 9) are characterized by their homology to yeast HDA1. They are considerably larger than class I proteins at an average of 1,069 amino acids (reviewed in [9]). The class IIa C-terminal catalytic deacetylase domain exhibits approximately 1,000-fold lower activity toward histone substrates than the class I subtypes [7]. Thus, regulation of transcription by histone deacetylation has been called into question for this class of enzymes. Indeed, to date there is no endogenous substrate identified for which class IIa enzymes are *bona fide* deacetylases. In HDAC4 preparations purified from mammalian cells, co-associated HDAC3 was proposed to confer all of the observable deacetylase activity [10]. In addition to the carboxy terminal "catalytic domain", however, class IIa enzymes express a conserved N-terminal extension of roughly 600 additional residues [11]. The N-terminus adopts various regulatory functions and is involved in protein–protein interactions, and notably transcription factor binding. For instance, binding sites for the family of MEF2 transcription factors are located within this domain. MEF2 transcription factors are important for muscle differentiation, synaptogenesis, and apoptosis, and there is good evidence

Table 1 Characteristics of individual HDAC isoforms: domains, lengths, cellular location, selected binding partners, and substrates. Catalytic domain ▢, MEF2-binding domain ■, chaperone 14-3-3-binding domain ▢, zinc finger domain □. Information from [5, 7, 8, 25–27, 29–32]

Class I	Isoform	Protein	Length	Cellular location	Example partners	Substrate
	1	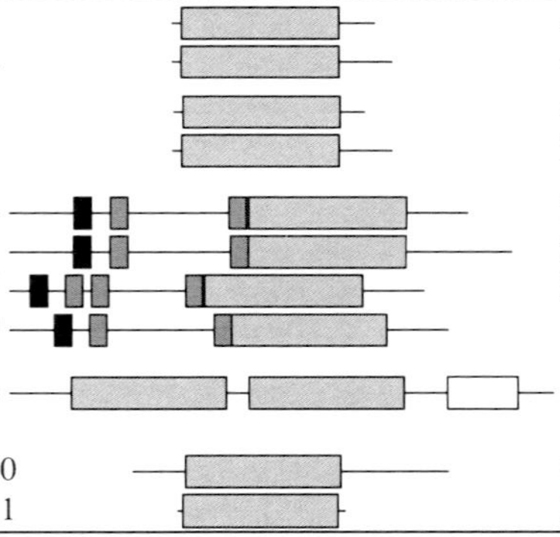	482	Nucleus	Sin3, NURD, CoREST	Histone
	2		488			
	3		428		NCOR,CPS2, HDACs4,5,7	
	8		377		CREB, PP1	–
Class IIa	4		1,084	Nucleus/cytoplasm	MEF2,SMRT, NCOR, other HDAC isoforms	–
	5		1,122			
	7		912			
	9		1,069			
Class IIb	6		1,215	Cytoplasm	Chaperone p97/VCP, ubiquitin	HSP90, α-tubulin, IFN α R, chaperone
	10		669	Nucleus/cytoplasm	NCOR2	HDAC2, SMRT
Class IV	11		347	Nucleus	–	Cdt1

to suggest that HDAC4 and 5 are critical repressors of MEF2 function *in vivo*. The role, if any, of the catalytic domain of class IIa enzymes on control of MEF2-controlled transcriptional repression is currently unclear. The MEF2-interaction transcription repressor (MITR) shares homology with the NH-terminal extensions of class IIa HDACs but lacks a deacetylase catalytic domain [12]. However, the repressive actions of MITR are mediated in part by its formation into macromolecular complexes with other class I and II enzymes, which may supply a necessary deacetylase function [13]. While there is no evidence for MEF2 deacetylation by HDAC4 [14], there are reports that blocking the catalytic site with HDAC inhibitors can influence MEF2-regulated transcription, possibly by interfering with the recruitment of transcriptional co-repressor complexes [15–17]. Of interest, mice harboring a viral insertion mutation that deletes the putative deacetylase domain of HDAC4, while preserving the N-terminal portion of the protein, are viable, have normal bone and muscle development and only subtle phenotypes [18]. This is in striking contrast to mice in which the complete gene is knocked out, which results in premature ossification and associated defects resulting in postnatal lethality [19].

All class IIa enzymes can shuttle between the cytoplasm and nucleus in a phosphorylation state-dependent manner. Phosphorylation by kinases such as calcium/calmodulin-dependent protein kinase and protein kinase D results in binding of chaperone protein 14-3-3 at the N-terminal domains and retention within the cytoplasm [20]. This regulated phosphorylation provides a mechanism whereby extracellular signal transduction can influence transcriptional modulation mediated by complexes containing class IIa enzymes by inducing relocation to nuclear compartments [3, 21]. Class IIa enzyme expression is more restricted according to cell type; HDAC4 is predominantly expressed in brain [22] and skeletal growth plates, HDAC5 and 9 are highly expressed in heart, muscle, and brain, and HDAC7 is enriched in endothelial cells and thymocytes [6, 22].

Class IIb HDACs (6 and 10) are also characterized by homology to yeast HDA1, but possess two deacetylase-like domains. However, only in HDAC6 are both functional. The C-terminal "catalytic" domain of HDAC10 is only partially present and does not retain activity [23, 24]. HDAC6 is ubiquitously expressed and cytoplasmic in location. It is promiscuous in its substrates which include chaperones, transmembrane proteins, α-tubulin, and cortactin [25–27]. HDAC10 has been identified as multiple splice variants. It is broadly expressed across cell types and has both a nuclear and cytosolic intracellular distribution. The C-terminal region contains putative retinoblastoma protein-binding domains [5].

Class IV HDAC11 is most closely related to the class I family but also displays common characteristics with class II HDACs. The low overall homology to either of these classes has resulted in a separate classification. HDAC11 is highly expressed in heart, brain, testis, muscle, and kidney cells and is predominantly located in the nucleus [28]. This deacetylase has short N- and C-terminal extensions; little is known about its function.

2.2 Structural Aspects of Zinc-Dependent Histone Deacetylases

The key to understanding the function of histone deacetylases lies in their three dimensional architecture. As outlined above, the class I, II, and IV enzymes are all metal ion dependent; in most cases, a zinc ion is essential for activation and hydrolysis of the amide group, which is located within the active site of the enzyme. However, it has been shown that other metal ions can efficiently adopt the role of the catalytic ion. For instance, the nature of the ion bound to the catalytic site influences the specific activity of HDAC8 in the following order: $Co^{2+} > Fe^{2+} > Zn^{2+} > Ni^{2+}$. These data suggest that Fe^{2+} rather than Zn^{2+} may be responsible for the *in vivo* activity of HDAC8 [33].

Several crystal structures of different HDAC subtypes and HDAC-like proteins have been published over the past decade. Related to the zinc-dependent enzymes, structural information is available for the class I enzymes HDAC2 [34] and HDAC8 [35, 36]. The recent HDAC2 structure revealed a foot pocket in proximity of the zinc ion, which can be accessed by small molecules [34]. This pocket contains multiple water molecules, which can be replaced by an inhibitor.

The catalytic domains of HDAC4 [37] and HDAC7 [38], members of the class IIa family, have also been solved. In addition, the structures of bacterial HDAC homologs HDLP [39] and HDAH [40, 41] have been elucidated. Of particular significance is the investigation of the HDAC4 catalytic domain as inhibitor-free and inhibitor-bound structures and a gain-of-function mutant protein (GOF HDAC4cd). This study revealed a likely structural explanation for the intrinsically low enzymatic activity of the class IIa enzymes toward acetylated lysines compared to the class I subtypes. In essence, the OH group of Tyr^{306}, which is conserved in all class I and class IIb subtypes, has been proposed to form a hydrogen bond to the tetrahedral anionic intermediate during the amide hydrolysis, thus accelerating the hydrolytic process. In all class IIa subtypes, the amino acid residue 306 is mutated to a histidine (His^{976}), which is rotated away from the active site. As a result, the amide hydrolysis is much less efficient. Supporting this finding is the fact that mutating the His^{976} of HDAC4cd to a tyrosine residue results in a gain of deacetylase function by roughly a 1,000-fold to levels similar to those of the class I enzymes [7]. In addition, the structural study of HDAC4cd has uncovered a second zinc ion besides the one that is critical for the catalytic process, at the bottom of the active site. This zinc ion is bound to a conformationally flexible domain that is present in all class IIa enzymes and appears to adopt a structural function. Structural information is also available for a glutamine-rich segment of the N-terminus of HDAC4 containing 19 Gln out of 68 residues [42]. The precise physiological function of this glutamine-rich stretch is not known; it has been proposed that it may be involved in protein–protein interactions, which is discussed later in relation to the CNS proteopathies.

2.3 General Overview of Known Class I, II, and IV HDAC Inhibitors

2.3.1 Introduction

In principle, lysine deacetylase inhibitors may be classified according to their structure or properties. The latter may be analyzed from several aspects, which may include parameters such as selectivity and pharmacokinetic and pharmacodynamic (potentially therapeutic) properties. Itoh *et al.* have published a review article, in which the inhibitors are classified based on their isoform selectivity [43]. A classification of lysine deacetylase inhibitors according to their pharmacokinetic properties would be interesting to better assess their drug-like properties; however, information about *in vitro* and *in vivo* metabolic stability and brain permeability is lacking for a lot of compounds, especially for structures that are in early development. Access to the CNS is of particular relevance to the topic of this article. Many methods and guidelines for the prediction of passive CNS permeability of small molecules have been described, which involve a combination of lipophilicity, topological polar surface area, molecular weight, and number of hydrogen bond donors. Besides, it is well known that active transport in both directions (uptake and efflux) is common for small molecules. Nevertheless, an accurate *in silico* prediction of CNS exposure of specific molecules remains challenging (for reviews, see, e.g. [44, 45]). Understanding the brain permeability of small molecules is further complicated by the finding that the blood–brain barrier, which adopts a neuroprotective role, may be compromised in patients with neurodegenerative diseases [46].

This section is intended to give an overview of the known structural chemotypes of Zn^{2+}-dependent N^{ε}-acetyl lysine deactylase inhibitors and will not be restricted to compounds with known brain-permeability and/or potential as a therapeutic agent for a neurodegenerative disease. Molecules with known activity related to neurodegenerative diseases will be discussed in the context of the specific neurological disorders further below. Complementary to this review, Wang and Dymock have recently published an extensive survey of the recent patent literature covering histone deacetylase inhibitors [47].

Most inhibitors of the class I, II, and IV enzymes known to date interact with the metal ion within the catalytic site preventing deacetylation of N^{ε}-acetyl Lys residues. Even though the molecules with such inhibitory property appear structurally very diverse, most of them share three common structural features, namely, a metal-binding group (also referred to as the molecular "warhead"), a linker region, and a surface recognition or capping domain as recognized by Schreiber and Grozinger [48]. In the following section, selected examples will be highlighted with the intention to illustrate the nature of the major groups of HDAC class I, II, and IV inhibitors. Sirtuin (class III) inhibitors will be discussed separately.

2.3.2 Hydroxamic Acids

The hydroxamic acids, also referred to as hydroxamates or hydroxy amides, are perhaps the most intensely investigated class of N^{ε}-acetyl lysine deacetylase inhibitors. From other areas of investigation, for instance the matrix metallo-proteases [49], it is well known that the hydroxamate function exhibits a strong affinity to Zn^{2+} and other metal ions and hence contributes significantly to the affinity of the ligands to their biological targets. Within the HDAC field, a large variety of hydroxamic acids have been studied aiming at therapeutic agents for oncology and other disease areas. These efforts culminated in the FDA approval of Zolinza (also suberoylanilide hydroxamic acid = SAHA or vorinostat, Fig. 1) for the treatment of cutaneous T-cell lymphoma. This breakthrough spurred a lot of efforts to develop therapeutic agents interfering with HDAC function that show improved efficacy and safety profiles [50].

The hydroxamic acid-based HDAC inhibitors can be further divided according to the nature of their linker and surface recognition elements. It is apparent from the literature that the linker region may be conformationally flexible and linear (e.g., Zolinza) or conformationally rigid as exemplified by the olefin TSA (**2**) and the cinnamide-type pyrrole **3** [51] shown below. Compounds in which the hydroxamate war head is directly attached to an aryl or heteroaryl group have been investigated as well.

Importantly, the structural differences within the linker and surface recognition portion will have profound effects on subtype selectivity and physicochemical/ ADME properties of the compounds. Structural changes within these areas will hence allow for optimization and fine-tuning of the drug-like parameters that are required for a successful therapy.

Dual inhibitors have been reported, in which structural features essential for HDAC inhibition are combined with elements that are known to interact with other target classes. The recently reported multi-acting HDAC and EGFR/HER2 inhibitor **4** (Fig. 2) serves as an example to demonstrate this concept [52].

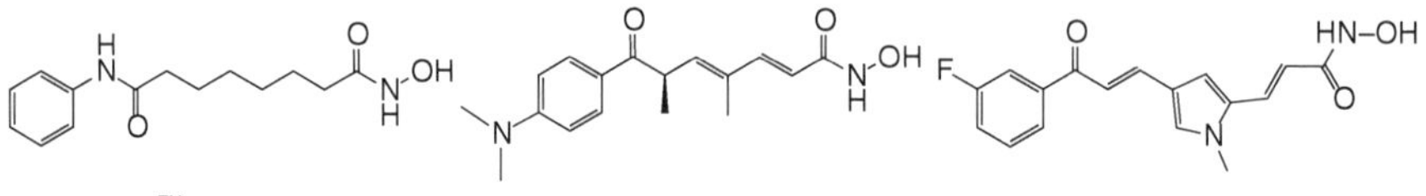

1 (Zolinza™, SAHA, vorinostat) **2** (TSA) **3** (aryl-pyrroyl-hydroxy-amide: APHA)

Fig. 1 Selected examples of hydroxamic acid-based HDAC inhibitors

4

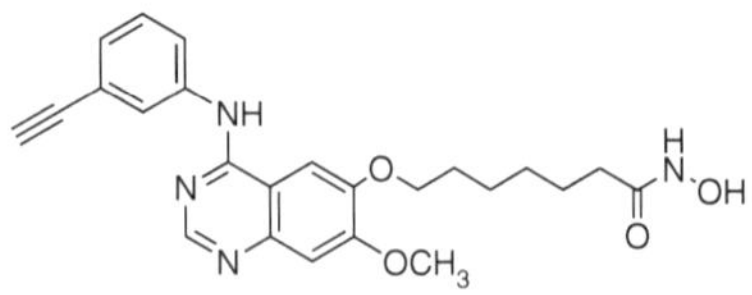

Fig. 2 Structure of a multi-acting HDAC and kinase inhibitor

2.3.3 *Ortho-N*-Acyl-Phenylene Diamines

The common feature of *ortho-N*-acyl-phenylene diamines, also referred to as benzamides, is an acylated *ortho*-phenylene diamine unit, which is thought to interact with the zinc ion of HDACs. Bressi *et al.* have recently published a crystal structure of compound **7** bound to HDAC2 (Fig. 3), in which the carbonyl oxygen and the amino group chelate the zinc ion in a bidentate fashion [34]. By analogy to the hydroxamic acids, there are family members comprised of a flexible linker and others that are conformationally rigid. Again, the linker and surface recognition portions provide excellent handles for fine-tuning the overall properties of the molecules. There is evidence that this class of molecules has the potential for subtype selectivity and drug-like properties. This is demonstrated by MS275 (compound **5**), which originally has been reported to selectively inhibit HDAC1 with an IC_{50} value of 181 nM [43, 54]. A subsequent publication suggested that this compound is essentially equipotent against HDAC1 and HDAC3 [36], but the compound did not show inhibitory activity for HDAC4, 6, 7, and 8 (>10 μM). MS-275 has progressed into phase II clinical trials for the treatment of cancer, demonstrating an acceptable PK profile following oral administration [55].

Notably, for some benzamides a time-dependent increase in affinity has been observed [34, 56, 57]. Bressi *et al.* proposed that disruption of an intramolecular hydrogen bond of the NH_2 group to the carbonyl oxygen is required for tight binding and may cause the slow binding kinetics [34].

2.3.4 *Ortho-N*-Acyl-Phenolamines

The *ortho-N*-acyl-phenolamines may be regarded as a structural variant of the above-mentioned benzamides. In this case, an *N*-acylated phenolamine function acts as the warhead chelating the zinc ion. The *ortho-N*-acyl-phenolamine **8** (MC1863; Fig. 4) serves as an example and exhibits inhibitory activity for HDAC1 and selectivity over HDAC4 that is similar to that of its corresponding benzamide or MS275 (Fig. 3).

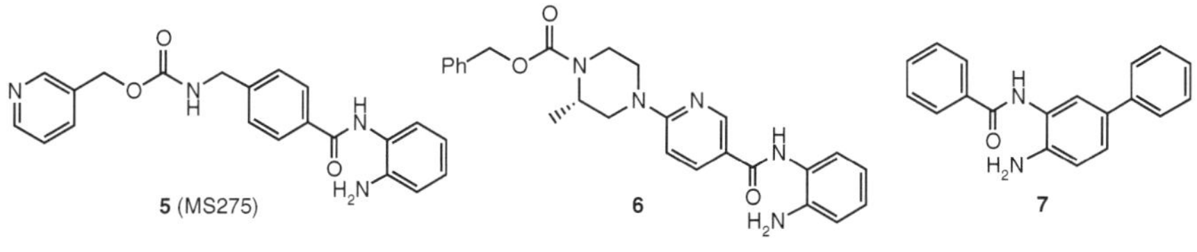

Fig. 3 Structures of selected benzamides (MS275, 6-amino-nicotinamide **6** [53], and HDAC2 inhibitor **7**)

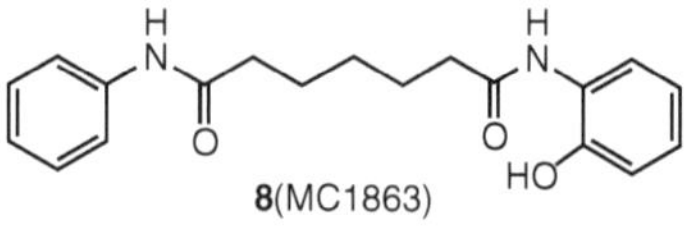

Fig. 4 Structure of *ortho*-acyl-*N*-phenolamine MC1863

2.3.5 Macrocyclic Natural Products

A variety of macrocyclic natural products, either peptides or peptide-mimetic structures with HDAC inhibitory activity and promising pharmacological effects, have been reported. Some of them are shown in Fig. 5 for illustration; for instance, romidepsin **9** (FK228) exhibits excellent class I inhibitory activity and has recently been approved by the FDA for the treatment of cutaneous T-cell lymphoma patients [58, 59]. Similarly, the marine natural product largazole **10** that demonstrates potent antiproliferative activity [60] and HC-toxin **11**, a fungal metabolite with immunosuppressant activity, have been investigated as cytostatic agents [61, 62].

At first glance, the structure of these macrocycles may appear very different from classical HDAC inhibitors. However, they share the same common features: the macrocyclic structure comprises the surface recognition element, which is connected to the metal-binding group via a relatively flexible linker moiety. Notably, reductive cleavage of the disulfide bond in romidepsin is required to liberate a mercapto group that can bind the zinc ion within the catalytic site.

The antiprotozoal agent apicidin **12** [63] and FR235222 **13** [64] share a keto group as the metal-binding group, which will be discussed in more detail below.

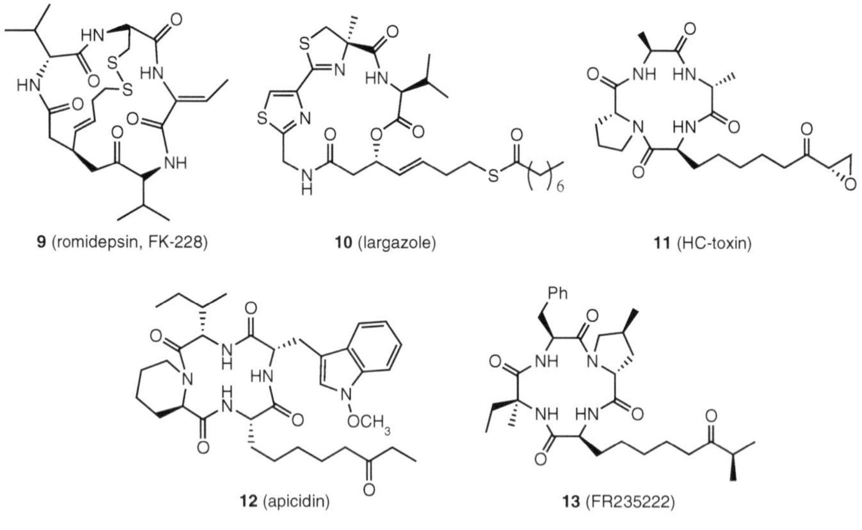

Fig. 5 Macrocyclic natural products with HDAC inhibitory activity

The creation of drug-like molecules capable of CNS penetration from these starting points would be a challenging task. This will be made harder as room for maneuver is limited by the likely demands for isoform selectivity. For a comprehensive review of delivery of peptide and protein drugs across the blood–brain barrier, see [65]. Notably, Ghadiri and co-workers exploited this structural class and have developed an efficient synthetic access to one-bead-one-compound combinatorial libraries of cyclic tetrapeptide analogs with promising subtype selectivity [66, 67].

2.3.6 Ketones and Trifluoromethyl Ketones

The natural product apicidin **12** bears an ethyl ketone as the metal-binding group and has inspired further work using a ketone function to engage with the zinc ion of HDACs. Examples of more selective class I inhibitors with improved drug-like properties are illustrated in Fig. 6. While the bisamide **14** [68] was metabolically not stable, the mono-amides **15** and **16** demonstrated efficacy in a colon cancer xenograft model [69, 70].

Trifluoromethyl ketones are more electrophilic than the alkyl ketones described above and typically exist in equilibrium with their hydrate forms. Potent inhibitory activity of trifluoromethyl ketones toward metal-dependent proteases has been well documented, which sparked the investigation of compounds bearing this war head in the context of inhibition of HDAC function [71]. Indeed, the alkyl-linked trifluoroketone **17** (Fig. 7) demonstrated submicromolar HDAC inhibitory activity and antiproliferative effects in HT1080 and MDA 435 cell lines. Molecules within

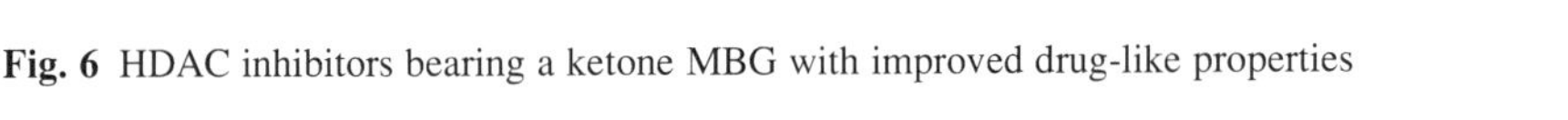

Fig. 6 HDAC inhibitors bearing a ketone MBG with improved drug-like properties

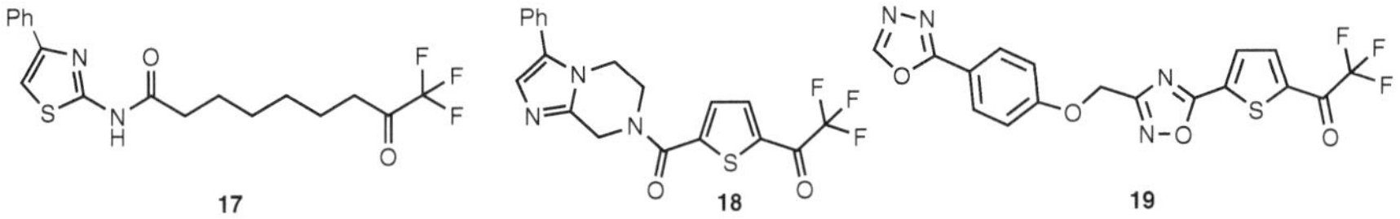

Fig. 7 Examples of trifluoromethyl ketones

this chemotype have been studied in more detail: selectivity of this class ranges from nonselective to compounds demonstrating good intra- and interclass selectivity. For instance, compound **19** has an IC_{50} against the class II HDAC4 (wt) of 7 nM with >100-fold selectivity over HDAC6 and >1,000-fold over the class I HDAC1 and 3 [72]. They have demonstrated cell permeability, but their promise as drug-like molecules has been compromised by high metabolic turnover. This is at least in part due to keto-reductase activity giving rise to the inactive alcohol form [69, 73]. Further optimization of this structural subclass has led to metabolically more stable compounds [74]. Although additional improvements will be required to achieve molecules with pharmacokinetic profiles that are suitable for further development, this effort yielded the disubstituted thiophene **18**. This molecule has been crystallized in combination with HDAC4cd [37], which confirmed the proposed chelation of Zn^{2+} by the hydrate form thus forming a four-membered ring.

2.3.7 Carboxylic Acids

The antiproliferative activity of sodium butyrate (Na^+ salt of **20**, Fig. 8) toward several types of carcinogenic cells has long been known, but it was not until much later that its anticancer activity was linked to HDAC inhibition [75, 76]. Pivanex **23** is a prodrug, which is metabolized *in vivo* to release butyric acid [77]. Other short-chain fatty acids such as valproic acid **21** and phenyl butyrate **22** have been investigated in the same context [78, 79]. These alkyl carboxylates show HDAC inhibitory activity typically in the low millimolar range and are much weaker than the strong chelators described above.

Interestingly methotrexate **24**, a well-known dihydrofolate reductase inhibitor that is clinically used for a number of indications such as leukemia and severe psoriasis, has recently been shown to interfere with HDAC activity [80]. Structurally, it may be regarded as a butyric acid derivative.

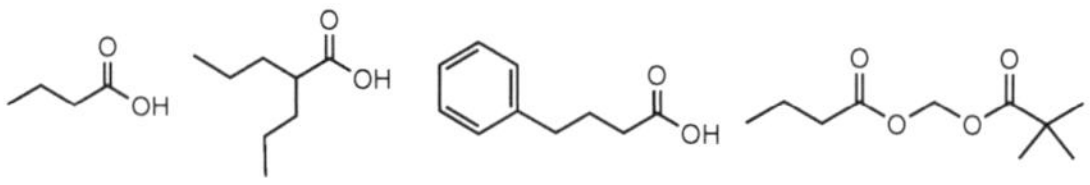

20 (butyric acid) **21** (valproic acid, VPA) **22** (phenyl butyrate, 4PBA) **23** (pivanex, AN-9)

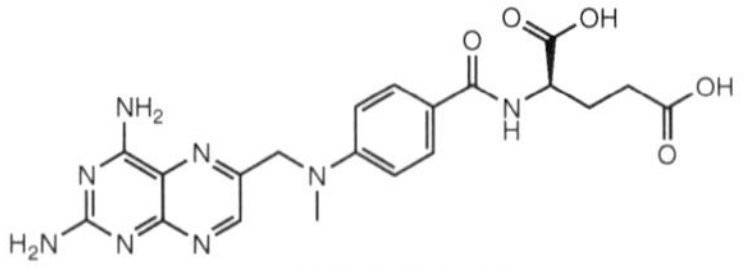

24 (methotrexate)

Fig. 8 Structure of short fatty acid inhibitors and methotrexate

2.3.8 Selectivity Determination

Determination of the selectivity of HDAC inhibitors toward isolated HDAC isoforms has been challenging. As a result, many publications describe inhibitory activity using cell lysates. Understanding in this area has been dogged by difficulties in the purification of individual isoforms, leading at times to contradictory results. This problem has subsequently been overcome by the isolation of protein from transfected *Escherichia coli*, which contain no endogenous HDACs [7].

Little is known about the natural substrate specificities of the HDAC isoforms, indeed, as stated earlier, it may be that class IIa HDACs have no meaningful catalytic deacetylation activity *in vivo* [7]; rather, they may act as N^{ε}-acetyl lysine recognition domains [81]. Various assays used to measure the inhibition of HDACs have been developed. These range from high resolution mass spectrometry profiling of whole cell systems [1] to isolated enzyme assays using a synthetic acetyl lysine substrate [82]. The low intrinsic catalytic activity of class IIa HDACs has been particularly problematic. To overcome this, gain-of-function mutants have been used [7]. These results have the caveat that they rely on modification of the active site and may not give an accurate reflection of the true binding event. Indeed, for some inhibitors, a marked difference is observed between gain-of-function mutant and wild-type HDAC4 [74].

Recently, an activated substrate has been designed which is successfully turned over by class IIa isoforms. This trifluoroacetate derivative is much more labile than the corresponding acetate and as such does not require the stabilizing tyrosine residue found in class I, IIb, and IV HDACs for the enzymatic conversion [7]. This substrate is efficiently deacetylated by the class IIa HDACs and intriguingly seems to show selectivity over class I [81]. This observed selectivity is likely to be a result of the more sterically demanding CF_3 group clashing with the more congested catalytic site of the class I isoforms.

The tools to de-convolute HDAC isoform activity have now been developed. Recent results highlighting the difficulty in purification of individual isoforms from mammalian systems suggest that a cautious interpretation of historical data is required.

In summary, the HDAC inhibitors known to date interact with the metal ion within the catalytic domain of the enzymes. However, besides recognition and deacetylation of substrates bearing N^{ε}-acetylated lysine residues, HDACs clearly adopt other cellular functions such as recruitment of other HDAC family members or transcription factors via protein–protein interactions. Molecules that selectively interfere with such a non-catalytic function would serve as excellent tools to further understand and dissect the function of these enzymes [83]. As compared to the currently investigated HDAC inhibitors, such molecules would likely exhibit a different selectivity profile.

In addition, the cellular localization of certain HDAC subtypes is governed by chaperone proteins (e.g., 14-3-3) which, depending on the phosphorylation state, transport the enzyme from the nucleus to the cytoplasm. Hence, inhibition of interaction with phosphatases, kinases, or chaperone molecules may also provide yet another avenue for regulation of HDAC activity with small-molecule ligands.

2.4 The Sirtuins (Class III HDACs)

2.4.1 Structure, Mechanism, and Function of Sirtuins

The class III histone deacetylases are referred to as the sirtuins. Seven mammalian sirtuins (Sirt1-7) are known to date and are classified by similarity to the Sir2 family from yeast. Most of them catalyze the deacetylation of N^{ε}-acetylated lysine side chains of histones as well as other protein substrates. Sirt1 alone is reported to have in excess of 30 substrates, which include p53, FOXO1, FOXO4, COUP-TF, NCOR, NF-κB-p65, and MEF2, respectively [84]. Sirt2 has been shown to be a tubulin deacetylase and an important regulator of cell division and myelinogenesis [85–87]. Sirt4 catalyzes ADP-ribosylation and Sirt6 accelerates both reaction types [88]. While sirtuins are expressed ubiquitously across tissue types [89], their intracellular localization varies: Sirt1, 6, and 7 are predominantly found in the nucleus, whereas Sirt1 and 2 are cytoplasmic, and isoforms 3, 4, and 5 are localized in mitochondria [90].

Various studies indicate that modulation of sirtuin activity (activation or inhibition) may lead to beneficial therapeutic effects, depending on the disease. There is ample evidence that overexpression of Sir2 (equivalent to the mammalian homolog Sirt1) leads to prolonged lifespan in various species, including yeast [91], fruit flies [92], and nematodes [93]. Furthermore, increased longevity due to a calorie restricted diet has been connected to upregulated sirtuin activity [92–94]. Enhancement of longevity and other health-promoting effects of sirtuins has frequently been attributed to regulation of metabolism. Since neuronal degeneration is a major pathophysiological aspect of aging, understanding the mechanisms of sirtuin-mediated neuroprotection promises novel strategies in clinical intervention of neurodegenerative diseases [95]. Inhibition of sirtuin function may also be beneficial in cancer therapy; for instance, prevention of Sirt1-mediated deacetylation of p53 might facilitate apoptosis in response to DNA damage and oxidative stress [96].

Compared to the zinc-dependent HDACs, the sirtuins act by a very different mechanism and require NAD^+ as a cofactor. Unsurprisingly, they show no sequence similarity with the other HDACs and are structurally very distinct [97]. The size of most sirtuins (Sirt2 to Sirt7) varies from 310 to 400 amino acid residues, while Sirt1 is larger (747 residues). Multiple crystal structures of eukaryotic and prokaryotic sirtuin proteins have been reported, which either are *apo*-forms or include ligands such as NAD^+ derivatives, N^{ε}-acetylated lysine substrates, and/or other small molecules [98–110]. These data have shed much light on the mode of action of this enzyme class.

As illustrated in Fig. 9, sirtuins convert one equivalent of NAD^+ to nicotinamide and 2′-*O*-acetyl-ADP-ribose (2′-OAADPr) to deacetylate an N^{ε}-acetyl lysine group [111]. This mechanism requires a conformational change of NAD^+ resulting in weakening of the C1′-N bond, which is induced upon binding of the substrate to the enzyme [109, 112]. A nucleophilic substitution at the anomeric

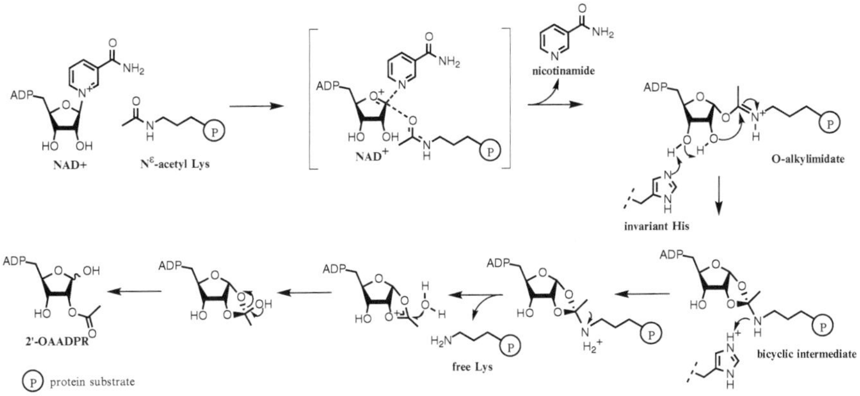

Fig. 9 Description of the catalytic cycle of sirtuins: N^{ε}-acetylated lysine substrate and NAD^{+} are converted to free lysine, nicotinamide, and 2′-OAADPR

center of the ribosyl unit leads to release of nicotinamide and formation of an *O*-alkylimidate intermediate. Importantly, the enzyme protects this activated intermediate from hydrolysis [113], which otherwise would revert the process back to N^{ε}-acetyl lysine. Instead, the invariant histidine functions as a general base assisting in a neighboring group participation of HO-C3′ and HO-C2′, which provides a bicyclic intermediate [114]. Hydrolysis of this intermediate, again supported by the invariant histidine, liberates the deacetylated substrate and 2′-*O*-acetylated ADP-ribose (2′-OAADPR).

All sirtuins share a catalytic NAD^{+} binding domain, which is fairly well conserved across the family [115] and a substrate-binding pocket. Structural data also provided insights to the substrate selectivity of sirtuins [108].

2.4.2 Sirtuin Inhibitors

Several structurally diverse sirtuin inhibitors have been reported, some of which are illustrated in Fig. 10. Nicotinamide is a product of NAD^{+} degradation that occurs during sirtuin-mediated catalytic process. Its inhibitory function at high concentrations is a result of a reaction with the ribosyl oxycarbenium intermediate formed as part of the mechanism, thus reversing the catalytic process and preventing deacetylation. Sirtinol **26** and salermide **27** [116], cambinol **28** [117], the tenovins **29** [118], and splitomycin **30** all show moderate inhibitory activity in the micromolar range.

The pseudo-spiro compound **31**, a moderately active Sirt2 inhibitor, has been identified by an *in silico* approach [119]. A crystal structure of the poly-sulfonylated symmetric urea suramin **32** bound to the catalytic site of Sirt5 has been elucidated [110] and demonstrated that the poly-aromatic compound covers the

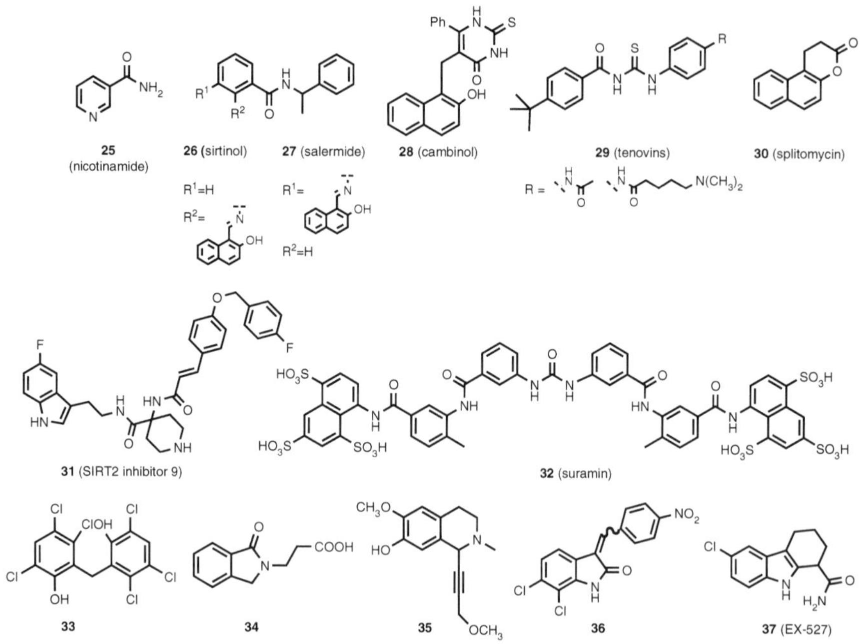

Fig. 10 Representative structures of known sirtuin inhibitors

NAD^+ binding site. Although suramin is unlikely to be brain permeable due to its size and highly negatively charged functionalities, the five last scaffolds display interesting characteristics: the diphenol **33**, the indanone **34,** and the tetrahydroisoquinoline **35** have been discovered by high throughput screening [120]. They exhibit moderate inhibitory activity for Sirt1 to Sirt3, and given their relatively low molecular weight, they may be suitable for further optimization for therapeutic purposes in neurological disorders. Indanone derivatives such as compound **36** with selective Sirt2 inhibitory activity have recently been published [121]. The fused indole EX-527 (**37**) and analogs remain the most potent sirtuin inhibitors described so far and show activity at well below micromolar levels in a fluorogenic assay with IC_{50} values of 98 nM for Sirt1, 19.6 μM for Sirt2, and 48.7 μM for Sirt3 [122]. Furthermore, neuroprotective properties have been reported for EX-527 in a variety of *in vitro* systems.

In a *Drosophila* model of neurodegeneration which overexpressed polyQ-expanded Htt Exon 1 (Httex1p Q93), both Sir2 (the *Drosophila* ortholog of mammalian Sirt1) and Sirt2 overexpression increased neuronal survival, as measured by the number of remaining photoreceptor neurons in the eye. However, none of these genetic mutations rescued the early lethality of the flies compared to wild-type lifespan [123]. Feeding Httex1p Q93-challenged flies on nicotinamide **25**- or sirtinol **26**-containing food also increased survival of photoreceptor neurons. Niacin, a vitamin supplement that is readily converted to nicotinamide, exhibited a comparable rescue. Similar results with nicotinamide feeding have been reported for a *Drosophila* model of SCA3 [124].

2.4.3 Sirtuin Activators

Several types of sirtuin activators have been reported. Resveratrol (**38**, Fig. 11), a triphenolic component of red wine, is one of the most intensely studied molecules; it has been used as a tool compound in several neurodegenerative models and there is ample data suggesting a positive impact in metabolic, neurodegenerative, and oncology models.

For instance, Parker *et al.* have demonstrated that resveratrol specifically rescued early neuronal dysfunction phenotypes induced by mutations of polyglutamines in transgenic *Caenorhabditis elegans*, indicating that activation of Sir2 (the *C. elegans* homolog of mammalian Sirt1) may result in a neuroprotective effect [125]. Separately, Kumar *et al.* studied resveratrol in a disease model in rodents, whereby i.p. administration of 3-nitropropionic acid (20 mg/kg for 4 days) caused significant loss of body weight, a decline in motor function, and poor retention of memory [126]. Repeated treatment with resveratrol significantly improved the motor and cognitive impairments.

However, these data have to be taken with caution as resveratrol is known to interfere with multiple pathways including Sirt1 activation, AMPK activation (which may have an indirect effect on sirtuins by changing the NAD^+/NADH equilibrium), nonspecific anti-oxidative properties, mitochondrial membrane polarization, and even AKT signaling. It should be noted that while resveratrol was shown to rescue neuronal degeneration in the Htt-challenged flies described above, it has been shown that resveratrol was equally effective in Htt-challenged flies homozygous for a Sir2 null mutation, indicating that the ability of resveratrol to suppress neurodegeneration did not depend on Sir2 [123]. Most importantly, it has recently been shown that resveratrol and compounds SRT1720 (**44**), SRT2183 (**45**), and SRT1460 (**46**; Fig. 12a), which have been reported as sirtuin activators [127, 128], are in fact not direct activators of Sirt1 [129]. Instead, it has been

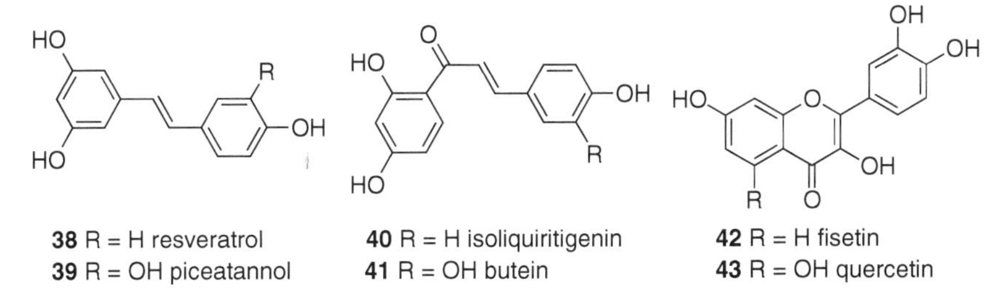

Fig. 11 Structure of resveratrol and other polyphenolic natural products with reported sirtuin-activating properties

unambiguously demonstrated that the Sirt1 activation observed by these agents is an artifact only observed when using a fluorescently labeled peptide substrate in the assay. All of these compounds have been shown to directly interact with the fluorophore moiety, thus resulting in a signal that is unrelated to Sirt1 activation.

A series of aza-benzimidazoles (Fig. 12b) with Sirt1-activating properties such as compound **47** have been described by the same research group who published the derivatives **44**, **45**, and **46** [130].

The pyrroloquinoxaline **48** (Fig. 13a [131]) and the dihydropyridines **49–51** (Fig. 13b [132]) are structurally diverse sirtuin activators.

Interestingly, the dihydropyridines have been discovered using a rational design approach starting from nicotinamide, and activators as well as inhibitors have been found within the same chemical series. In cell-based functional assays, compounds **49–51** have indeed induced a phenotype that could be a result of Sirt1 activation.

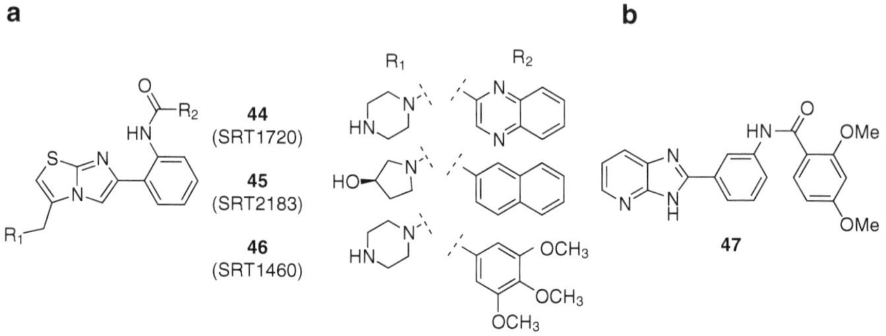

Fig. 12 (**a**) Structures of non-natural Sirt1 "activators" identified by a high throughput screen. (**b**) Reported aza-benzimidazole Sirt1 activator

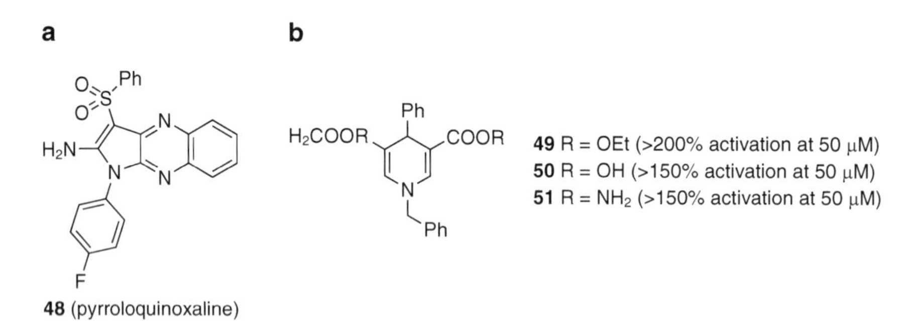

Fig. 13 Structures of sirtuin activators: (**a**) pyrroloquinoxaline **48**; (**b**) Structure and Sirt1 activation data for dihydropyridines **49–51**

3 Potential Treatment of Neurodegenerative Disorders with HDAC Inhibitors

3.1 Introduction

The remaining sections are dedicated to the potential role that HDAC inhibition may play in the amelioration of various pathway dysfunctions common to neurodegenerative disorders. Key biological signatures common to neurodegenerative disorders and the influence of HDAC activity on such processes are discussed. The use of specific HDAC inhibitors in the treatment of the specific disease domains will be illustrated.

3.2 HDACs, Chromatin Remodeling, and Control of the Epigenome

Epigenetics can be defined as persistent phenotypic changes which occur without alteration of primary DNA sequence. Dynamic changes in chromatin architecture have evolved as a key factor in regulating the epigenome [133]. While the genetic code in each cell of the body contains identical information at the level of the primary DNA sequence, activation and repression of genes by different combinations of transcriptional activators and repressors are responsible for orchestrating the diversity in cell type specification and function that is required for any living organism. Chromatin is the complex of genomic DNA, histone, and non-histone proteins that condenses and organizes genomic DNA. The fundamental unit of chromatin is the nucleosome: 147 base pairs of DNA spooled twice around an octamer of histones; composed of two copies each of the "core" histones (H2A, H2B, H3, and H4), along with a linker histone H1. This lower order or open structure resembles beads on a string with the individual beads about 10 nm apart from each other. However, chromatin adopts many higher order structures, moving up in complexity whereby short range inter-nucleosomal interactions condense the chromatin further [133, 134]. This *meta*-structure of chromatin governs DNA access of the transcriptional co-repressors, activators, and enzyme complexes that will regulate gene expression in a given cell type. It is generally accepted that loosely packaged "euchromatin" enables transcriptional activation, while more compact "heterochromatin" is thought to be more transcriptionally inactive or silent.

There are a number of different ways by which chromatin remodeling can occur. Removal, destabilization, or mobilization of nucleosomes may regulate access of transcription factors to genomic DNA for transcriptional activation. Posttranslational modification of histone proteins can also affect chromatin structure, which occurs most usually at the N-terminal tails of the histones and occasionally in the globular domains. Such modifications include acetylation, methylation, phosphorylation, sumoylation, biotinylation, and ubiquitination, or a combination thereof. Each of these biochemical events affects the DNA–histone interactions and may result in specific functional consequences. For example, modifications of histones in gene promoter regions have the potential to activate or repress binding of transcription factors and thus regulate gene transcription. Of pertinence to this review, Histone acetyl transferases (HATs) acetylate the amino group of lysine residues in the histone N-terminal region, which renders their side chains nonbasic. Hence, salt-bridges between the histone proteins and the highly negatively charged DNA are perturbed, and as a result the histones favor an open and accessible euchromatin conformation. These stable acetamido groups can be converted back to the free N^{ε}-amino-lysine by HDACs. Consequently, the interactions between the histone proteins and DNA are stronger, and the chromatin adopts a more compact and transcriptionally silent heterochromatin state. Importantly, a fine balance of the HATs and HDACs activities is required to maintain the equilibrium between open (accessible) and closed (silent) chromatin structure. Of particular interest is the fact that altered histone acetylation states play a part in many of the CAG repeat and other disorders, discussed further below [135].

3.3 Control and Dysregulation of Gene Transcription in Neurodegeneration

Transcriptional regulation depends on a complex molecular machine consisting of more than 100 proteins, which function in a highly synchronized fashion. There is now a substantial body of evidence that dysfunction in transcriptional mechanisms plays a central role in neurodegenerative diseases. Analysis of brain tissue from postmortem AD [136, 137] and PD patients [138–141] as well as AD [142] and PD [143] animal models have implicated transcriptional dysregulation as a strong feature of these diseases [144]. The same holds true for the polyQ diseases: in HD, transcriptional profiling of human tissue derived from HD-affected individuals shows an aberrant expression of a large number of genes and proteins [145–147]. Characteristic gene expression changes have also been observed in various mouse models designed to replicate certain aspects of the disease [147–151]. Similarly, mice overexpressing polyQ-expanded atrophin (a model of DRPLA) [152], expanded ataxin-7 [152, 153], and the mutant androgen receptor [152] also show aberrant gene expression.

What underlies transcriptional dysregulation? In the case of PD, mutations in transcriptional factors themselves are an emerging theme. Two recent genetic association studies in a screening sample of large cohorts of individuals with idiopathic PD have revealed evidence for a novel association of PITX3 promoter single nucleotide polymorphisms with PD, suggesting that an allele-dependent dysregulation of PITX3 expression might contribute to the susceptibility to PD [154, 155]. PITX3 is a homeodomain protein and transcription factor which is important for the differentiation and maintenance of midbrain dopaminergic neurons during development and the long-term survival of these neurons [156]. An isolated report of one individual with a mutation in the orphan nuclear receptor NURR1 has also been reported. This mutation markedly attenuated NURR1-induced transcriptional activation when tested *in vitro*, suggesting a role for this mutation and subsequent transcriptional dysregulation in the predisposition to idiopathic PD [157]. In familial forms of PD, mutations in the DJ-1 gene cause a rare early-onset autosomal recessive PD. DJ-1 adopts a transcriptional coactivator function, and pathogenic mutations impair that function and render dopaminergic neurons vulnerable to apoptosis [158]. Similarly mutations of the ubiquitin ligase parkin account for most autosomal recessive forms of juvenile PD (AR-JP). Parkin has been shown to act as a transcriptional repressor of p53 independently of its ubiquitin ligase function and downregulates the p53 pathway both *in vitro* and *in vivo* [159]. A main cell survival molecular pathway involves phosphorylation of Akt/PKB mediated by phosphatidylinositol-3-kinase. Several studies have consistently documented a molecular cascade linking Akt and NFκB that ultimately leads to p53 inhibition and cell survival [160]. Parkin mutations associated with familial AR-JP abolish the parkin-mediated control of p53, enhancing p53 expression in human brains affected by juvenile PD [159, 161]. P53 upregulation has also been implicated as a factor contributing to neurodegeneration in HD brain and mouse models [162]. More recently, huntingtin protein itself has been shown to facilitate the activity of the multi-subunit epigenetic silencer polycomb repressive complex 2 (PRC2), a function that is augmented in a polyQ tract length-dependent manner, providing another mechanism by which aberrant transcriptional repression could occur in this disease due to a pathogenic polyQ expansion [163].

3.4 Sequestration of HATs and Transcriptional Factors in Neurodegeneration

In terms of the underlying causes of transcriptional dysregulation, the polyQ diseases are possibly the best characterized pathologies. In these cases, epigenetic mechanisms and the sequestration of transcriptional cofactors have been clearly implicated. Along with the propensity for CAG expanded proteins to self aggregate, there are now multiple lines of evidence to suggest that proteins with expanded polyglutamine stretches are capable of forming abnormal interactions with other proteins containing short polyglutamine tracts (reviewed in [164, 165]). One of the strongest candidates to emerge thus far is the HAT CREB-binding protein (CBP), first identified as a coactivator for the transcription factor CREB

[166]. More recently, CBP has been shown to bind and modulate the activity of many different transcription factors [167–169]. Both CBP and the closely related P300 HAT/transcription factor contain a compactly folded 46 residue domain (named IBiD domain). Structural determination by NMR showed a helical framework containing a flexible polyglutamine loop that participates in ligand binding to multiple DNA-bound transcription factors [170]. It has been shown that poly-L-glutamine stretches aggregate into β-pleated sheets by forming hydrogen bonds between the side chain and the backbone amides, turning them into "polar zippers." This is also reflected in the propensity of Htt Exon 1 to self aggregate [171]. An interaction between short polyQ stretches in nonpathogenic proteins with expanded polyQ proteins is thus suspected to sequester normally soluble proteins into pathogenic inclusions, which are a hallmark of polyQ diseases. However, other lines of evidence underplay the role of polyQ expansion, suggesting that this may be a more general response to the presence of misfolded proteins in the nucleus [172]. Nevertheless, sequestration of CBP into insoluble polyQ aggregates has been demonstrated with expanded forms of the androgen receptor (SBMA) [173], atrophin-1 (DRPLA) [174], huntingtin (HD) [174, 175], and ataxin-3 (SCA-3) [173, 176]. In fact in HD, polyQ-expanded Exon 1 Htt has been shown to inhibit the acetyltransferase activity of a least three HATs: CBP, P300, and the p300/CBP-associated factor (P/CAF) [177].

The consequence of a reduction in HAT-mediated gene transcription is exemplified by Rubenstein–Taybi syndrome (RTS). RTS is caused by mutations in the CBP gene, which leads to an insufficient amount of produced functional CBP. The disease is characterized by developmental abnormalities and mental retardation. A number of mouse models with CBP mutations have been developed. These mice exhibit histone hypo-acetylation due to impaired CBP function, transcriptional repression, and memory impairment, while homozygous knockouts are embryonic lethal (reviewed in [178]). Thus, sequestration of CBP into insoluble protein aggregates may well be expected to phenocopy some aspects of RTS pathology.

CBP and P300 are not the only HATs that have been shown to be compromised in neurodegenerative diseases. Unsurprisingly, early evidence also came from investigation into SCA-17, whereby the pathogenic polyQ expansion occurs in a transcription factor itself, the TBP. TBP is part of a larger multi-subunit complex [179], similar to the TFTC-type GCN5 HAT-containing complexes, of which ataxin-7 (both wild-type and expanded) is also a component [180]. SCA-17 is characterized by late-onset neurological symptoms that are very similar to those of Huntington's disease and is often referred to as Huntington's disease-like 4 (HDL4).

TBP is a key transcriptional factor required for transcriptional initiation by the three major RNA polymerases (RNAP I, II, and III) and is involved in gene expression of most eukaryotic genes. Expanding the polyQ stretch of TBP from 31Q into the pathogenic range of 71 Q reduced *in vitro* binding of TBP to the TATA box DNA [181]. In a SCA-17 mouse model, N-terminal TBP fragments are present, which harbor the expanded polyQ tract but lack an intact C-terminal

DNA-binding domain. This polyQ-expanded TBP, incapable of binding DNA, formed nuclear inclusions and caused a severe neurological phenotype in transgenic mice. Together, these results indicate that polyQ-expanded TBP is inhibitory to TATA-dependent transcription as it is unable to bind DNA productively [181, 182]. PolyQ-expanded Htt exon 1 protein has also been shown to bind and sequester TBP [183]. In the case of mutant ataxin-7, HAT activity of the STAGA complex is compromised, and this has been directly linked to the retinal degeneration common in this disease [184, 185]. Transcriptional abnormalities have also been detected in ALS patients and mouse models thereof [186–188].

3.5 HDAC Inhibitors Ameliorate Transcriptional Dysregulation

Given the strong evidence for the sequestration of HAT complexes into insoluble nuclear aggregates in the polyQ diseases, it is not surprising that there are several lines of evidence to suggest that the acetylation status of histones are altered. In PC12 cells induced to express mutant huntingtin (Htt) Exon 1 protein with either 25Q or 103Q, a Q-length dose-dependent reduction in the level of the acetylated histones H3 and H4 was demonstrated, which could be reversed by treatment with SAHA (**1**, Fig. 1), TSA (**2**, Fig. 1), or sodium butyrate (**20**, Fig. 8) [177], a feature that was shared when polyQ peptides alone were expressed in cells [189]. This also proved to be the case in other cell models of HD, using both immortalized striatal cell lines derived from transgenic mice overexpressing full-length human expanded Htt and neuronal progenitor cells containing Htt with polyQ expansions in the pathological (but not wild-type) range [190]. Model organisms, such as *Drosophila* or *C. elegans*, which have been engineered to express either (Htt) Exon 1 protein or overexpressed polyQ-expanded peptides, exhibit histone hypo-acetylation, neurodegeneration, and compromised survival [191, 192]. In *Drosophila*, overexpression of CBP or treatment with the HDAC inhibitors sodium butyrate (**20**, Fig. 8) and SAHA (**1**, Fig. 1) has been shown to reverse histone hypo-acetylation and resulted in ameliorated pathology and extended survival [177, 192].

In the R6/2 model of HD, mice carry a transgenic Htt exon 1 fragment with an expanded polyQ repeat and exhibit a very aggressive phenotype consisting of motor and cognitive impairment, dramatic weight loss, and premature death occurring at approximately 4 months of age. Modest global hypo-acetylation of both histone H3 and H4 has been reported in R6/2 mice as compared to wild-type littermates; interperitoneal injection of sodium butyrate (**20**, 0.2–1.2 g/kg/day) resulted in brain histone hyper-acetylation as well as partial amelioration of the symptoms [193]. In a later study, Sadri-Vakili *et al.* demonstrated by *in vivo* chromatin immunoprecipitation that histone H3 was hypo-acetylated. In this instance, they did not find convincing evidence of global histone hypo-acetylation, instead showing significantly reduced AcH3 association within the promoter regions of known downregulated genes [190]. Furthermore, these investigators showed that treatment with sodium phenyl butyrate (**20**), either *in vivo* (0.4 g/kg/day for 7 days) or in HD cell models (10 μM), rescued the hypo-acetylation and partially reversed the associated gene transcript downregulation.

The finding that broad spectrum HDAC inhibitors do not universally increase gene expression is not surprising; earlier studies have shown that treatment with HDAC inhibitors change the expression of only ~2–10% of human genes significantly [194]. It should also be noted that following this treatment, almost equal numbers of genes are downregulated rather than upregulated, which underscores the complexity by which HDAC inhibitors affect gene expression [195, 196].

In "atro-118Q" transgenic mice, neuronal expression of the mutant human atrophin-1 protein containing an expanded stretch of 118 polyQ results in several neurodegenerative phenotypes that are commonly seen in DRPLA patients. Symptoms include ataxia, tremors, and other motor defects. Biochemical analysis of these mice also revealed histone H3 hypo-acetylation in brain tissue [197]. Furthermore, histone hypo-acetylation has also been demonstrated in transgenic ALS mice [198].

Friedreich's ataxia (FRDA) is the result of a GAA·TTC triplet hyper-expansion in an intron of the frataxin (FXN) gene that leads to transcriptional silencing. Frataxin is an essential mitochondrial protein and the resultant FXN insufficiency results in progressive spinocerebellar neurodegeneration and cardiomyopathy, leading to a progressive lack of motor coordination, incapacity, and death, usually in early adulthood. Interference with transcription due to the high GAA content of the mutated gene as well as the ability of expanded GAA·TTC regions to favor a heterochromatin assembly has been implicated as the reason for the observed transcriptional silencing [199]. Histones H3 and H4 of the FXN gene were noted to be hypo-acetylated in transformed lymphoid cell lines taken from an FRDA patient and a concomitant upregulation of trimethylated H3K9 has been observed. These findings imply a repressed heterochromatin state [200]. The effects on both H3 and H4 acetylation and FXN mRNA levels were assessed using valproic acid (**21**, Fig. 8), TSA (**2**, Fig. 1), SAHA (**1**, Fig. 1), and suberoyl bishydroxamic acid (**52**, Fig. 14) with variable results that were confounded by the cellular toxicity of these compounds. However, the benzamide derivative BML-210 (**53**, Fig. 14) did indeed increase FXN mRNA without showing cytotoxicity at the concentration tested. Furthermore, treatment with an analog of BML-210, pimelic

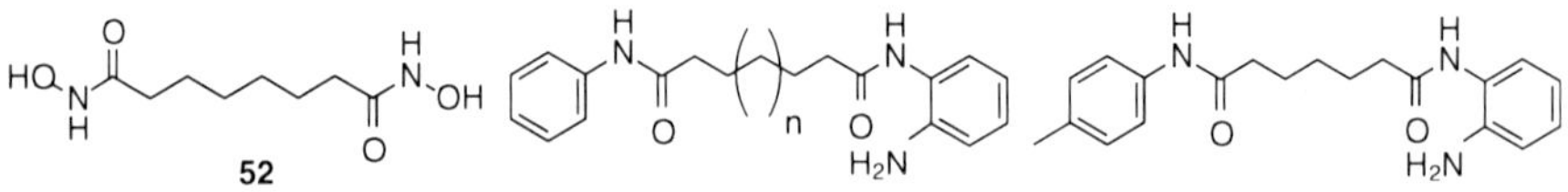

52 (suberoyl bishydroxamic acid, SBHA)

53 n = 2 (BML-210)
54 n = 1 (compound 4b)

55 (compound 106)

Fig. 14 Structure of SBHA (**52**), BML-210 (**53**), and analogs **54** and **55**

diphenylamide **54**, resulted in a 2.5-fold enhancement of FXN mRNA (5 μM), acetylation of H3K14, H4K5, and H4K12 in the chromatin region immediately upstream of the GAA repeats, and an approximate 3.5-fold increase in FXN protein levels (2.5 μM) [200] .

A subsequent short pharmacodynamic study showed that a very close analog of **54**, the tolyl derivative **55**, corrected the frataxin deficiency in a Friedreich's ataxia mouse model [201]. These mice carry a homozygous $(GAA)_{230}$ expansion in the first intron of the mouse FXN gene (KI/KI mice). Biochemical analysis revealed that these mice carry the same heterochromatin marks, close to the GAA repeat as those detected in patient cell lines and have mildly but significantly reduced frataxin mRNA and protein levels. However, they show no overt phenotype. Once a day treatment with compound **55** at 150 mg/kg subcutaneously for 3 days increased global brain tissue histone acetylation as well as histone acetylation close to the GAA repeat and restored frataxin levels in the nervous system and heart (determined by qPCR and semiquantitative western blot analysis). Reversion of other differentially expressed genes toward wild-type levels was also observed. The compound showed no apparent toxicity.

HDAC inhibitor **54** has also demonstrated a therapeutic effect in the R6/2 Huntington's disease model, which expressed the Exon 1 Htt protein with an expanded polyglutamine region of ~300 repeats, and shows a more delayed phenotype than the R6/2 model with shorter polyQ expansions [202]. Again, a short pharmacodynamic trial (once a day subcutaneous treatment with 150 mg/kg for 3 days) successfully ameliorated gene expression abnormalities as detected by microarray analysis in these mice and showed increased histone H3 acetylation in association with selected downregulated genes. For a chronic efficacy trial, the TFA salt of compound **54** was solubilized in 2-hydroxypropyl-β-cyclodextrin and dissolved in drinking water to an estimated dosage of 150 mg/kg/day and administered to mice from 4 months of age. While the expected differences in oral versus parenteral administration preclude a direct correlation between the pharmacodynamic and efficacy trial, these mice exhibited improved motor performance, improvement in overall appearance, and an amelioration of body weight loss, which are features of this mouse model. Brain weight and striatal atrophy were also improved.

The successful use of benzamide **54** in treating R6/2 mice loosely correlates with an earlier report, whereby SAHA (**1**) was administered in drinking water to R6/2 mice that harbor a smaller polyQ repeat (~250 Q) and exhibit a more aggressive phenotype [203]. In this study, the authors also showed significant improvement in the motor dysfunction in R6/2 as assessed by rotarod performance and grip strength, but this improvement was offset by the failure of both wild-type and R6/2 mice to gain weight at the maximum tolerated dose (0.67 g/L in drinking water), which is suggestive of a narrow therapeutic window. Increasing this dose further showed overt toxicity in R6/2. Toxicity of broad spectrum HDAC inhibitors in the clinic is a general concern, especially when considering neurodegenerative indications which requires long-term treatment. This underscores a necessity to narrow the therapeutic focus by targeting the specific HDAC isoforms that would have most impact in ameliorating the pathophysiology and symptoms of the disease, while minimizing potential adverse effects resulting from redundant inhibition of other isoforms.

4 Effect of HDAC Inhibitors on Reversing Protein Accumulation in Neurodegeneration

4.1 The Role of the Ubiquitin–Proteosome System and Autophagy Pathways

The two main catabolic pathways responsible for degrading proteins are the ubiquitin–proteosome system (UPS) and the autophagy–lysosomal system. Ubiquitin is a small 8.5 kD protein composed of 76 amino acids and can be covalently attached to lysine moieties by a peptide bond. This is mediated by three types of enzymes known as E1, E2 (Ubc), and E3, which have the ability to activate, conjugate, and transfer the ubiquitin moiety to a target protein. The E3 ligases catalyze the formation of an isopeptide bond between a Lys residue of the target protein and the C-terminal Gly of ubiquitin. In many cases, additional ubiquitin monomers are added to the first ubiquitin by the subsequent action of E4 ligases to form poly-ubiquitin chains, consisting of four to seven ubiquitin monomers. The type of linkage conferred by E4 ligases affects the fate of the ubiquitinated protein. K48-linked polyubiquitinated proteins are generally degraded by the 26S proteosome, while K63-linked poly-ubiquitinated proteins are ultimately degraded in the lysosome.

The aggregated protein inclusions, which are a hallmark of neurodegenerative proteopathies, are often heavily ubiquitinated. This finding may suggest that the cell is attempting to dispose of these abnormal proteins. Thus, a unifying concept has emerged regarding an underlying mechanism that contributes to these classes of diseases: certain proteins, which are vulnerable to misfolding to pathological conformations, assemble into aggregates as the capacity of the cell to dispose of them is exceeded. In other cases, mutation of UPS components may directly cause a pathological accumulation of proteins. For instance, a frame shift mutation of ubiquitin UBB+1, found in some sporadic and hereditary AD patients, leads to an inhibition of the UPS and an enhancement of toxic protein aggregation in a yeast model [204, 205]. As mentioned previously, in familial juvenile PD, there is a defect in the E3 ubiquitin ligase activity of parkin, and accumulation of which is found in the Lewy body aggregates along with its target substrates [206]. Either boosting the UPS proteosome pathway or increasing autophagy has been proposed as viable therapeutic approaches for the treatment of many neurodegenerative disorders. The contribution of impairment in the UPS system to neuropathological conditions has recently been reviewed in depth [207].

4.2 Regulation of Protein Turnover by HATS and HDACs

It is now clear that lysine acetylation by HATs and deacetylation by HDACs can also occur in non-histone proteins, implicating their involvement in a variety of cellular processes aside from transcription [208]. The regulation of protein stability is an important example (reviewed in [209]).

Three general mechanisms link lysine acetylation to protein stability. Lysine acetylation of proteins can create docking sites favoring protein–protein interactions, or conversely interfere with the binding of specific partners, and hence stabilize or destabilize particular protein complexes. A striking example of this is the control of the chaperone activity of HSP90 by acetylation, with HDAC6 emerging as a key regulator [27]. Proteosome dysfunction and accumulation of protein aggregates leads to the activation of the major heat shock transcription factor, HSF1, which in turn induces the accumulation of the cellular heat shock proteins (HSPs). Correct folding of proteins by these chaperones has a major impact on protein stability and in safeguarding stressed cells. In yeast, Hsp90 mutants that cannot be acetylated at K294 have reduced viability and chaperone function. Reduction in Hsp90 function has also been observed on knockdown of HDAC6 [210]. Importantly however, Hsp90 can be acetylated at multiple sites, and it has been shown that HDAC6 is not capable of deacetylating all sites, suggesting that other HDACs may play a role. Another example of the acetylation of a protein specifically influencing degradation came from Huntington's disease research. Jeong *et al.* showed that acetylation of mutant Htt at Lys9 and Lys444 can promote clearance of the mutant protein by autophagy in both primary neurons and in a *C. elegans* model of the disease, whereas a mutant version of huntingtin that cannot be acetylated accumulates leading to neurodegeneration [211]. The HDACs responsible for this activity have yet to be determined.

Second, in several reported cases, the lysine "locking" activity of an acetylation event hinders subsequent protein ubiquitination and leads to increased protein stabilization [212]. For instance, the stability of p53 can be regulated in this way: the activity of CBP/p300-mediated acetylation of p53 increases the stability of the protein, which is counterbalanced by the action of a HDAC1–MDM2 (E3-ligase) complex enhancing p53 degradation [213]. Sirt1 has also been implicated as a major p53 deacetylase in mammalian cells [96]. In a SCA-7 model, CBP-dependent acetylation at Lys257 of ataxin-7 prevents autophagy-mediated turnover of an N-terminal caspase-7 cleavage fragment, a process which can be replicated by selective HDAC7 knockdown [214].

Interestingly, four independent HATs, namely CBP, p300, PCAF, and TAF1, and one HDAC (HDAC6) have been shown to also possess intrinsic ubiquitin-linked functions in addition to their regular HAT/HDAC activities. P300 and CBP possess intrinsic E4 ligase activity by means of a specific domain distinct from their HAT activity [215], and p53 is a target for this activity. PCAF exhibits E3 ligase activity, which is only partially independent of its HAT activity [216, 217]. For HDAC6, the ubiquitin-binding activity has been shown to arise from a conserved zinc finger-containing domain named ZnF-UBP (PAZ domain). This ability for binding ubiquitinated proteins correlates with its dramatic relocalization into aggresomes on proteosome inhibition [29, 218, 219].

4.3 Effect of SUMO E3 Ligase Activity of Class IIa HDACs

More recently, a family of ubiquitin-like modifiers known as small ubiquitin-like modifier (SUMO) 1–4 has been described. These modifiers are small proteins similar in size to ubiquitin. Through a process analogous to ubiquitination known as sumoylation, SUMO monomers are conjugated to target proteins. In some cases mono-ubiquitination or sumoylation may act to protect proteins from ubiquitin-dependent degradation, and in other cases they appear to trigger poly-ubiquitination. Several studies indicate that lysine sumoylation negatively regulates transcription factors. Sumoylation can also alter the subcellular localization of a protein, as is the case for HDAC4. Sumoylation of HDAC4 at K559 by SUMO-1 is coupled to its nuclear import by nature of the interaction with RanBP2, a SUMO E3 ligase that comprises part of the nuclear pore complex [220]. MITR, HDAC1, and HDAC6 are similarly SUMO-modified [220], indicating that sumoylation may be an important regulatory mechanism for the control of transcriptional repression and protein stability by HDACs.

Of particular interest is the fact that HDAC4 and HDAC5 bind the universal E2 ligase Ubc9, by nature of the N-terminal coiled-coil domain of class IIa enzymes. Furthermore, both HDAC4 and HDAC5 appear to have intrinsic SUMO E3 ligase activity (regardless of their own sumoylation state), potently stimulating MEF2 sumoylation at Lys424 both *in vitro* and *in vivo*. In the case of MEF2, the HDAC4-induced SUMOylation does not appear to involve prior deacetylation of the Lys by HDAC4. Instead, this is performed by Sirt1, which forms a complex with HDAC4 and MEF2 [14].

The finding that HATs and HDACs are also associated directly with E3 ligases in multi-subunit complexes adds another layer of complexity to the regulation of protein degradation pathways. The implication of this emerging role of class IIa HDACs for neurodegenerative diseases is currently unclear, but it raises two important points. First, endogenous class IIa deacetylase activity has yet to be demonstrated against a native substrate and may in fact not be functionally relevant; transcriptional control and protein–protein interactions may be largely mediated through the MITR domain. The effect of catalytic site inhibition on the regulation of class IIa MITR-domain-derived activity is not known. Second, identification of Ubc9 binding and SUMO E3 ligase activity of HDAC4 and 5 may have important consequences for protein clearance and stability in the CNS proteopathies, especially given the fact that HDAC4 has been shown to segregate into poly-ubiquitinated inclusions in neurodegenerative mouse and cell models [221]. There is emerging interest in SUMO posttranslational modification in regulating the stability and clearance of the polyQ-expanded proteins, and this field is likely to expand in the coming years [222–224].

4.4 HDAC6: A Master Regulator of Cell Response to Cytotoxic Insults

In a *Drosophila* model of SBMA, flies express a polyQ-expanded human androgen receptor and exhibit a hormone-dependent degeneration of the eye. A genetic screen in this model identified HDAC6 depletion as an enhancer of neurodegeneration, which was confirmed by HDAC6 RNAi knockdown experiments [225, 226]. It was also shown that upregulation of HDAC6 suppressed the degeneration, which was absolutely dependent on its catalytic activity; a catalytically inactive mutant failed to suppress the degeneration. The neuroprotective properties of overexpressed HDAC6 and the enhanced neurodegeneration following HDAC6 knockout were not exclusive to the polyQ-expanded AR in this model: this was also demonstrated using expanded ataxin-3, a 127 polyQ fragment and the pathogenic Aβ1-42 fragment of APP, but interestingly no effect of HDAC6 was noted in flies expressing expanded ataxin-1.

The mechanism by which HDAC6 can affect misfolded protein stress arose from studying the aggresome pathway. Aggresomes are ubiquitinated inclusions that form when the proteosome is impaired or when misfolded proteins are overexpressed. In this pathway, a microtubule-organizing center (MTOC) transports misfolded proteins to lysosomes which are degraded through autophagy. Thus, it is suggested that aggresome formation is a protective response that provides an alternative route for the clearance of substrates that are resistant to proteosomal degradation. HDAC6 is a microtubule-associated HDAC and the main deacetylase of α-tubulin, a component of the MTOC. The HDAC6 inhibitor tubacin (**60**, Fig. 16) prevents deacetylation of α-tubulin, which resulted in the accumulation of poly-ubiquitinated proteins and apoptosis [227]. In addition, via its previously mentioned ubiquitin-linking domain and its direct interaction with microtubule motor complexes containing p150 (glued) [25], HDAC6 may function in part by providing a physical link between the poly-ubiquinated cargo and dynein motors, which permits the transport of the cargo to the lysosome [228]. Indeed, in the SBMA fly model, overexpression of HDAC6 accelerated the degradation of polyQ protein in an autophagy-dependent manner, which is consistent with earlier reports of HDAC6-dependent autophagic clearance [229, 230]. Analysis of brain microtubule protein from AD patients has also identified that α-tubulin levels decreased along with increased acetylation of α-tubulin, mainly in neurons containing neurofibrillary tau pathology. In an *in vitro* study, tau was shown to bind HDAC6, which decreased its activity and resulted in increased tubulin acetylation and impaired the autophagic pathway [231].

The results described above indicate that HDAC6 inhibition would likely be detrimental and result in exacerbation of protein misfolding and cellular stress in neurodegenerative proteopathies. However, despite the convincing literature suggesting that HDAC6 is a critical sensor of cellular stress via autophagic and proteosomal pathways, it was unexpected that HDAC6 (−/−) mice are viable and show no overt phenotype (discussed in [232]). Perhaps even more surprisingly, selective HDAC6 inhibition has been proposed as a viable treatment for HD, by nature of accelerating anterograde transport of kinesin-1 cargo on acetylated microtubules [233]. Htt associates with molecular motors and activates the microtubule-dependent transport of vesicles containing brain-derived neurotrophic factor (BDNF). Wild-type htt enhances the velocity of vesicle transport. With polyQ-expanded Htt, the intracellular transport of BDNF-containing vesicles is altered, resulting in reduced trophic support to neurons and their death. This effect can be overcome *in vitro* by HDAC6 inhibition [234, 235]. Others have also demonstrated that selective inhibition of HDAC6 by two HDAC6-selective inhibitors (MA-I and MA-II) can exert a neuroprotective function in response to oxidative stress *in vitro* [236].

The structures of the mercaptoacetamides MA-I and MA-II are shown in Fig. 15. Other mercaptanes with selective impact on HDAC6 activity have been described. For instance, compound **58**, a thioester, is thought to be hydrolyzed to the corresponding mercaptan **59** within cells, thus acting as a prodrug [237]. Under cell-free conditions, thiol **59** demonstrated inhibitory activity for HDAC6 with an IC_{50} value of 29 nM and 42- and 36-fold selectivity over HDAC1 and HDAC4, respectively. In agreement with this, the prodrug **58** led to increased α-tubulin acetylation in a cellular context without affecting the acetylation state of histone H4.

Several selective hydroxamate-based HDAC6 inhibitors have been documented (examples shown in Fig. 16). Schreiber and coworkers discovered the first selective HDAC6 inhibitor by screening: the high molecular weight compound tubacin (**60**, M_w 721) [238]. Tubacin binds one of the two catalytic domains of HDAC6 and blocks its function in α-tubulin deacetylation. Jung *et al.* investigated hydroxamates with variable spacer length (C_6 and C_7) and various surface recognition elements [239, 240]. This study yielded compounds like bromophenyl alanine **61**, which showed HDAC6 inhibitory potency in the low micromolar range and modest selectivity over HDAC1.

Similarly, Kozikowski *et al.* used the hydroxamate metal-binding group and a flexible C_6 alkyl linker and developed a [2+3] cycloaddition approach to optimize the surface recognition element [241]. This work culminated in the isoxazole **62**, for which an IC_{50} value of 2 pM against HDAC6 and significant selectivity over several

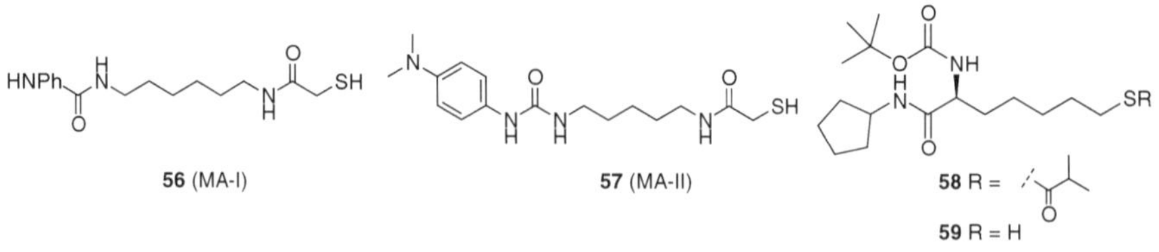

56 (MA-I) **57** (MA-II) **58** R = **59** R = H

Fig. 15 Structures of selective HDAC6 inhibitors MA-I, MA-II, and thiol **59** and its prodrug **58**

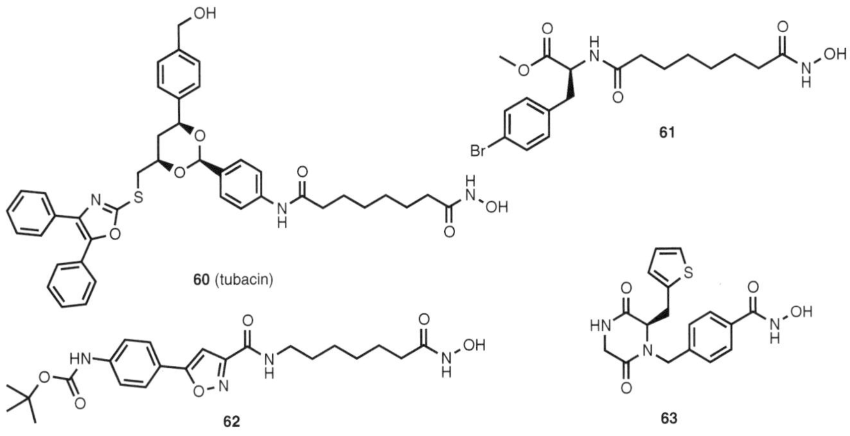

Fig. 16 Examples of selective HDAC6 inhibitors

other isoforms has been reported. The same compound also demonstrated antiproliferative activity between 0.1 and 1 μM in various pancreatic cancer cell lines.

Smil *et al.* studied hydroxamates bearing a phenylene linker and various chiral diketo-piperazine derivatives as the capping group. This work yielded compounds inhibitory activity for HDAC6 in the low nanomolar range, selectivity of up to 40-fold over other class I and IIa isoforms, and a remarkably low molecular weight (e.g., **63**, M_w 359, [242]). All of these compounds may serve as chemical tools to investigate HDAC6 function. Given the disparate hypotheses of HDAC6 overexpression versus inhibition, genetic cross of the HDAC6 null mice with neurodegenerative mouse models or *in vivo* interrogation with a selective brain-penetrant HDAC6 inhibitor will be important experiments to elucidate the role of HDAC6 in protein clearance pathways and neuroprotection.

4.5 HDACs and Sirtuins Regulate Autophagy Pathways

HDAC1 inhibition has been shown to be an effective stimulator of the autophagic pathway in cell systems, using either class I inhibition by FK228 (**9**, Fig. 5) or RNAi of HDAC1 [243]. The NAD^+-dependent deacetylase Sirt1 also appears to be an important regulator of autophagy. Sirt1 is capable of forming a molecular complex with several essential components of the autophagy machinery, including Atg5, Atg7, and Atg8, and can deacetylate these proteins. In contrast to HDAC1, a transient *increase* in expression of Sirt1 is sufficient to stimulate basal rates of autophagy. Furthermore, Sirt1(−/−) mouse embryonic fibroblasts show impaired autophagy under starved conditions, and Sirt1(−/−) mice demonstrate an accumulation of damaged organelles, disruption of energy homeostasis, and early perinatal mortality [244].

5 Neuroprotection Through HDAC Inhibition

5.1 Introduction

Given that the earliest therapeutic effects of HDAC inhibitors in oncology arose from the propensity of these compounds to kill rapidly proliferating cells, it may seem somewhat incongruent that these compounds protect compromised neuronal cells. Indeed, it is well known that some HDAC inhibitors, such as TSA (**2**, Fig. 1), exhibit basal toxicity, and prolonged treatment often induce neuronal death. However, a number of studies indicate that HDAC inhibition can show direct neuroprotective properties to neuronal cells *in vitro* under particular insults [245]. TSA can rescue cortical neurons from oxidative stress when applied for a short period of time [246]. In cultured cortical neurons, Ryu and colleagues showed that treatment with TSA (**2**, Fig. 1), SAHA (**1**, Fig. 1), or sodium butyrate (**20**, Fig. 8) protected against glutathione depletion induced oxidative stress, which involved acetylation and activation of the DNA-binding activity of Sp1 [247]. As previously discussed, selective HDAC6 inhibitors have also demonstrated neuroprotective potential against oxidative stress *in vitro* [236]. Class I/II inhibitors can also block BAX-dependent apoptosis of mouse cortical neurons by p53-dependent and -independent mechanisms [248] and protect against excessive glutamate challenge *in vitro* [249].

Emerging evidence also supports the notion that HDAC inhibition in microglia may play a significant role in mediating anti-inflammatory effects. HDAC inhibitors protect against dopaminergic neuronal death and neuroinflammation induced by exposure to lipopolysaccharide (LPS) [250, 251], in part through the induction of microglial apoptosis [252]. Partially contributing to their neuroprotective effect, HDAC inhibitors have been shown to increase the expression of pro-survival neurotrophins. Upregulation of BDNF was shown to underlie neuroprotection in rat cortical neurons [253]; GDNF and BDNF induction upon HDAC inhibition has been demonstrated in primary cultures of astrocytes [250, 254].

In vivo neuroprotection by HDAC inhibition has been linked to upregulation of transcription of antioxidant and growth factor proteins, stimulation of neurogenesis [255], and anti-inflammatory effects [256–258]. An anti-inflammatory effect has been achieved by suppression of microglial activation [259], inhibition of pro-inflammatory cytokine expression [260], or NFκB-mediated inflammatory responses. Treatment with HDAC inhibitors also markedly inhibited ischemia-induced p53 overexpression [261, 262]. In an animal model of multiple sclerosis (experimental autoimmune encephalomyelitis, EAE), treatment with TSA (**2**, Fig. 1) activated a transcriptional program that culminated in decreased caspase 3 activity [263]. In HD, treatment of *Drosophila* mutants expressing Htt with the HDAC inhibitors SAHA or TSA (**1** or **2**, Fig. 1) suppressed neuronal photoreceptor generation [177].

The precise HDAC isoforms involved in HDAC inhibitor-mediated neuroprotection are unclear. Extrapolating from the cardiac field, it is noteworthy that knockdown of HDAC4 reduced infarct size following myocardial ischemia-induced reperfusion injury [264]. Intracellular trafficking of HDAC4 from the cytoplasm to the nucleus was shown to be a critical component of low-potassium or excitotoxic glutamate-induced cell death in cerebellar granule cells [265]. Interestingly, Paroni and colleagues showed that during UV irradiation to cells (to trigger apoptosis), HDAC4 is cleaved by both caspase 2 and 3, which separates the carboxy terminal and N-terminal fragments; the C-terminus becomes localized in the cytoplasm, whereas the N-terminal fragment accumulates in the nucleus. They demonstrated that it was the N-terminal portion of HDAC4 that triggered cell death, which correlated with strong repressive MEF2 activity [2]. Similarly, inactivation of a MEF2D/HDAC5 complex by depolarization-mediated calcium influx promoted cerebellar granule cell survival, and overexpression of HDAC5 induced apoptosis [266]. An increase of nuclear HDAC4 in granule neurons is also observed in weaver mice, which harbor a mutation that promotes CGN apoptosis [265]. While these data collectively suggest that HDAC4 inhibition or its cytoplasmic retention may be neuroprotective, one study described the opposite scenario; here, HDAC4 overexpression protected cerebellar granule cells from oxidative stress, which is mediated by nuclear HDAC4 [267]. HDAC4 also appears to regulate neuronal survival in the retina, with a reduction in HDAC4 expression during retinal development leading to apoptosis of bipolar inter-neurons and rod photoreceptors. Conversely, HDAC4 overexpression in a mouse model of retinal degeneration prolonged photoreceptor survival [268]. In this instance, the survival effect was attributed to cytoplasmic HDAC4.

5.2 *Neuroprotection by Sirtuins*

A growing body of evidence implicates Sirt1 and Sirt2 as important regulators of neurodegeneration [269, 270]. Overexpression of Sirt1 prevents neuronal death in tissue culture models of AD, amyotropic lateral sclerosis, and polyglutamine toxicity and reduces hippocampal degeneration in a mouse model of AD [271, 272]. It has been suggested that some of the neuroprotection stems from the regulation of the peroxisome proliferator-activated receptor γ (PPARγ) coactivator-1α (PGC-1α). PGC-1α is a master orchestrator of mitochondrial function that integrates signals regulating mitochondrial biogenesis and respiration, detoxification of ROS, energy metabolism, and thermogenesis [273]. PGC-1α interacts with a number of transcription factors including PPARγ of the PPAR family, which regulates adipogenesis and lipid metabolism, and the nuclear respiratory factor-1/2 (NRF-1/2), which play a pivotal role in mitochondrial respiration. Expression of PGC-1α has been shown to be repressed in both *in vitro* and *in vivo* models of HD, partially due to downregulation of the CREB/TAF4 signaling pathway which is a predominant regulator of PGC-1α expression [274]. Primary striatal neurons are significantly protected from mHtt-induced toxicity by exogenous expression of PGC-1α, and lentiviral delivery of PGC-1α into the striatum of HD mice was shown to attenuate brain atrophy [274]. Sirt1 is known to deacetylate PGC-1α leading to activation [275–278], which may in part be responsible for its neuroprotective effect.

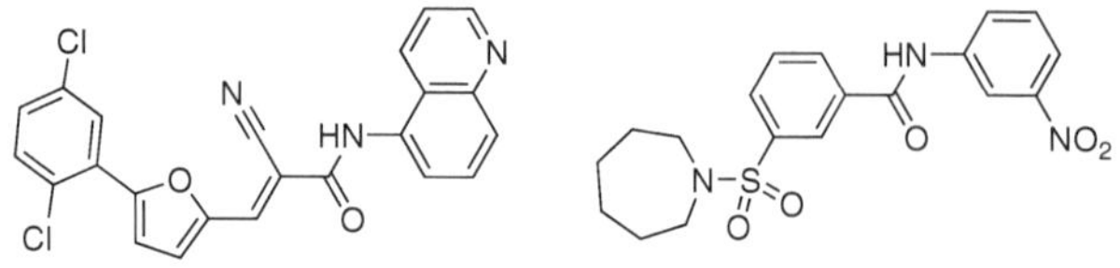

64 (AGK2) 65 (AK−1)

Fig. 17 Structures of Sirt2 inhibitors AGK-2 and AK-1

Paradoxically, Sirt2 inhibition has also been reported to exert neuroprotective effects in both cell and invertebrate models of neurodegeneration, which is in line with the early findings of Pallos *et al.* [123]. In particular, the use of the two Sirt2 inhibitors AGK2 and AK-1 (**64** and **65**, Fig. 17) has underlined this hypothesis. The furan AGK2 and the sulfonamide AK-1 have been reported as selective Sirt2 inhibitors with IC_{50} values for Sirt2 of 3.5 and 12.5 μM, respectively, and no inhibitory activity for Sirt1 up to at least 50 μM [279]. The compounds are thought to block the NAD^+ site of Sirt2.

Outeiro *et al.* demonstrated that both inhibitors ameliorated α-synuclein-mediated dopaminergic cell death *in vitro*, with AGK-2 displaying more potency and dose-dependency in this response. Subsequently, the same authors showed that treatment of *Drosophila* with 0.5–1 mM AGK-2 for 20 days was neuroprotective in a model of neurodegeneration resulting from α-synuclein overexpression. Although the molecular mechanism of action is not fully understood, the authors postulated that these compounds may function by promoting the formation of enlarged inclusion bodies, which were suggested to provide a cell survival advantage [280]. A recent study using these same inhibitors in cellular and invertebrate models of HD showed equivalent neuroprotection. In an *in vitro* primary striatal neuron model overexpressing polyQ-expanded Htt fragment, both AK-1 and AGK-2 rescued neuronal toxicity. In this instance, the authors reported a decrease in Htt inclusion number, but no effect on inclusion size or morphology. These inhibitors also rescued neuronal dysfunction associated with expression of N-terminal Htt in *C. elegans* touch receptor neurons, as well as in the HD flies previously described by Pallos *et al.* [123]. Interestingly, this study revealed a unique role for Sirt2 in the control of neuronal metabolism and in particular sterol biosynthesis. While Sirt2 inhibition did not alter the global transcriptional dysfunction associated with mHtt, it decreased sterol levels by decreasing the nuclear trafficking of the sterol response element binding protein 2 (SREBP-2). This regulation of SREBP-2 by Sirt2 was shown to happen via an extranuclear mechanism [281]. These data raise the intriguing possibility that negative regulation of sterol production might be the cellular neuroprotective mechanism of selective Sirt2 inhibition.

Further evaluation of the role of the sirtuin isoforms in a mammalian context will be necessary to dissect the apparent paradox of the neuroprotective properties of sirtuins and the necessity for isotype selective inhibition. In regard to this, Pfister and colleagues overexpressed each of the seven sirtuin proteins in healthy cerebellar granule neurons *in vitro* or in neurons that have been induced to die by low potassium (LK) treatment, which provided the first analysis of the role of sirtuin isoforms 3–7 in neuronal survival functions [282]. Lysine acetylation is a very abundant posttranslational modification in mitochondria [1]. As Sirt3, Sirt4, and Sirt5 localize in the mitochondria, they are thought to play a role in energy metabolism and responses to oxidative stress [90]. Although their role has not been well studied in neurodegeneration, manipulation of these enzymes may have important consequences, as alterations of mitochondrial function have been demonstrated in many neurodegenerative conditions (recently reviewed in [283–286]). In the study by Pfister *et al.*, Sirt1 overexpression protected neurons from LK-induced cell death, while Sirt2, 3, and 6 overexpression induced apoptosis in otherwise healthy neurons. Ectopic Sirt5 overexpression showed differential effects based on its subcellular location: if localized to either nuclear or cytoplasmic compartments, Sirt5 was protective, but induced apoptosis when localized exclusively to mitochondria (as it is endogenously). Of importance, the rescue by Sirt1 overexpression was also observed on transfection of either of the two catalytically dead Sirt1 isoforms (H363Y and H355A), suggesting that the Sirt1 protective effect did not rely on the deacetylase activity of the enzyme. This has important consequences for any conceived therapy with potential Sirt1 activators. Indeed, addition of the Sirt inhibitors nicotinamide (**25**, Fig. 10), sirtinol (**26**, Fig. 10), or splitomycin (**30**, Fig. 10) failed to prevent the Sirt1 overexpression rescue of cell death, and resveratrol (**38**, Fig. 11) failed to mimic Sirt1 overexpression.

6 Treating Cognitive Impairment and Depression with HDAC Inhibitors

6.1 Introduction

Despite the plethora of causal insults and molecular mechanisms that underlie neurodegeneration in CNS proteopathies, the resultant symptoms are commonly manifested (with different severity depending on each disease) as disturbances in motor control, dementia, cognitive impairment, depression, and sleep disturbances. This last section briefly summarizes the data that support a therapeutically beneficial role for HDAC inhibition on these different "disease domains."

6.2 Procognitive Effects of HDAC Inhibition

As an epigenetic mechanism of transcriptional control, chromatin modification has been shown to participate in maintaining cellular "memory" and may underlie the strengthening and maintenance of synaptic connections required for long-term changes in behavior [287]. This was demonstrated elegantly in the CBP (+/−) HAT haplo-insufficiency model of the Rubenstein–Taybi syndrome, whereby long-term memory impairment was correlated both with chromatin hypo-acetylation and deficits in synaptic plasticity, known as long-term potentiation (LTP) [288]. LTP is a phenomenon whereby synaptic connections and hence synaptic transmission are strengthened in response to a brief increase in neural activity. This strengthening outlasts the original induction stimulus and has been suggested as one of many means in which neurons maintain a "memory" of their previous activity. Molecularly, the transcriptional and translational dependency of the late phase of this process on the generation of new synaptic proteins has been well established [289]. Synaptic "plasticity" is bidirectional and can be induced by different stimuli. Although the mechanism is extremely complex, there is ample evidence suggesting that experience-dependent alterations in synaptic plasticity underpin learning, memory, and cognition (reviewed in [290–292]). In CBP (+/−) mice, global histone H2B acetylation was reduced, the late phase of LTP measured in the hippocampal CA3-CA1 pathway was significantly impaired compared to wild-type mice, and the mice demonstrated reduced long-term memory for fear and object recognition [288]. Treatment of acutely prepared brain slices with the HDAC inhibitor SAHA (**1**, Fig. 1) rescued the deficit in late phase-LTP back to wild-type levels and increased H2B acetylation. Intraventricular infusion of SAHA also significantly improved the deficits in contextual fear conditioning in these mice.

Aside from motor impairment, reduced cognition is the major debilitating symptom of many neurodegenerative disorders [293–296]. There are few drugs approved to treat this disease aspect, which are at best only partially effective [297]. The finding that HDAC inhibition may positively impact cognition may thus be applied as a favorable therapeutic strategy. Indeed, subsequent to Alarcon's findings in RTS models, the positive impact of HDAC inhibition in improving memory and reversing synaptic dysfunction has been reported by many different investigators (reviewed in [178]). Sodium butyrate (**20**, Fig. 8) was used in conjunction with a cognitive training paradigm to improve memory performance in brain-injured mice [298] and in the enhancement of long-term memory in a novel object recognition test in wild-type mice [299]. While Alarcon's and Levenson's results suggested that HDAC inhibition may affect global gene expression to modulate memory and synaptic plasticity through epigenetic means [287, 288], other investigators have shown that HDAC inhibitors enhance memory processes by the activation of selected key genes. Bredy and Barad demonstrated that conditioned fear in mice, an experimental model used to assess therapeutics for human anxiety disorders, could be improved using valproic acid (**21**, Fig. 8). Valproic acid was shown to act through a mechanism that depended on the epigenetic regulation of BDNF expression by enhanced histone acetylation at BDNF promoter regions [300, 301]. Other groups have demonstrated the key dependence of HDAC inhibition (using TSA **2** and sodium butyrate **20**) on the CREB:CBP transcriptional complex in mediating improvement in memory and enhancement in LTP [302, 303] or the association of NF-κB:p65:CBP complex in the amygdala by p65 acetylation in mediating enhancement of fear conditioning [304].

Deficits in both synaptic plasticity and memory formation have been reported for numerous mouse models of neurodegenerative diseases, especially for AD, PD, and HD models [305–314]. Despite this, there are few published reports directly assessing the role of HDAC inhibitors in treating this aspect of disease pathology. In the APP/PS1 double transgenic mouse model of AD, systemic injection of sodium butyrate (**20**, Fig. 8), valproate (**21**, Fig. 8), or SAHA (**1**, Fig. 1) completely reversed the contextual memory deficits in these mice, an action that was attributed to specifically blocking class I HDACs [315]. TSA (**2**, Fig. 1) and sodium butyrate (**20**) also rescued the cognitive deficits induced by Kainate administration and accelerated aging in SAMP-8 mice [316]. In an elegant study, Fischer and colleagues used the CK-p25 transgenic mouse model, which allows temporally and spatially restricted induction of neuronal loss through controlled p25 expression, to investigate the role of HDAC inhibition on cognitive impairment arising from neurodegeneration in this model. They demonstrated that long-term treatment with sodium butyrate (**20**) could reinstate learning and access to long-term memories in injured mice, a process which was accompanied by a concomitant increase in synaptic density [317]. The same group also showed that HDAC1 inactivation by p25 is part of the mechanism underlying the ability of p25 to elicit double-stranded DNA breaks that precede neurotoxicity [318].

To date, the published work in respect to the procognitive effects of HDAC inhibitors has relied on using sodium butyrate (**20**), valproate (**21**), SAHA (**1**), or TSA (**2**), which are not selective inhibitors; hence, there is little understanding of the specific HDAC isoforms that are the critical regulators of procognitive processes. Further exploration of this aspect will be critical for the rational design of compounds with maximum therapeutic benefit. EnVivo Pharmaceuticals reported the development of a small molecule, which is a CNS-penetrant and orally bioavailable HDAC inhibitor (EVP-0334) for the treatment of the cognitive deficits associated with neurological disorders. Successful completion of a phase I clinical trial has been reported in spring of 2010. To the best of our knowledge, neither the structure nor a clinical trajectory for EVP-0334 has been disclosed so far.

In the absence of truly isoform selective HDAC compounds, it is likely that genetic experiments to either overexpress or knockout particular HDAC brain isoforms will provide much needed information about the impact of individual HDAC isoforms. The first report on this approach clearly implicated HDAC2, but not HDAC1, as a major regulator. Neuron-specific overexpression of HDAC2 decreased dendritic spine density, synapse number and synaptic plasticity in the hippocampus, and impaired memory formation. Conversely, HDAC2 conditional knockout mice showed increased synapse number and memory facilitation [319]. Notably, reduced synapse number and learning impairment of HDAC2-overexpressing mice were ameliorated by chronic treatment with SAHA (**1**), while treatment with SAHA failed to further facilitate memory formation in HDAC2-deficient mice. These observations suggest that HDAC2 may be the major, if not exclusive, target of SAHA in enhancing hippocampal memory formation. These data encourage the development of HDAC2-selective inhibitors for human diseases associated with memory impairment [319].

Table 2 Isoform selectivity of representative HDAC inhibitors (μM)

Isoform	SAHA (**1**)	TSA (**2**)	APHA (**3**)	MS275 (**5**)	FK288 (**9**)	CF_3CO (**19**)	VPA (**21**)
HDAC1	0.0013	0.0002	0.055	0.022	0.0000015	4.8	700
HDAC2	0.0016	0.00065	0.125	0.065	0.000038	–	800
HDAC3	0.005	0.0005	0.25	0.36	0.00015	>1	1,000
HDAC4	>10	1.4	17.5	>10	0.0205	0.07	1,500
HDAC5	3.6	0.26	11.5	>10	0.55	–	1,000
HDAC6	0.0016	0.001	0.03	>10	0.0095	0.76	–
HDAC7	>10	0.195	7	>10	1.25	–	1,300
HDAC8	0.48	0.045	0.6	>10	0.00015	–	–
HDAC9	>10	0.8	10	>10	1.1	–	–

Data from [43, 72, 81]

SAHA (**1**) is a class I/IIb selective HDAC inhibitor, with little observable activity against the class IIa enzymes (Table 2), and thus the role of the class IIa enzymes in synaptic plasticity and memory has yet to be investigated. There is reason to believe that these enzymes are well poised to potentially affect these processes. As previously described, class IIa enzymes can shuttle between the cytoplasm and nuclear compartments of neurons. This process is governed by the phosphorylation-dependent binding of class IIa HDACs to 14-3-3 scaffolding proteins, resulting in their cytoplasmic sequestration [20]. The calcium–calmodulin-dependent kinase is one of the enzymes responsible for this phosphorylation [20, 320]. As calcium signaling is a major signal transducer at the synapse, the calcium dependence of nucleo-cytoplasmic shuttling of the class IIa enzymes provides a means for the activation-state of a neuron to drive transcriptional programs [320, 321]. Although investigation of this hypothesis in the CNS is limited at present, the effect of nuclear-cytoplasmic redistribution of HDAC4 on neuronal activity has been demonstrated in cultures [322].

By immunohistochemistry of mouse brain, the CNS distribution of HDAC4 has been shown to demonstrate a mixed nuclear and cytoplasmic location, which has been postulated to be a result of the activity state of the neuron at the time of killing. In addition, HDAC4 accumulated at the postsynaptic density of some synapses, placing it at the right location to sense and respond to synaptic calcium transients [323]. Adding to this hypothesis, it has recently been reported that the activity-dependent regulation of the transcription factor MEF2, a well-known target of HDAC4/5-mediated transcriptional repression, can influence synapse number, spine density, learning, and memory [324, 325]. Furthermore, regulation of transcriptional reprogramming by HDAC4 in response to activity at the neuromuscular junction has been demonstrated in detail [326–328], and it is plausible that similar parallels exist in the CNS. To this note, it has recently been documented that a skeletal, muscle-specific microRNA (miR-206) is dramatically induced in a mouse model of ALS; miR-206 induction delayed ALS progression and promoted the regeneration of neuromuscular synapses following acute nerve injury [329]. HDAC4 mRNA is among the strongest computationally predicted targets of miR-206 and was shown to repress HDAC4 translation. Thus, inhibition of peripheral (muscular) HDAC4 may offer an attractive strategy for the treatment of ALS.

6.3 *Antidepressant Effects of HDAC Inhibition*

Depressive symptoms often accompany neurodegenerative disorders, especially in PD and HD. Although the "monoamine hypothesis" of depression has long been proposed, the pathologies and mechanisms for depressive disorders remain only partially understood. A number of proposed mechanisms for depression such as diminishing neurotrophic factors and neuroinflammation appear to be similar to those implicated in neurodegenerative diseases [330, 331]. Many patients suffering from these disorders are treated with conventional antidepressants.

In recent years, a role for chromatin remodeling in the treatment of depression has been proposed; chronic exposure to antidepressant drugs alters histone methylation and acetylation in specific brain regions [332, 333]. Interestingly, HDAC5 inhibition appears to impact antidepressant actions [332, 334]. Tsankova and colleagues showed that the antidepressant imipramine reversed BDNF downregulation and increased histone acetylation of BDNF promoters in a model of chronic social defeat stress. This action was correlated with a selective downregulation of hippocampal HDAC5. Conversely, virally mediated overexpression of HDAC5 blocked the antidepressive effect of imipramine. Furthermore, class I HDAC inhibitors have been shown to have antidepressant-like effects. Direct injection of either SAHA (**1**, Fig. 1) or MS275 (**5**, Fig. 3) into the nucleus accumbens of mice resulted in robust antidepressant-like effects in the same chronic social defeat stress paradigm. This was accompanied by increased histone acetylation and a change of global patterns of gene expression [335].

In peripheral leukocytes of patients with major depression, HDAC5 mRNA is elevated, which is decreased after an 8-week treatment with paroxetine, a selective serotonin uptake inhibitor [336]. In peripheral white blood cells from patients suffering from major depressive disorder (MDD), other groups have found an increase in both HDAC2 and HDAC5 mRNA, with HDAC4 elevated in bipolar disorder (BPD) [337]. In this study, a striking correlation of a reduction in the mRNA of HDAC2 and the class II HDACs (4,5, 7, and 9) was observed in MDD and BPD patients when in remission, although normalization of HDAC levels was not achieved upon treatment with standard antidepressants. These data suggest that aberrant transcriptional regulation caused by the altered expression of HDACs is associated with the pathology of mood disorders. Consequently, specific inhibitors of HDAC2 and/or class IIa enzymes may provide a new therapeutic avenue [337].

7 Conclusion

The understanding of the biological function and role of individual HDAC isoforms in the context of CNS disorders is still in its infancy, but recent work has clearly shed light on the relevance of this protein class. A number of isoform and class selective inhibitors of HDACs/sirtuins are now available, which will serve as tools to interrogate the function of these enzymes in a chemo-genomic fashion. The development of potent, selective inhibitors with suitable CNS permeability and stability now appears an attainable goal. Another avenue of research may aim at developing molecules that interrupt protein–protein interactions, especially as it becomes more and more apparent that HDACs exhibit many functions such as recruiting other proteins and are highly regulated through the interaction with other partners (chaperones, kinases, phosphatases, ubiquitin ligases, etc.).

Clearly, the clinical toxicity and side effects of the first generation of HDAC inhibitors such as SAHA will need to be thoroughly evaluated before embarkation on a chronic long-term dosing strategy.

Nevertheless, HDAC inhibition is a propitious avenue to pursue the clinical treatment of neurodegenerative disorders: it is already apparent that their realm of influence in treating neurodegenerative disorders is far more reaching than the original conception of reversing transcriptional dysregulation.

References

1. Choudhary C et al (2009) Lysine acetylation targets protein complexes and co-regulates major cellular functions. Science 325(5942):834–840
2. Paroni G et al (2004) Caspase-dependent regulation of histone deacetylase 4 nuclear-cytoplasmic shuttling promotes apoptosis. Mol Biol Cell 15(6):2804–2818
3. McKinsey TA et al (2000) Signal-dependent nuclear export of a histone deacetylase regulates muscle differentiation. Nature 408(6808):106–111
4. Glaser KB et al (2003) Gene expression profiling of multiple histone deacetylase (HDAC) inhibitors: defining a common gene set produced by HDAC inhibition in T24 and MDA carcinoma cell lines. Mol Cancer Ther 2(2):151–163
5. Fischer DD et al (2002) Isolation and characterization of a novel class II histone deacetylase, HDAC10. J Biol Chem 277(8):6656–6666
6. Haberland M, Montgomery RL, Olson EN (2009) The many roles of histone deacetylases in development and physiology: implications for disease and therapy. Nat Rev Genet 10(1):32–42
7. Lahm A et al (2007) Unraveling the hidden catalytic activity of vertebrate class IIa histone deacetylases. Proc Natl Acad Sci USA 104(44):17335–17340
8. de Ruijter AJ et al (2003) Histone deacetylases (HDACs): characterization of the classical HDAC family. Biochem J 370(Pt 3):737–749
9. Verdin E, Dequiedt F, Kasler HG (2003) Class II histone deacetylases: versatile regulators. Trends Genet 19(5):286–293
10. Fischle W et al (2002) Enzymatic activity associated with class II HDACs is dependent on a multiprotein complex containing HDAC3 and SMRT/N-CoR. Mol Cell 9(1):45–57
11. Fischle W et al (1999) A new family of human histone deacetylases related to Saccharomyces cerevisiae HDA1p. J Biol Chem 274(17):11713–11720
12. Zhang CL, McKinsey TA, Olson EN (2001) The transcriptional corepressor MITR is a signal-responsive inhibitor of myogenesis. Proc Natl Acad Sci USA 98(13):7354–7359
13. Zhang CL et al (2001) Association of COOH-terminal-binding protein (CtBP) and MEF2-interacting transcription repressor (MITR) contributes to transcriptional repression of the MEF2 transcription factor. J Biol Chem 276(1):35–39
14. Zhao X et al (2005) Regulation of MEF2 by histone deacetylase 4- and SIRT1 deacetylase-mediated lysine modifications. Mol Cell Biol 25(19):8456–8464

15. Nebbioso A et al (2009) Selective class II HDAC inhibitors impair myogenesis by modulating the stability and activity of HDAC-MEF2 complexes. EMBO Rep 10(7):776–782
16. Matsuoka H et al (2007) Disruption of HDAC4/N-CoR complex by histone deacetylase inhibitors leads to inhibition of IL-2 gene expression. Biochem Pharmacol 74(3):465–476
17. Huang EY et al (2000) Nuclear receptor corepressors partner with class II histone deacetylases in a Sin3-independent repression pathway. Genes Dev 14(1):45–54
18. Rajan I et al (2009) Loss of the putative catalytic domain of HDAC4 leads to reduced thermal nociception and seizures while allowing normal bone development. PLoS ONE 4(8):e6612
19. Vega RB et al (2004) Histone deacetylase 4 controls chondrocyte hypertrophy during skeletogenesis. Cell 119(4):555–566
20. Grozinger CM, Schreiber SL (2000) Regulation of histone deacetylase 4 and 5 and transcriptional activity by 14-3-3-dependent cellular localization. Proc Natl Acad Sci USA 97 (14):7835–7840
21. Wang AH et al (2000) Regulation of histone deacetylase 4 by binding of 14-3-3 proteins. Mol Cell Biol 20(18):6904–6912
22. Broide RS et al (2007) Distribution of histone deacetylases 1-11 in the rat brain. J Mol Neurosci 31(1):47–58
23. Kao HY et al (2002) Isolation and characterization of mammalian HDAC10, a novel histone deacetylase. J Biol Chem 277(1):187–193
24. Yang XJ, Gregoire S (2005) Class II histone deacetylases: from sequence to function, regulation, and clinical implication. Mol Cell Biol 25(8):2873–2884
25. Hubbert C et al (2002) HDAC6 is a microtubule-associated deacetylase. Nature 417 (6887):455–458
26. Bali P et al (2005) Inhibition of histone deacetylase 6 acetylates and disrupts the chaperone function of heat shock protein 90: a novel basis for antileukemia activity of histone deacetylase inhibitors. J Biol Chem 280(29):26729–26734
27. Kovacs JJ et al (2005) HDAC6 regulates Hsp90 acetylation and chaperone-dependent activation of glucocorticoid receptor. Mol Cell 18(5):601–607
28. Gao L et al (2002) Cloning and functional characterization of HDAC11, a novel member of the human histone deacetylase family. J Biol Chem 277(28):25748–25755
29. Boyault C et al (2006) HDAC6-p97/VCP controlled polyubiquitin chain turnover. EMBO J 25(14):3357–3366
30. Gao J et al (2009) Inactivation of CREB mediated gene transcription by HDAC8 bound protein phosphatase. Biochem Biophys Res Commun 379(1):1–5
31. Bertrand P (2010) Inside HDAC with HDAC inhibitors. Eur J Med Chem 45(6):2095–2116
32. Glozak MA, Seto E (2009) Acetylation/deacetylation modulates the stability of DNA replication licensing factor Cdt1. J Biol Chem 284(17):11446–11453
33. Gantt SL, Gattis SG, Fierke CA (2006) Catalytic activity and inhibition of human histone deacetylase 8 is dependent on the identity of the active site metal ion. Biochemistry 45 (19):6170–6178
34. Bressi JC et al (2010) Exploration of the HDAC2 foot pocket: synthesis and SAR of substituted N-(2-aminophenyl)benzamides. Bioorg Med Chem Lett 20(10):3142–3145
35. Somoza JR et al (2004) Structural snapshots of human HDAC8 provide insights into the class I histone deacetylases. Structure 12(7):1325–1334
36. Vannini A et al (2004) Crystal structure of a eukaryotic zinc-dependent histone deacetylase, human HDAC8, complexed with a hydroxamic acid inhibitor. Proc Natl Acad Sci USA 101 (42):15064–15069
37. Bottomley MJ et al (2008) Structural and functional analysis of the human HDAC4 catalytic domain reveals a regulatory structural zinc-binding domain. J Biol Chem 283(39): 26694–26704
38. Schuetz A et al (2008) Human HDAC7 harbors a class IIa histone deacetylase-specific zinc binding motif and cryptic deacetylase activity. J Biol Chem 283(17):11355–11363

39. Finnin MS et al (1999) Structures of a histone deacetylase homologue bound to the TSA and SAHA inhibitors. Nature 401(6749):188–193
40. Nielsen TK et al (2005) Crystal structure of a bacterial class 2 histone deacetylase homologue. J Mol Biol 354(1):107–120
41. Nielsen TK et al (2007) Complex structure of a bacterial class 2 histone deacetylase homologue with a trifluoromethylketone inhibitor. Acta Crystallogr F Struct Biol Cryst Commun 63(Pt 4):270–273
42. Guo L et al (2007) Crystal structure of a conserved N-terminal domain of histone deacetylase 4 reveals functional insights into glutamine-rich domains. Proc Natl Acad Sci USA 104(11):4297–4302
43. Itoh Y, Suzuki T, Miyata N (2008) Isoform-selective histone deacetylase inhibitors. Curr Pharm Des 14(6):529–544
44. Hitchcock SA, Pennington LD (2006) Structure-brain exposure relationships. J Med Chem 49(26):7559–7583
45. Klon AE (2009) Computational models for central nervous system penetration. Curr Comput Aided Drug Des 5(2):71–89
46. Carvey PM, Hendey B, Monahan AJ (2009) The blood-brain barrier in neurodegenerative disease: a rhetorical perspective. J Neurochem 111(2):291–314
47. Wang H, Dymock BW (2009) New patented histone deacetylase inhibitors. Expert Opin Ther Pat 19(12):1727–1757
48. Grozinger CM, Schreiber SL (2002) Deacetylase enzymes: biological functions and the use of small-molecule inhibitors. Chem Biol 9(1):3–16
49. Musso DL et al (2001) N-hydroxyformamide peptidomimetics as TACE/matrix metalloprotease inhibitors: oral activity via P1' isobutyl substitution. Bioorg Med Chem Lett 11 (16):2147–2151
50. Kattar SD et al (2009) Parallel medicinal chemistry approaches to selective HDAC1/HDAC2 inhibitor (SHI-1:2) optimization. Bioorg Med Chem Lett 19(4):1168–1172
51. Mai A et al (2005) Class II (IIa)-selective histone deacetylase inhibitors. 1. Synthesis and biological evaluation of novel (aryloxopropenyl)pyrrolyl hydroxyamides. J Med Chem 48(9):3344–3353
52. Cai X et al (2010) Discovery of 7-(4-(3-ethynylphenylamino)-7-methoxyquinazolin-6-yloxy)-N-hydroxyheptanam ide (CUDc-101) as a potent multi-acting HDAC, EGFR, and HER2 inhibitor for the treatment of cancer. J Med Chem 53(5):2000–2009
53. Hamblett CL et al (2007) The discovery of 6-amino nicotinamides as potent and selective histone deacetylase inhibitors. Bioorg Med Chem Lett 17(19):5300–5309
54. Suzuki T et al (1999) Synthesis and histone deacetylase inhibitory activity of new benzamide derivatives. J Med Chem 42(15):3001–3003
55. Ryan QC et al (2005) Phase I and pharmacokinetic study of MS-275, a histone deacetylase inhibitor, in patients with advanced and refractory solid tumors or lymphoma. J Clin Oncol 23(17):3912–3922
56. Chou CJ, Herman D, Gottesfeld JM (2008) Pimelic diphenylamide 106 is a slow, tight-binding inhibitor of class I histone deacetylases. J Biol Chem 283(51):35402–35409
57. Xu C et al (2009) Chemical probes identify a role for histone deacetylase 3 in Friedreich's ataxia gene silencing. Chem Biol 16(9):980–989
58. Ueda H et al (1994) FR901228, a novel antitumor bicyclic depsipeptide produced by Chromobacterium violaceum No. 968. I. Taxonomy, fermentation, isolation, physico-chemical and biological properties, and antitumor activity. J Antibiot (Tokyo) 47(3):301–310
59. Campas-Moya C (2009) Romidepsin for the treatment of cutaneous T-cell lymphoma. Drugs Today (Barc) 45(11):787–795
60. Taori K, Paul VJ, Luesch H (2008) Structure and activity of largazole, a potent antiproliferative agent from the Floridian marine cyanobacterium Symploca sp. J Am Chem Soc 130(6):1806–1807
61. Walton JD (2006) HC-toxin. Phytochemistry 67(14):1406–1413

62. Shute RE, Dunlap B, Rich DH (1987) Analogues of the cytostatic and antimitogenic agents chlamydocin and HC-toxin: synthesis and biological activity of chloromethyl ketone and diazomethyl ketone functionalized cyclic tetrapeptides. J Med Chem 30(1):71–78
63. Darkin-Rattray SJ et al (1996) Apicidin: a novel antiprotozoal agent that inhibits parasite histone deacetylase. Proc Natl Acad Sci USA 93(23):13143–13147
64. Mori H et al (2003) FR235222, a fungal metabolite, is a novel immunosuppressant that inhibits mammalian histone deacetylase (HDAC). I. Taxonomy, fermentation, isolation and biological activities. J Antibiot (Tokyo) 56(2):72–79
65. Brasnjevic I et al (2009) Delivery of peptide and protein drugs over the blood-brain barrier. Prog Neurobiol 87(4):212–251
66. Montero A et al (2009) Design, synthesis, biological evaluation, and structural characterization of potent histone deacetylase inhibitors based on cyclic alpha/beta-tetrapeptide architectures. J Am Chem Soc 131(8):3033–3041
67. Olsen CA, Ghadiri MR (2009) Discovery of potent and selective histone deacetylase inhibitors via focused combinatorial libraries of cyclic α3β-tetrapeptides. J Med Chem 52 (23):7836–7846
68. Jones P et al (2006) A series of novel, potent, and selective histone deacetylase inhibitors. Bioorg Med Chem Lett 16(23):5948–5952
69. Jones P et al (2008) A novel series of potent and selective ketone histone deacetylase inhibitors with antitumor activity *in vivo*. J Med Chem 51(8):2350–2353
70. Kinzel O et al (2009) Discovery of a potent class I selective ketone histone deacetylase inhibitor with antitumor activity *in vivo* and optimized pharmacokinetic properties. J Med Chem 52(11):3453–3456
71. Frey RR et al (2002) Trifluoromethyl ketones as inhibitors of histone deacetylase. Bioorg Med Chem Lett 12(23):3443–3447
72. Muraglia E et al (2008) 2-Trifluoroacetylthiophene oxadiazoles as potent and selective class II human histone deacetylase inhibitors. Bioorg Med Chem Lett 18(23):6083–6087
73. Scarpelli R et al (2008) Studies of the metabolic stability in cells of 5-(trifluoroacetyl) thiophene-2-carboxamides and identification of more stable class II histone deacetylase (HDAC) inhibitors. Bioorg Med Chem Lett 18(23):6078–6082
74. Ontoria JM et al (2009) Identification of novel, selective, and stable inhibitors of class II histone deacetylases. Validation studies of the inhibition of the enzymatic activity of HDAC4 by small molecules as a novel approach for cancer therapy. J Med Chem 52(21): 6782–6789
75. Chen JS, Faller DV, Spanjaard RA (2003) Short-chain fatty acid inhibitors of histone deacetylases: promising anticancer therapeutics? Curr Cancer Drug Targets 3(3):219–236
76. Riggs MG et al (1977) n-Butyrate causes histone modification in HeLa and Friend erythroleukaemia cells. Nature 268(5619):462–464
77. Blank-Porat D et al (2007) The anticancer prodrugs of butyric acid AN-7 and AN-9, possess antiangiogenic properties. Cancer Lett 256(1):39–48
78. Kothari V et al (2010) HDAC inhibitor valproic acid enhances tumor cell kill in adenovirus-HSVtk mediated suicide gene therapy in HNSCC xenograft mouse model. Int J Cancer 126 (3):733–742
79. Ryningen A, Stapnes C, Bruserud O (2007) Clonogenic acute myelogenous leukemia cells are heterogeneous with regard to regulation of differentiation and effect of epigenetic pharmacological targeting. Leuk Res 31(9):1303–1313
80. Yang PM et al (2010) Inhibition of histone deacetylase activity is a novel function of the antifolate drug methotrexate. Biochem Biophys Res Commun 391(3):1396–1399
81. Bradner JE et al (2010) Chemical phylogenetics of histone deacetylases. Nat Chem Biol 6(3):238–243
82. Riester D et al (2004) Members of the histone deacetylase superfamily differ in substrate specificity towards small synthetic substrates. Biochem Biophys Res Commun 324(3): 1116–1123

83. Su H, Altucci L, You Q (2008) Competitive or noncompetitive, that's the question: research toward histone deacetylase inhibitors. Mol Cancer Ther 7(5):1007–1012
84. Anastasiou D, Krek W (2006) SIRT1: linking adaptive cellular responses to aging-associated changes in organismal physiology. Physiology (Bethesda) 21:404–410
85. North BJ, Verdin E (2007) Interphase nucleo-cytoplasmic shuttling and localization of SIRT2 during mitosis. PLoS ONE 2(8):e784
86. Werner HB et al (2007) Proteolipid protein is required for transport of sirtuin 2 into CNS myelin. J Neurosci 27(29):7717–7730
87. Li W et al (2007) Sirtuin 2, a mammalian homolog of yeast silent information regulator-2 longevity regulator, is an oligodendroglial protein that decelerates cell differentiation through deacetylating alpha-tubulin. J Neurosci 27(10):2606–2616
88. Ahuja N et al (2007) Regulation of insulin secretion by SIRT4, a mitochondrial ADP-ribosyltransferase. J Biol Chem 282(46):33583–33592
89. Jin D et al (2009) Molecular cloning and characterization of porcine sirtuin genes. Comp Biochem Physiol B Biochem Mol Biol 153(4):348–358
90. Michishita E et al (2005) Evolutionarily conserved and nonconserved cellular localizations and functions of human SIRT proteins. Mol Biol Cell 16(10):4623–4635
91. Lin SJ, Defossez PA, Guarente L (2000) Requirement of NAD and SIR2 for life-span extension by calorie restriction in Saccharomyces cerevisiae. Science 289(5487):2126–2128
92. Rogina B, Helfand SL (2004) Sir2 mediates longevity in the fly through a pathway related to calorie restriction. Proc Natl Acad Sci USA 101(45):15998–16003
93. Tissenbaum HA, Guarente L (2001) Increased dosage of a sir-2 gene extends lifespan in Caenorhabditis elegans. Nature 410(6825):227–230
94. Cohen HY et al (2004) Calorie restriction promotes mammalian cell survival by inducing the SIRT1 deacetylase. Science 305(5682):390–392
95. Tang BL, Chua CE (2008) SIRT1 and neuronal diseases. Mol Aspects Med 29(3):187–200
96. Luo J et al (2001) Negative control of p53 by Sir2alpha promotes cell survival under stress. Cell 107(2):137–148
97. Smith BC, Denu JM (2006) Sirtuins caught in the act. Structure 14(8):1207–1208
98. Finnin MS, Donigian JR, Pavletich NP (2001) Structure of the histone deacetylase SIRT2. Nat Struct Biol 8(7):621–625
99. Min J et al (2001) Crystal structure of a SIR2 homolog-NAD complex. Cell 105(2):269–279
100. Chang JH et al (2002) Structural basis for the NAD-dependent deacetylase mechanism of Sir2. J Biol Chem 277(37):34489–34498
101. Avalos JL et al (2002) Structure of a Sir2 enzyme bound to an acetylated p53 peptide. Mol Cell 10(3):523–535
102. Zhao K et al (2003) Structure and autoregulation of the yeast Hst2 homolog of Sir2. Nat Struct Biol 10(10):864–871
103. Zhao K, Chai X, Marmorstein R (2003) Structure of the yeast Hst2 protein deacetylase in ternary complex with 2'-O-acetyl ADP ribose and histone peptide. Structure 11(11): 1403–1411
104. Zhao K, Chai X, Marmorstein R (2004) Structure and substrate binding properties of cobB, a Sir2 homolog protein deacetylase from Escherichia coli. J Mol Biol 337(3):731–741
105. Zhao K et al (2004) Structural basis for nicotinamide cleavage and ADP-ribose transfer by NAD(+)-dependent Sir2 histone/protein deacetylases. Proc Natl Acad Sci USA 101(23): 8563–8568
106. Avalos JL, Bever KM, Wolberger C (2005) Mechanism of sirtuin inhibition by nicotinamide: altering the NAD(+) cosubstrate specificity of a Sir2 enzyme. Mol Cell 17(6):855–868
107. Avalos JL, Boeke JD, Wolberger C (2004) Structural basis for the mechanism and regulation of Sir2 enzymes. Mol Cell 13(5):639–648
108. Cosgrove MS et al (2006) The structural basis of sirtuin substrate affinity. Biochemistry 45(24):7511–7521

109. Hoff KG et al (2006) Insights into the sirtuin mechanism from ternary complexes containing NAD^+ and acetylated peptide. Structure 14(8):1231–1240
110. Schuetz A et al (2007) Structural basis of inhibition of the human NAD^+ -dependent deacetylase SIRT5 by suramin. Structure 15(3):377–389
111. Sauve AA et al (2001) Chemistry of gene silencing: the mechanism of NAD^+ -dependent deacetylation reactions. Biochemistry 40(51):15456–15463
112. Hawse WF et al (2008) Structural insights into intermediate steps in the Sir2 deacetylation reaction. Structure 16(9):1368–1377
113. Smith BC, Denu JM (2006) Sir2 protein deacetylases: evidence for chemical intermediates and functions of a conserved histidine. Biochemistry 45(1):272–282
114. Sauve AA et al (2006) The biochemistry of sirtuins. Annu Rev Biochem 75:435–465
115. Yamamoto H, Schoonjans K, Auwerx J (2007) Sirtuin functions in health and disease. Mol Endocrinol 21(8):1745–1755
116. Grozinger CM et al (2001) Identification of a class of small molecule inhibitors of the sirtuin family of NAD-dependent deacetylases by phenotypic screening. J Biol Chem 276(42): 38837–38843
117. Heltweg B et al (2006) Antitumor activity of a small-molecule inhibitor of human silent information regulator 2 enzymes. Cancer Res 66(8):4368–4377
118. Lain S et al (2008) Discovery, *in vivo* activity, and mechanism of action of a small-molecule p53 activator. Cancer Cell 13(5):454–463
119. Tervo AJ et al (2006) Discovering inhibitors of human sirtuin type 2: novel structural scaffolds. J Med Chem 49(24):7239–7241
120. Sanders BD et al (2009) Identification and characterization of novel sirtuin inhibitor scaffolds. Bioorg Med Chem 17(19):7031–7041
121. Huber K et al (2010) Novel 3-arylideneindolin-2-ones as inhibitors of NAD + -dependent histone deacetylases (sirtuins). J Med Chem 53(3):1383–1386
122. Napper AD et al (2005) Discovery of indoles as potent and selective inhibitors of the deacetylase SIRT1. J Med Chem 48(25):8045–8054
123. Pallos J et al (2008) Inhibition of specific HDACs and sirtuins suppresses pathogenesis in a Drosophila model of Huntington's disease. Hum Mol Genet 17(23):3767–3775
124. Ghosh S, Feany MB (2004) Comparison of pathways controlling toxicity in the eye and brain in Drosophila models of human neurodegenerative diseases. Hum Mol Genet 13(18):2011–2018
125. Parker JA et al (2005) Resveratrol rescues mutant polyglutamine cytotoxicity in nematode and mammalian neurons. Nat Genet 37(4):349–350
126. Kumar P et al (2006) Effect of resveratrol on 3-nitropropionic acid-induced biochemical and behavioural changes: possible neuroprotective mechanisms. Behav Pharmacol 17(5–6):485–492
127. Milne JC et al (2007) Small molecule activators of SIRT1 as therapeutics for the treatment of type 2 diabetes. Nature 450(7170):712–716
128. Vu CB et al (2009) Discovery of Imidazo[1,2-b]thiazole Derivatives as Novel SIRT1 Activators. J Med Chem 52(5):1275–1283
129. Pacholec M et al (2010) SRT1720, SRT2183, SRT1460, and resveratrol are not direct activators of SIRT1. J Biol Chem 285(11):8340–8351
130. Bemis JE et al (2009) Discovery of oxazolo[4, 5-b]pyridines and related heterocyclic analogs as novel SIRT1 activators. Bioorg Med Chem Lett 19(8):2350–2353
131. Nayagam VM et al (2006) SIRT1 modulating compounds from high-throughput screening as anti-inflammatory and insulin-sensitizing agents. J Biomol Screen 11(8):959–967
132. Mai A et al (2009) Study of 1, 4-dihydropyridine structural scaffold: discovery of novel sirtuin activators and inhibitors. J Med Chem 52(17):5496–5504
133. Berger SL (2007) The complex language of chromatin regulation during transcription. Nature 447(7143):407–412

134. Horn PJ, Peterson CL (2002) Molecular biology. Chromatin higher order folding – wrapping up transcription. Science 297(5588):1824–1827
135. Keppler BR, Archer TK (2008) Chromatin-modifying enzymes as therapeutic targets–Part 1. Expert Opin Ther Targets 12(10):1301–1312
136. Colangelo V et al (2002) Gene expression profiling of 12633 genes in Alzheimer hippocampal CA1: transcription and neurotrophic factor down-regulation and up-regulation of apoptotic and pro-inflammatory signaling. J Neurosci Res 70(3):462–473
137. Kitamura Y et al (1997) Alteration of transcription factors NF-κB and STAT1 in Alzheimer's disease brains. Neurosci Lett 237(1):17–20
138. Bossers K et al (2009) Analysis of gene expression in Parkinson's disease: possible involvement of neurotrophic support and axon guidance in dopaminergic cell death. Brain Pathol 19 (1):91–107
139. Sutherland GT et al (2009) A cross-study transcriptional analysis of Parkinson's disease. PLoS ONE 4(3):e4955
140. Mandel S et al (2005) Gene expression profiling of sporadic Parkinson's disease substantia nigra pars compacta reveals impairment of ubiquitin-proteasome subunits, SKP1A, aldehyde dehydrogenase, and chaperone HSC-70. Ann NY Acad Sci 1053:356–375
141. Vogt IR et al (2006) Transcriptional changes in multiple system atrophy and Parkinson's disease putamen. Exp Neurol 199(2):465–478
142. Iivonen S et al (2002) Seladin-1 transcription is linked to neuronal degeneration in Alzheimer's disease. Neuroscience 113(2):301–310
143. Yacoubian TA et al (2008) Transcriptional dysregulation in a transgenic model of Parkinson disease. Neurobiol Dis 29(3):515–528
144. Robakis NK (2003) An Alzheimer's disease hypothesis based on transcriptional dysregulation. Amyloid 10(2):80–85
145. Hodges A et al (2006) Regional and cellular gene expression changes in human Huntington's disease brain. Hum Mol Genet 15(6):965–977
146. Strand AD et al (2005) Gene expression in Huntington's disease skeletal muscle: a potential biomarker. Hum Mol Genet 14(13):1863–1876
147. Becanovic K et al (2010) Transcriptional changes in Huntington disease identified using genome-wide expression profiling and cross-platform analysis. Hum Mol Genet 19(8): 1438–1452
148. Mazarei G et al (2010) Expression analysis of novel striatal-enriched genes in Huntington disease. Hum Mol Genet 19(4):609–622
149. Kuhn A et al (2007) Mutant huntingtin's effects on striatal gene expression in mice recapitulate changes observed in human Huntington's disease brain and do not differ with mutant huntingtin length or wild-type huntingtin dosage. Hum Mol Genet 16(15):1845–1861
150. Luthi-Carter R et al (2000) Decreased expression of striatal signaling genes in a mouse model of Huntington's disease. Hum Mol Genet 9(9):1259–1271
151. Chiang MC et al (2007) Systematic uncovering of multiple pathways underlying the pathology of Huntington disease by an acid-cleavable isotope-coded affinity tag approach. Mol Cell Proteomics 6(5):781–797
152. Luthi-Carter R et al (2002) Polyglutamine and transcription: gene expression changes shared by DRPLA and Huntington's disease mouse models reveal context-independent effects. Hum Mol Genet 11(17):1927–1937
153. Huen NY, Wong SL, Chan HY (2007) Transcriptional malfunctioning of heat shock protein gene expression in spinocerebellar ataxias. Cerebellum 6(2):111–117
154. Fuchs J et al (2009) The transcription factor PITX3 is associated with sporadic Parkinson's disease. Neurobiol Aging 30(5):731–738
155. Le W et al (2009) Transcription factor PITX3 gene in Parkinson's disease. Neurobiol Aging [Epub ahead of print]
156. Li J, Dani JA, Le W (2009) The role of transcription factor Pitx3 in dopamine neuron development and Parkinson's disease. Curr Top Med Chem 9(10):855–859

157. Jacobsen KX et al (2008) A Nurr1 point mutant, implicated in Parkinson's disease, uncouples ERK1/2-dependent regulation of tyrosine hydroxylase transcription. Neurobiol Dis 29(1):117–122
158. Xu J et al (2005) The Parkinson's disease-associated DJ-1 protein is a transcriptional co-activator that protects against neuronal apoptosis. Hum Mol Genet 14(9):1231–1241
159. da Costa CA, Checler F (2010) A novel parkin-mediated transcriptional function links p53 to familial Parkinson's disease. Cell Cycle 9(1):16–17
160. Jeong SJ et al (2005) Activated AKT regulates NF-kappaB activation, p53 inhibition and cell survival in HTLV-1-transformed cells. Oncogene 24(44):6719–6728
161. da Costa CA et al (2009) Transcriptional repression of p53 by parkin and impairment by mutations associated with autosomal recessive juvenile Parkinson's disease. Nat Cell Biol 11(11):1370–1375
162. Ryan AB, Zeitlin SO, Scrable H (2006) Genetic interaction between expanded murine Hdh alleles and p53 reveal deleterious effects of p53 on Huntington's disease pathogenesis. Neurobiol Dis 24(2):419–427
163. Seong IS et al (2010) Huntingtin facilitates polycomb repressive complex 2. Hum Mol Genet 19(4):573–583
164. Hughes RE (2002) Polyglutamine disease: acetyltransferases awry. Curr Biol 12(4):R141–R143
165. McCampbell A, Fischbeck KH (2001) Polyglutamine and CBP: fatal attraction? Nat Med 7(5):528–530
166. Kwok RP et al (1994) Nuclear protein CBP is a coactivator for the transcription factor CREB. Nature 370(6486):223–226
167. Janknecht R (2002) The versatile functions of the transcriptional coactivators p300 and CBP and their roles in disease. Histol Histopathol 17(2):657–668
168. McManus KJ, Hendzel MJ (2001) CBP, a transcriptional coactivator and acetyltransferase. Biochem Cell Biol 79(3):253–266
169. Rouaux C, Loeffler JP, Boutillier AL (2004) Targeting CREB-binding protein (CBP) loss of function as a therapeutic strategy in neurological disorders. Biochem Pharmacol 68 (6):1157–1164
170. Lin CH et al (2001) A small domain of CBP/p300 binds diverse proteins: solution structure and functional studies. Mol Cell 8(3):581–590
171. Perutz MF et al (2002) Aggregation of proteins with expanded glutamine and alanine repeats of the glutamine-rich and asparagine-rich domains of Sup35 and of the amyloid beta-peptide of amyloid plaques. Proc Natl Acad Sci USA 99(8):5596–5600
172. Fu L, Gao YS, Sztul E (2005) Transcriptional repression and cell death induced by nuclear aggregates of non-polyglutamine protein. Neurobiol Dis 20(3):656–665
173. McCampbell A et al (2000) CREB-binding protein sequestration by expanded polyglutamine. Hum Mol Genet 9(14):2197–2202
174. Nucifora FC Jr et al (2001) Interference by huntingtin and atrophin-1 with cbp-mediated transcription leading to cellular toxicity. Science 291(5512):2423–2428
175. Jiang H et al (2006) Depletion of CBP is directly linked with cellular toxicity caused by mutant huntingtin. Neurobiol Dis 23(3):543–551
176. Li F et al (2002) Ataxin-3 is a histone-binding protein with two independent transcriptional corepressor activities. J Biol Chem 277(47):45004–45012
177. Steffan JS et al (2001) Histone deacetylase inhibitors arrest polyglutamine-dependent neurodegeneration in Drosophila. Nature 413(6857):739–743
178. Barrett RM, Wood MA (2008) Beyond transcription factors: the role of chromatin modifying enzymes in regulating transcription required for memory. Learn Mem 15(7):460–467
179. Hardy S et al (2002) TATA-binding protein-free TAF-containing complex (TFTC) and p300 are both required for efficient transcriptional activation. J Biol Chem 277(36): 32875–32882
180. Helmlinger D et al (2004) Ataxin-7 is a subunit of GCN5 histone acetyltransferase-containing complexes. Hum Mol Genet 13(12):1257–1265

181. Friedman MJ et al (2008) Polyglutamine expansion reduces the association of TATA-binding protein with DNA and induces DNA binding-independent neurotoxicity. J Biol Chem 283(13):8283–8290
182. Friedman MJ et al (2007) Polyglutamine domain modulates the TBP-TFIIB interaction: implications for its normal function and neurodegeneration. Nat Neurosci 10(12):1519–1528
183. Schaffar G et al (2004) Cellular toxicity of polyglutamine expansion proteins: mechanism of transcription factor deactivation. Mol Cell 15(1):95–105
184. Palhan VB et al (2005) Polyglutamine-expanded ataxin-7 inhibits STAGA histone acetyltransferase activity to produce retinal degeneration. Proc Natl Acad Sci USA 102(24): 8472–8477
185. Helmlinger D et al (2006) Both normal and polyglutamine- expanded ataxin-7 are components of TFTC-type GCN5 histone acetyltransferase-containing complexes. Biochem Soc Symp 73:155–163
186. Malaspina A, Kaushik N, de Belleroche J (2001) Differential expression of 14 genes in amyotrophic lateral sclerosis spinal cord detected using gridded cDNA arrays. J Neurochem 77(1):132–145
187. Ishigaki S et al (2002) Differentially expressed genes in sporadic amyotrophic lateral sclerosis spinal cords–screening by molecular indexing and subsequent cDNA microarray analysis. FEBS Lett 531(2):354–358
188. Yoshihara T et al (2002) Differential expression of inflammation- and apoptosis-related genes in spinal cords of a mutant SOD1 transgenic mouse model of familial amyotrophic lateral sclerosis. J Neurochem 80(1):158–167
189. McCampbell A et al (2001) Histone deacetylase inhibitors reduce polyglutamine toxicity. Proc Natl Acad Sci USA 98(26):15179–15184
190. Sadri-Vakili G et al (2007) Histones associated with downregulated genes are hypo-acetylated in Huntington's disease models. Hum Mol Genet 16(11):1293–1306
191. Bates EA et al (2006) Differential contributions of Caenorhabditis elegans histone deacetylases to huntingtin polyglutamine toxicity. J Neurosci 26(10):2830–2838
192. Taylor JP et al (2003) Aberrant histone acetylation, altered transcription, and retinal degeneration in a Drosophila model of polyglutamine disease are rescued by CREB-binding protein. Genes Dev 17(12):1463–1468
193. Ferrante RJ et al (2003) Histone deacetylase inhibition by sodium butyrate chemotherapy ameliorates the neurodegenerative phenotype in Huntington's disease mice. J Neurosci 23(28):9418–9427
194. Van Lint C, Emiliani S, Verdin E (1996) The expression of a small fraction of cellular genes is changed in response to histone hyperacetylation. Gene Expr 5(4–5):245–253
195. Gray SG et al (2004) Microarray profiling of the effects of histone deacetylase inhibitors on gene expression in cancer cell lines. Int J Oncol 24(4):773–795
196. Smith CL (2008) A shifting paradigm: histone deacetylases and transcriptional activation. Bioessays 30(1):15–24
197. Ying M et al (2006) Sodium butyrate ameliorates histone hypoacetylation and neurodegenerative phenotypes in a mouse model for DRPLA. J Biol Chem 281(18):12580–12586
198. Ryu H et al (2005) Sodium phenylbutyrate prolongs survival and regulates expression of anti-apoptotic genes in transgenic amyotrophic lateral sclerosis mice. J Neurochem 93(5):1087–1098
199. Marmolino D, Acquaviva F (2009) Friedreich's Ataxia: from the (GAA)n repeat mediated silencing to new promising molecules for therapy. Cerebellum 8(3):245–259
200. Herman D et al (2006) Histone deacetylase inhibitors reverse gene silencing in Friedreich's ataxia. Nat Chem Biol 2(10):551–558
201. Rai M et al (2008) HDAC inhibitors correct frataxin deficiency in a Friedreich ataxia mouse model. PLoS ONE 3(4):e1958
202. Thomas EA et al (2008) The HDAC inhibitor 4b ameliorates the disease phenotype and transcriptional abnormalities in Huntington's disease transgenic mice. Proc Natl Acad Sci USA 105(40):15564–15569

203. Hockly E et al (2003) Suberoylanilide hydroxamic acid, a histone deacetylase inhibitor, ameliorates motor deficits in a mouse model of Huntington's disease. Proc Natl Acad Sci USA 100(4):2041–2046
204. Tank EM, True HL (2009) Disease-associated mutant ubiquitin causes proteasomal impairment and enhances the toxicity of protein aggregates. PLoS Genet 5(2):e1000382
205. Hol EM, van Leeuwen FW, Fischer DF (2005) The proteasome in Alzheimer's disease and Parkinson's disease: lessons from ubiquitin B + 1. Trends Mol Med 11(11):488–495
206. Staropoli JF et al (2003) Parkin is a component of an SCF-like ubiquitin ligase complex and protects postmitotic neurons from kainate excitotoxicity. Neuron 37(5):735–749
207. Lehman NL (2009) The ubiquitin proteasome system in neuropathology. Acta Neuropathol 118(3):329–347
208. Spange S et al (2009) Acetylation of non-histone proteins modulates cellular signalling at multiple levels. Int J Biochem Cell Biol 41(1):185–198
209. Sadoul K et al (2008) Regulation of protein turnover by acetyltransferases and deacetylases. Biochimie 90(2):306–312
210. Scroggins BT et al (2007) An acetylation site in the middle domain of Hsp90 regulates chaperone function. Mol Cell 25(1):151–159
211. Jeong H et al (2009) Acetylation targets mutant huntingtin to autophagosomes for degradation. Cell 137(1):60–72
212. Caron C, Boyault C, Khochbin S (2005) Regulatory cross-talk between lysine acetylation and ubiquitination: role in the control of protein stability. Bioessays 27(4):408–415
213. Ito A et al (2002) MDM2-HDAC1-mediated deacetylation of p53 is required for its degradation. EMBO J 21(22):6236–6245
214. Mookerjee S et al (2009) Posttranslational modification of ataxin-7 at lysine 257 prevents autophagy-mediated turnover of an N-terminal caspase-7 cleavage fragment. J Neurosci 29(48):15134–15144
215. Grossman SR et al (2003) Polyubiquitination of p53 by a ubiquitin ligase activity of p300. Science 300(5617):342–344
216. Kass EM et al (2009) Mdm2 and PCAF increase Chk2 ubiquitination and degradation independently of their intrinsic E3 ligase activities. Cell Cycle 8(3):430–437
217. Linares LK et al (2007) Intrinsic ubiquitination activity of PCAF controls the stability of the oncoprotein Hdm2. Nat Cell Biol 9(3):331–338
218. Seigneurin-Berny D et al (2001) Identification of components of the murine histone deacetylase 6 complex: link between acetylation and ubiquitination signaling pathways. Mol Cell Biol 21(23):8035–8044
219. Hook SS et al (2002) Histone deacetylase 6 binds polyubiquitin through its zinc finger (PAZ domain) and copurifies with deubiquitinating enzymes. Proc Natl Acad Sci USA 99(21):13425–13430
220. Kirsh O et al (2002) The SUMO E3 ligase RanBP2 promotes modification of the HDAC4 deacetylase. EMBO J 21(11):2682–2691
221. Takahashi-Fujigasaki J, Fujigasaki H (2006) Histone deacetylase (HDAC) 4 involvement in both Lewy and Marinesco bodies. Neuropathol Appl Neurobiol 32(5):562–566
222. Steffan JS et al (2004) SUMO modification of Huntingtin and Huntington's disease pathology. Science 304(5667):100–104
223. Terashima T et al (2002) SUMO-1 co-localized with mutant atrophin-1 with expanded polyglutamines accelerates intranuclear aggregation and cell death. NeuroReport 13(17): 2359–2364
224. Riley BE, Zoghbi HY, Orr HT (2005) SUMOylation of the polyglutamine repeat protein, ataxin-1, is dependent on a functional nuclear localization signal. J Biol Chem 280(23): 21942–21948
225. Pandey UB et al (2007) HDAC6 at the intersection of autophagy, the ubiquitin-proteasome system and neurodegeneration. Autophagy 3(6):643–645

226. Pandey UB et al (2007) HDAC6 rescues neurodegeneration and provides an essential link between autophagy and the UPS. Nature 447(7146):859–863
227. Simms-Waldrip T et al (2008) The aggresome pathway as a target for therapy in hematologic malignancies. Mol Genet Metab 94(3):283–286
228. Kawaguchi Y et al (2003) The deacetylase HDAC6 regulates aggresome formation and cell viability in response to misfolded protein stress. Cell 115(6):727–738
229. Kopito RR (2003) The missing linker: an unexpected role for a histone deacetylase. Mol Cell 12(6):1349–1351
230. Iwata A et al (2005) HDAC6 and microtubules are required for autophagic degradation of aggregated huntingtin. J Biol Chem 280(48):40282–40292
231. Perez M et al (2009) Tau – an inhibitor of deacetylase HDAC6 function. J Neurochem 109(6):1756–1766
232. Matthias P, Yoshida M, Khochbin S (2008) HDAC6 a new cellular stress surveillance factor. Cell Cycle 7(1):7–10
233. Reed NA et al (2006) Microtubule acetylation promotes kinesin-1 binding and transport. Curr Biol 16(21):2166–2172
234. Gauthier LR et al (2004) Huntingtin controls neurotrophic support and survival of neurons by enhancing BDNF vesicular transport along microtubules. Cell 118(1):127–138
235. Dompierre JP et al (2007) Histone deacetylase 6 inhibition compensates for the transport deficit in Huntington's disease by increasing tubulin acetylation. J Neurosci 27(13): 3571–3583
236. Rivieccio MA et al (2009) HDAC6 is a target for protection and regeneration following injury in the nervous system. Proc Natl Acad Sci USA 106(46):19599–19604
237. Itoh Y et al (2007) Design, synthesis, structure–selectivity relationship, and effect on human cancer cells of a novel series of histone deacetylase 6-selective inhibitors. J Med Chem 50(22):5425–5438
238. Haggarty SJ et al (2003) Domain-selective small-molecule inhibitor of histone deacetylase 6 (HDAC6)-mediated tubulin deacetylation. Proc Natl Acad Sci USA 100(8):4389–4394
239. Schafer S et al (2008) Phenylalanine-containing hydroxamic acids as selective inhibitors of class IIb histone deacetylases (HDACs). Bioorg Med Chem 16(4):2011–2033
240. Schafer S et al (2009) Pyridylalanine-containing hydroxamic acids as selective HDAC6 inhibitors. ChemMedChem 4(2):283–290
241. Kozikowski AP et al (2008) Use of the nitrile oxide cycloaddition (NOC) reaction for molecular probe generation: a new class of enzyme selective histone deacetylase inhibitors (HDACIs) showing picomolar activity at HDAC6. J Med Chem 51(15):4370–4373
242. Smil DV et al (2009) Novel HDAC6 isoform selective chiral small molecule histone deacetylase inhibitors. Bioorg Med Chem Lett 19(3):688–692
243. Oh M, Choi IK, Kwon HJ (2008) Inhibition of histone deacetylase1 induces autophagy. Biochem Biophys Res Commun 369(4):1179–1183
244. Lee IH et al (2008) A role for the NAD-dependent deacetylase Sirt1 in the regulation of autophagy. Proc Natl Acad Sci USA 105(9):3374–3379
245. Jeong MR et al (2003) Valproic acid, a mood stabilizer and anticonvulsant, protects rat cerebral cortical neurons from spontaneous cell death: a role of histone deacetylase inhibition. FEBS Lett 542(1–3):74–78
246. Langley B et al (2008) Pulse inhibition of histone deacetylases induces complete resistance to oxidative death in cortical neurons without toxicity and reveals a role for cytoplasmic p21 (waf1/cip1) in cell cycle-independent neuroprotection. J Neurosci 28(1):163–176
247. Ryu H et al (2003) Histone deacetylase inhibitors prevent oxidative neuronal death independent of expanded polyglutamine repeats via an Sp1-dependent pathway. Proc Natl Acad Sci USA 100(7):4281–4286
248. Uo T, Veenstra TD, Morrison RS (2009) Histone deacetylase inhibitors prevent p53-dependent and p53-independent Bax-mediated neuronal apoptosis through two distinct mechanisms. J Neurosci 29(9):2824–2832

249. Leng Y, Chuang DM (2006) Endogenous alpha-synuclein is induced by valproic acid through histone deacetylase inhibition and participates in neuroprotection against glutamate-induced excitotoxicity. J Neurosci 26(28):7502–7512
250. Chen PS et al (2006) Valproate protects dopaminergic neurons in midbrain neuron/glia cultures by stimulating the release of neurotrophic factors from astrocytes. Mol Psychiatry 11(12):1116–1125
251. Peng GS et al (2005) Valproate pretreatment protects dopaminergic neurons from LPS-induced neurotoxicity in rat primary midbrain cultures: role of microglia. Brain Res Mol Brain Res 134(1):162–169
252. Chen PS et al (2007) Valproic acid and other histone deacetylase inhibitors induce microglial apoptosis and attenuate lipopolysaccharide-induced dopaminergic neurotoxicity. Neuroscience 149(1):203–212
253. Yasuda S et al (2009) The mood stabilizers lithium and valproate selectively activate the promoter IV of brain-derived neurotrophic factor in neurons. Mol Psychiatry 14(1):51–59
254. Wu X et al (2008) Histone deacetylase inhibitors up-regulate astrocyte GDNF and BDNF gene transcription and protect dopaminergic neurons. Int J Neuropsychopharmacol 11 (8):1123–1134
255. Kim HJ, Leeds P, Chuang DM (2009) The HDAC inhibitor, sodium butyrate, stimulates neurogenesis in the ischemic brain. J Neurochem 110(4):1226–1240
256. Adcock IM (2007) HDAC inhibitors as anti-inflammatory agents. Br J Pharmacol 150(7): 829–831
257. Glauben R et al (2009) HDAC inhibitors in models of inflammation-related tumorigenesis. Cancer Lett 280(2):154–159
258. Halili MA et al (2009) Histone deacetylase inhibitors in inflammatory disease. Curr Top Med Chem 9(3):309–319
259. Zhang B et al (2008) HDAC inhibitor increases histone H3 acetylation and reduces microglia inflammatory response following traumatic brain injury in rats. Brain Res 1226:181–191
260. Dinarello CA (2006) Inhibitors of histone deacetylases as anti-inflammatory drugs. Ernst Schering Res Found Workshop 56:45–60
261. Ren M et al (2004) Valproic acid reduces brain damage induced by transient focal cerebral ischemia in rats: potential roles of histone deacetylase inhibition and heat shock protein induction. J Neurochem 89(6):1358–1367
262. Faraco G et al (2006) Pharmacological inhibition of histone deacetylases by suberoylanilide hydroxamic acid specifically alters gene expression and reduces ischemic injury in the mouse brain. Mol Pharmacol 70(6):1876–1884
263. Camelo S et al (2005) Transcriptional therapy with the histone deacetylase inhibitor trichostatin A ameliorates experimental autoimmune encephalomyelitis. J Neuroimmunol 164(1–2):10–21
264. Granger A et al (2008) Histone deacetylase inhibition reduces myocardial ischemia-reperfusion injury in mice. FASEB J 22(10):3549–3560
265. Bolger TA, Yao TP (2005) Intracellular trafficking of histone deacetylase 4 regulates neuronal cell death. J Neurosci 25(41):9544–9553
266. Linseman DA et al (2003) Inactivation of the myocyte enhancer factor-2 repressor histone deacetylase-5 by endogenous Ca^{2+} calmodulin-dependent kinase II promotes depolarization-mediated cerebellar granule neuron survival. J Biol Chem 278(42):41472–41481
267. Majdzadeh N et al (2008) HDAC4 inhibits cell-cycle progression and protects neurons from cell death. Dev Neurobiol 68(8):1076–1092
268. Chen B, Cepko CL (2009) HDAC4 regulates neuronal survival in normal and diseased retinas. Science 323(5911):256–259
269. Outeiro TF, Marques O, Kazantsev A (2008) Therapeutic role of sirtuins in neurodegenerative disease. Biochim Biophys Acta 1782(6):363–369
270. Westphal CH, Dipp MA, Guarente L (2007) A therapeutic role for sirtuins in diseases of aging? Trends Biochem Sci 32(12):555–560

271. Kim D et al (2007) SIRT1 deacetylase protects against neurodegeneration in models for Alzheimer's disease and amyotrophic lateral sclerosis. EMBO J 26(13):3169–3179
272. Li Y et al (2007) Bax-inhibiting peptide protects cells from polyglutamine toxicity caused by Ku70 acetylation. Cell Death Differ 14(12):2058–2067
273. Houten SM, Auwerx J (2004) PGC-1alpha: turbocharging mitochondria. Cell 119(1):5–7
274. Cui L et al (2006) Transcriptional repression of PGC-1alpha by mutant huntingtin leads to mitochondrial dysfunction and neurodegeneration. Cell 127(1):59–69
275. Dominy JE Jr et al (2010) Nutrient-dependent regulation of PGC-1alpha's acetylation state and metabolic function through the enzymatic activities of Sirt1/GCN5. Biochim Biophys Acta 1804(8):1676–1683
276. Rodgers JT et al (2008) Metabolic adaptations through the PGC-1 alpha and SIRT1 pathways. FEBS Lett 582(1):46–53
277. Rodgers JT et al (2005) Nutrient control of glucose homeostasis through a complex of PGC-1alpha and SIRT1. Nature 434(7029):113–118
278. Sugden MC, Caton PW, Holness MJ (2010) PPAR control: it's SIRTainly as easy as PGC. J Endocrinol 204(2):93–104
279. Garske AL, Smith BC, Denu JM (2007) Linking SIRT2 to Parkinson's disease. ACS Chem Biol 2(8):529–532
280. Outeiro TF et al (2007) Sirtuin 2 inhibitors rescue alpha-synuclein-mediated toxicity in models of Parkinson's disease. Science 317(5837):516–519
281. Luthi-Carter R et al (2010) SIRT2 inhibition achieves neuroprotection by decreasing sterol biosynthesis. Proc Natl Acad Sci USA 107(17):7927–7932
282. Pfister JA et al (2008) Opposing effects of sirtuins on neuronal survival: SIRT1-mediated neuroprotection is independent of its deacetylase activity. PLoS ONE 3(12):e4090
283. Bonda DJ et al (2010) Mitochondrial dynamics in Alzheimer's disease: opportunities for future treatment strategies. Drugs Aging 27(3):181–192
284. Burchell VS et al (2010) Targeting mitochondrial dysfunction in neurodegenerative disease: Part I. Expert Opin Ther Targets 14(4):369–385
285. Moreira PI et al (2010) Mitochondria: a therapeutic target in neurodegeneration. Biochim Biophys Acta 1802(1):212–220
286. Su B et al (2010) Abnormal mitochondrial dynamics and neurodegenerative diseases. Biochim Biophys Acta 1802(1):135–142
287. Levenson JM et al (2004) Regulation of histone acetylation during memory formation in the hippocampus. J Biol Chem 279(39):40545–40559
288. Alarcon JM et al (2004) Chromatin acetylation, memory, and LTP are impaired in CBP+/− mice: a model for the cognitive deficit in Rubinstein-Taybi syndrome and its amelioration. Neuron 42(6):947–959
289. Bailey CH, Bartsch D, Kandel ER (1996) Toward a molecular definition of long-term memory storage. Proc Natl Acad Sci USA 93(24):13445–13452
290. Martin SJ, Grimwood PD, Morris RG (2000) Synaptic plasticity and memory: an evaluation of the hypothesis. Annu Rev Neurosci 23:649–711
291. Kandel ER (2001) The molecular biology of memory storage: a dialogue between genes and synapses. Science 294(5544):1030–1038
292. Maren S, Baudry M (1995) Properties and mechanisms of long-term synaptic plasticity in the mammalian brain: relationships to learning and memory. Neurobiol Learn Mem 63(1):1–18
293. Petersen RC et al (2009) Mild cognitive impairment: ten years later. Arch Neurol 66(12):1447–1455
294. Gabelle A et al (2010) Neurodegenerative dementia and parkinsonism. J Nutr Health Aging 14(1):37–44
295. Levy JA, Chelune GJ (2007) Cognitive-behavioral profiles of neurodegenerative dementias: beyond Alzheimer's disease. J Geriatr Psychiatry Neurol 20(4):227–238
296. Bourne C et al (2006) Cognitive impairment and behavioural difficulties in patients with Huntington's disease. Nurs Stand 20(35):41–44

297. Buccafusco JJ (2009) Emerging cognitive enhancing drugs. Expert Opin Emerg Drugs 14(4):577–589
298. Dash PK, Orsi SA, Moore AN (2009) Histone deactylase inhibition combined with behavioral therapy enhances learning and memory following traumatic brain injury. Neuroscience 163(1):1–8
299. Stefanko DP et al (2009) Modulation of long-term memory for object recognition via HDAC inhibition. Proc Natl Acad Sci USA 106(23):9447–9452
300. Bredy TW, Barad M (2008) The histone deacetylase inhibitor valproic acid enhances acquisition, extinction, and reconsolidation of conditioned fear. Learn Mem 15(1):39–45
301. Bredy TW et al (2007) Histone modifications around individual BDNF gene promoters in prefrontal cortex are associated with extinction of conditioned fear. Learn Mem 14(4): 268–276
302. Vecsey CG et al (2007) Histone deacetylase inhibitors enhance memory and synaptic plasticity via CREB:CBP-dependent transcriptional activation. J Neurosci 27(23):6128–6140
303. Korzus E, Rosenfeld MG, Mayford M (2004) CBP histone acetyltransferase activity is a critical component of memory consolidation. Neuron 42(6):961–972
304. Yeh SH, Lin CH, Gean PW (2004) Acetylation of nuclear factor-kappaB in rat amygdala improves long-term but not short-term retention of fear memory. Mol Pharmacol 65(5): 1286–1292
305. Yamin G (2009) NMDA receptor-dependent signaling pathways that underlie amyloid beta-protein disruption of LTP in the hippocampus. J Neurosci Res 87(8):1729–1736
306. Viola KL, Velasco PT, Klein WL (2008) Why Alzheimer's is a disease of memory: the attack on synapses by A beta oligomers (ADDLs). J Nutr Health Aging 12(1):51S–57S
307. Rowan MJ et al (2007) Synaptic memory mechanisms: Alzheimer's disease amyloid beta-peptide-induced dysfunction. Biochem Soc Trans 35(Pt 5):1219–1223
308. Rowan MJ et al (2003) Synaptic plasticity in animal models of early Alzheimer's disease. Philos Trans R Soc Lond B Biol Sci 358(1432):821–828
309. Lynch G et al (2008) The substrates of memory: defects, treatments, and enhancement. Eur J Pharmacol 585(1):2–13
310. Di Filippo M et al (2007) Plastic abnormalities in experimental Huntington's disease. Curr Opin Pharmacol 7(1):106–111
311. Smith R, Brundin P, Li JY (2005) Synaptic dysfunction in Huntington's disease: a new perspective. Cell Mol Life Sci 62(17):1901–1912
312. Kung VW et al (2007) Dopamine-dependent long term potentiation in the dorsal striatum is reduced in the R6/2 mouse model of Huntington's disease. Neuroscience 146(4):1571–1580
313. Lynch G et al (2007) Brain-derived neurotrophic factor restores synaptic plasticity in a knock-in mouse model of Huntington's disease. J Neurosci 27(16):4424–4434
314. Simmons DA et al (2009) Up-regulating BDNF with an ampakine rescues synaptic plasticity and memory in Huntington's disease knockin mice. Proc Natl Acad Sci USA 106(12):4906–4911
315. Kilgore M et al (2010) Inhibitors of class 1 histone deacetylases reverse contextual memory deficits in a mouse model of Alzheimer's disease. Neuropsychopharmacology 35 (4):870–880
316. Fontan-Lozano A et al (2008) Histone deacetylase inhibitors improve learning consolidation in young and in KA-induced-neurodegeneration and SAMP-8-mutant mice. Mol Cell Neurosci 39(2):193–201
317. Fischer A et al (2007) Recovery of learning and memory is associated with chromatin remodelling. Nature 447(7141):178–182
318. Kim D et al (2008) Deregulation of HDAC1 by p25/Cdk5 in neurotoxicity. Neuron 60(5): 803–817
319. Guan JS et al (2009) HDAC2 negatively regulates memory formation and synaptic plasticity. Nature 459(7243):55–60

320. McKinsey TA, Zhang CL, Olson EN (2000) Activation of the myocyte enhancer factor-2 transcription factor by calcium/calmodulin-dependent protein kinase-stimulated binding of 14-3-3 to histone deacetylase 5. Proc Natl Acad Sci USA 97(26):14400–14405
321. Hardingham GE et al (1997) Distinct functions of nuclear and cytoplasmic calcium in the control of gene expression. Nature 385(6613):260–265
322. Chawla S et al (2003) Neuronal activity-dependent nucleocytoplasmic shuttling of HDAC4 and HDAC5. J Neurochem 85(1):151–159
323. Darcy MJ et al (2010) Regional and subcellular distribution of HDAC4 in mouse brain. J Comp Neurol 518(5):722–740
324. Tian X et al (2010) MEF-2 regulates activity-dependent spine loss in striatopallidal medium spiny neurons. Mol Cell Neurosci 44(1):94–108
325. Barbosa AC et al (2008) MEF2C, a transcription factor that facilitates learning and memory by negative regulation of synapse numbers and function. Proc Natl Acad Sci USA 105(27): 9391–9396
326. Cohen TJ et al (2009) The deacetylase HDAC4 controls myocyte enhancing factor-2-dependent structural gene expression in response to neural activity. FASEB J 23(1):99–106
327. Cohen TJ et al (2007) The histone deacetylase HDAC4 connects neural activity to muscle transcriptional reprogramming. J Biol Chem 282(46):33752–33759
328. Zhang CL, McKinsey TA, Olson EN (2002) Association of class II histone deacetylases with heterochromatin protein 1: potential role for histone methylation in control of muscle differentiation. Mol Cell Biol 22(20):7302–7312
329. Williams AH et al (2009) MicroRNA-206 delays ALS progression and promotes regeneration of neuromuscular synapses in mice. Science 326(5959):1549–1554
330. Wuwongse S, Chang RC, Law AC (2010) The putative neurodegenerative links between depression and Alzheimer's disease. Prog Neurobiol 91(4):362–375
331. Covington HE III, Vialou V, Nestler EJ (2010) From synapse to nucleus: novel targets for treating depression. Neuropharmacology 58(4–5):683–693
332. Tsankova NM et al (2006) Sustained hippocampal chromatin regulation in a mouse model of depression and antidepressant action. Nat Neurosci 9(4):519–525
333. Tsankova N et al (2007) Epigenetic regulation in psychiatric disorders. Nat Rev Neurosci 8(5):355–367
334. Renthal W et al (2007) Histone deacetylase 5 epigenetically controls behavioral adaptations to chronic emotional stimuli. Neuron 56(3):517–529
335. Covington HE III et al (2009) Antidepressant actions of histone deacetylase inhibitors. J Neurosci 29(37):11451–11460
336. Iga J et al (2007) Altered HDAC5 and CREB mRNA expressions in the peripheral leukocytes of major depression. Prog Neuropsychopharmacol Biol Psychiatry 31(3):628–632
337. Hobara T et al (2010) Altered gene expression of histone deacetylases in mood disorder patients. J Psychiatr Res 44(5):263–270

Top Med Chem 6: 57–90
DOI: 10.1007/7355_2010_8

Published online: 26 August 2010

Phosphodiesterase Inhibition to Target the Synaptic Dysfunction in Alzheimer's Disease

Kelly R. Bales, Niels Plath, Niels Svenstrup, and Frank S. Menniti

Abstract Alzheimer's Disease (AD) is a disease of synaptic dysfunction that ultimately proceeds to neuronal death. There is a wealth of evidence that indicates the final common mediator of this neurotoxic process is the formation and actions on synaptotoxic b-amyloid (Aβ). The premise in this review is that synaptic dysfunction may also be an initiating factor in for AD and promote synaptotoxic Aβ formation. This latter hypothesis is consistent with the fact that the most common risk factors for AD, apolipoprotein E (ApoE) allele status, age, education, and fitness, encompass suboptimal synaptic function. Thus, the synaptic dysfunction in AD may be both *cause* and *effect*, and remediating synaptic dysfunction in AD may have acute effects on the symptoms present at the initiation of therapy and also slow disease progression. The cyclic nucleotide (cAMP and cGMP) signaling systems are intimately involved in the regulation of synaptic homeostasis. The phosphodiesterases (PDEs) are a superfamily of enzymes that critically regulate spatial and temporal aspects of cyclic nucleotide signaling through metabolic inactivation of cAMP and cGMP. Thus, targeting the PDEs to promote improved synaptic function, or 'synaptic resilience', may be an effective and facile approach to new symptomatic and disease modifying therapies for AD. There continues to be a significant drug discovery effort aimed at discovering PDE inhibitors to treat a variety of neuropsychiatric disorders. Here we review the current status of those efforts as they relate to potential new therapies for AD.

K.R. Bales
Neuroscience Biology, Pfizer Global Research & Development, Groton, CT 06355 USA

N. Plath
In Vivo Neuropharmacology, H. Lundbeck A/S, 2500 Valby, Denmark

N. Svenstrup
Medicinal Chemistry Research, H. Lundbeck A/S2500 Valby, Denmark

F.S. Menniti (✉)
Mnemosyne Pharmaceuticals, Inc., 3 Davol Sq., Providence, RI 02903
e-mail: mennitifs@gmail.com

Keywords Alzheimer's disease, Phosphodiesterase, Cyclic nucleotide, Synaptic plasticity, PDE2A, PDE4, PDE5A, PDE7, PDE8B, PDE9A

Contents

1 Introduction

Alzheimer's disease (AD) is the most common form of chronic neurodegeneration, affecting as many as 5.3 million people in the USA alone. The major risk factor for AD is aging. Consequently, as the USA and most other countries continue to enjoy increased longevity, the prevalence of AD is projected to increase dramatically. AD is portended by deficits in short-term memory. Mild cognitive impairment (MCI), greater than expected cognitive deficiency in the elderly, is believed to be the earliest antecedent of this aspect of the disease, with those suffering from the amnesic variant of MCI having a high conversion to AD [1]. Progression of AD is accompanied by greater impairment in both declarative and nondeclarative memory domains, along with disruption of reasoning, abstraction, and language, and the emergence of disturbing behavioral problems including anxiety and excessive emotionality, aggression, and wandering [2]. These pervasive cognitive and behavioral symptoms are devastating to patients and place a tremendous burden on caregivers in the home and in care giving institutions. Thus, there is an aggressive effort to develop therapies that may alleviate the symptoms of AD. Two such therapies are currently available, the acetylcholinesterase inhibitors [3] and the NMDA receptor antagonist memantine [4]. These therapies offer symptomatic relief and slow down clinical progression;

however, they have little or no effect on disease modification and so have only a limited window of therapeutic benefit within the long-term course of the disease. Thus, the goal is discovery of new therapies that both alleviate symptoms and substantially slow or halt the progression of AD.

The symptoms of AD are the result of a progressive loss of neuronal function, beginning in the temporal lobes and then spreading to a widening network of interconnected cortical regions. The key to identifying approaches to slow disease progression is to understand the underlying cause of this neuronal dysfunction and the reason for the characteristic pattern of progressive pathology. A seminal finding was the discovery in 1991 that mutation in the gene encoding for the amyloid precursor protein (APP) results in autosomal dominant inheritance of AD [5, 6]. APP is metabolized to a family of small peptides, the β-amyloids (Aβ), which form the core of one of the hall mark pathological markers in the AD brain, the amyloid plaque. These observations focused attention on the Aβ peptides as somehow being the key causal toxic agent in AD. However, the putative causal role of Aβ, including the underlying toxic mechanism and primary target of toxicity, has yet to be definitively established. The hypotheses are many: compelling arguments have been made that Aβ, in either soluble form or as higher order aggregates, in various intra- and extracellular compartments, is directly toxic to neurons, is disruptive to the cerebral vasculature, and/or induces a deleterious inflammatory response. There are a myriad of corresponding therapeutic strategies currently under development that target these putative toxic mechanisms. Although it may seem chaotic, rigorous testing of these multiple hypotheses is precisely what is needed to reach a definitive understanding of the role of Aβ toxicity in AD.

A second seminal set of findings is that AD is a disease of synaptic failure [7]. A striking feature of the end stage AD brain is the tremendous loss of neurons. Consequently, there has been considerable focus on identifying therapies to prevent neuronal death in AD. However, it is becoming increasingly recognized that synaptic dysfunction is the more proximal pathological event. Synaptic pathology is responsible for the cognitive decline characteristic of the earliest phases of the disease. In addition, it is highly likely that the loss of synaptic interconnectivity contributes significantly or is directly responsible for the ultimate death of neurons in AD.

The novel premise from which we are working is that synaptic dysfunction may also be an initiating factor in AD in that it may promote synaptotoxic Aβ formation [8]. This latter hypothesis is particularly intriguing in that it may account for the neuroanatomical progression of AD pathology as well as the most common risk factors: apolipoprotein E (apoE) allele status, age, education, and fitness. Thus, the synaptic dysfunction in AD may encompass both *cause* and *effect*. Given this premise, remediating synaptic dysfunction in AD may be predicted to have acute effects on the symptoms present at the initiation of therapy and, significantly, may also slow disease progression.

The mechanistic approach we are pursuing to remediate the synaptic pathology of AD is the use of cyclic nucleotide phosphodiesterase (PDE) inhibitors. The cAMP and cGMP signaling systems are intimately involved in the regulation of synaptic homeostasis. The PDEs are the enzymes responsible for the metabolic inactivation of cAMP and cGMP and, as such, are critical regulators of cyclic nucleotide signaling [9]. Furthermore, among all of the classes of molecular targets in the cyclic nucleotide signaling cascades, the PDEs are the most highly amenable to pharmaceutical development. Thus, targeting the PDEs to promote "synaptic resilience" may be an effective and facile approach to new symptomatic and disease-modifying therapies for AD. We briefly provide additional context for this therapeutic approach and then present an analysis of the potential uses of inhibitors of PDE2A, PDE4A, B and D, PDE5A, PDE7A and B, PDE8B, and PDE9A to treat the synaptic dysfunction of AD.

2 AD as a Disease of Synaptic Dysfunction

2.1 Synapse Loss in AD

Synapse loss has been established as the strongest correlate of cognitive dysfunction in MCI and early AD [10, 11] and is apparent as a decreased synapse density in ultrastructural studies as well as decreased expression of synaptic proteins [12]. The significant reduction in the number of presynaptic boutons precedes frank pyramidal neuron loss. An illuminating finding has been that many of these synaptic changes also precede development of the diagnostic pathologies of the disease, parenchymal amyloid deposition, and intraneuronal neurofibrillary tangles (NFT) [13].

In individuals exhibiting the behavioral symptoms of AD, the diagnosis is formally confirmed at autopsy by the presence of two neuropathological features: the presence within brain of parenchymal plaques containing aggregated Aβ, and intraneuronal NFT arising from hyperphosphorylated fibrils of the microtubule-associated protein tau [14]. Much of modern AD research has focused on divining the underlying cause of the disease from these pathological markers. Neurofibrillary tau pathology in AD begins in the entorhinal cortex and spreads in a hierarchical manner into the hippocampus proper and cortex. Tau pathology increases as memory impairments become more severe and other cognitive and behavioral symptoms develop [15, 16]. In fact, the hierarchical progression of tau pathology "maps" the progressive deterioration of cortical systems reflected in the progression of symptoms. Given the importance of microtubules in intraneuronal transport, axonal growth, and maintenance of dendritic architecture, it is reasonable to suspect a role for tau dysregulation in the synaptic dysfunction of AD. While tau pathology may be an *effector* for synaptic toxicity, there is no compelling evidence to suggest that tau hyperphosphorylation and aggregation is the principal *causative* factor in the disease. In contrast, there is strong genetic evidence to suggest such a causative role for Aβ.

2.2 Aβ and Synapse Function

The term Aβ encompasses a small family of 39–43 amino acid peptides derived from the intramembranous cleavage of the APP by the sequential action of β- and γ-secretases [17]. Inheritance of the rare autosomal dominant early onset forms of AD (EOAD) is caused by mutations within the APP or presenilin genes. The latter encode for proteins that, together with three other proteins, form the γ-secretase complex. Various mutations in these different genes all result in an increase in the ratio of formation of Aβ42:Aβ40 [18]. Thus, this genetic evidence strongly suggests that aberrant over-production and/or mis-metabolism of APP/Aβ causes EOAD. The much more common form of AD is late in onset (LOAD) and of idiopathic etiology. Significantly, the clinical presentation, disease course, and neuropathology are nearly identical between EOAD and LOAD. This suggests a common underlying pathological mechanism and, thus, implicates a causal role for Aβ in LOAD as well [19]. Indeed, imaging studies using a ligand that binds to amyloid (thioflavin S β-pleated sheet material composed of deposited Aβ) have now documented an increase in brain amyloid burden in asymptomatic EOAD patients as well as patients diagnosed with "probable" LOAD [20]. Based on this compelling data, candidate compounds that inhibit the production or enhance clearance of Aβ are now entering late stages of clinical testing.

2.3 Synapse Loss as Both Cause and Effect in AD

The findings reviewed above beg a critical question – what causes aberrant over-production and/or mis-metabolism of APP to Aβ in the common, idiopathic form of AD? A plausible explanation linking APP processing to the cause of AD was proposed in 1993 based on two considerations [8]. First, the entorhinal cortex, the area of brain that demonstrates the earliest neurofibrillary pathology [15, 16], also has the highest levels of APP in brain [8, 21]. Second, this region undergoes an adaptive upregulation of APP turnover late in life in response to a life-long progressive loss of synaptic connectivity [8]. In some individuals, this response is hypothesized to cross a threshold resulting in the formation of neuropathological toxic products [8]. Thus, instead of promoting compensatory synaptic connectivity, the increased APP turnover results in synaptic toxicity. This synaptic toxicity disconnects the entorhinal projection from its postsynaptic targets [22], decreasing excitatory drive on the targets and thereby setting up a recurrent cycle of synaptic disconnection/APP upregulation/toxicity [8]. This cycle cascades in a hierarchical progression that is marked by NFT formation within a neuronal circuitry that mediates normal learning and memory processes in the anatomical progression of hyperphosphorylated tangle pathology defined by Braak and Braak [15, 16]. The earliest enunciation of such an "amyloid cascade hypothesis" of AD pathogenesis posited that accumulation of Aβ-containing plaques was causative to disease pathogenesis [19]. However, individuals who, in life, experienced no pathological memory impairment may be found, at post mortem, to have fulfilled the neuropathological criteria for amyloid plaque burden [23]. This implies that plaques per se are not directly causative in disease onset and/or progression. Instead, evidence is converging on soluble forms of Aβ (i.e., Aβ that is not sequestered in plaque) as the "synaptotoxic" agent [7].

Since the discovery that mutations in the gene encoding for APP result in autosomal dominant inheritance of AD, there has been considerable research into the physiological functions of APP and related proteins [24]. This research is consistent with the above hypothesis in indicating that APP is regulated by, and involved in the regulation of, synaptic activity at multiple levels. Evidence suggests a role for the APP holoprotein in axonal transport and extracellular cell/cell interactions and adhesion [24]. Furthermore, proteolytic fragments of APP are suggested to have distinct signaling functions. Processing by α- or β-secretase releases soluble N-terminal fragments (the sAPPα or sAPPβ) into the extracellular space, where these peptides appear to have neurotrophin-like signaling properties. The C-terminal fragment released following cleavage by γ-secretase is suggested to be transported to the nucleus, where it functions to regulate transcription. However, the most enigmatic aspect of APP processing is the minor (<10%) component comprising the sequential action of β- and γ-secretases to form Aβ. Initially, the Aβ peptides were considered as simple by-products of the formation of the other signaling fragments of APP. Instead, it is becoming increasingly clear that Aβ, too, has distinct roles in regulating synaptic function. Aβ formation is regulated by neuronal activity [25, 26]. In some studies, very low (pM) levels of soluble, cell-derived Aβ were found to reduce synaptic potentials and spine density when applied to primary neuronal cultures [27, 28]. When these same soluble Aβ species were administered intrathecally to rats, cognition was impaired [29]. However, in other experimental systems, synthetic Aβ42 positively modulated synaptic plasticity and enhanced hippocampal-dependent memory [30], and Aβ monomers were found to be neuroprotective [31]. Recently, Tampellini et al. provide evidence to suggest that there are two pools of Aβ, intra- and extracellular, that interact to impact synaptic function in different ways [25]. Taken together, these data suggest that Aβ peptides are formed in response to synaptic activity to impact normal neurophysiological function.

The key point of understanding is exactly why, in some individuals, formation of Aβ at excitatory synapses crosses from physiological to pathological. Our premise is that this is related to properties intrinsic to the synaptic physiology of these at risk individuals. All of the major risk factors for idiopathic AD are associated with reduced synaptic function. In addition to age, the major environmental risk factors for idiopathic LOAD are lower native intelligence (operationally defined as education level) and reduced overall physical health. Each of these factors has a negative impact on synaptic function. Particularly illuminating may be the emerging data suggesting that synaptic function is also impacted by apolipoprotein E4 (apoE4) status, the most significant genetic risk factor for AD [32, 33]. The E4 allele of the apoE gene is a well-characterized risk factor for AD, with E4 carriers having an increased probability of suffering AD, at an earlier age of onset [34]. In humans, E4 carriers exhibit reduced cognitive capacity, reductions in resting brain glucose metabolism, and a distinct pattern of brain activity that is observed well before onset of AD symptoms [35–37]. Furthermore, the apoE4 allele is positively linked to subclinical epileptiform activity, which is remarkable in light of the recent compelling evidence showing that aberrant excitatory neuronal activity is a primary upstream mechanism for cognitive decline in AD [38–41]. In mice that express human apoE4, dendritic architecture, spine number, and electrophysiological parameters are significantly reduced when compared to age- and background-matched mice expressing human apoE3 [42]. These findings suggest that the E4 allele may reduce overall synaptic function and that this occurs well before frank neurodegeneration.

Taking into account all of the factors discussed above, we hypothesize that reduced synaptic function is the key "initiating" factor in LOAD. This reduced synaptic function is hypothesized to be responsible for the susceptibility to a change in Aβ processing from physiological to pathological and/or an increase in susceptibility to Aβ toxicity. Reduced synaptic function is also hypothesized to be a key facilitatory factor in the progression of synaptic disconnection that initiates in entorhinal cortex and progresses throughout interconnected cortical networks. "Synaptic resilience" is the inverse of this reduced synaptic function. Thus, therapies that promote synaptic resilience may reduce the risk and/or slow AD progression; that is, such therapies may have a true disease-modifying effect.

Unfortunately, at present we do not have a complete understanding of the molecular underpinnings of the reduced synaptic function that is hypothesized to be causal to AD. There is, however, a tremendously expanding understanding of fundamental processes that mediate physiological synaptic function and plasticity. It seems reasonable to assume that we will want to manipulate some of these fundamental processes to get at the synaptic dysfunction of AD. Thus, this body of knowledge serves as the logical starting point to explore such therapies. The basis for our interest in the potential of PDE inhibitors in this regard is outlined below.

3 Cyclic Nucleotides and Synaptic Plasticity

Synaptic plasticity is a term that encompasses a wide range of complex processes. At the level of the individual synapse, synaptic plasticity is the process by which the architecture and complement of signaling molecules are adjusted in response to recent activity, in preparation for future activity. In the simplest terms, the past predicts the future, and so recent activity increases synaptic strength, whereas lack of activity leads to synapse deconstruction. A critical modulator of this general rule is the coordination (i.e., timing) of events between the pre- and postsynaptic sides of individual synapses. At the level of the neuron and neuronal circuit, synaptic plasticity is the means of encoding information. That is, changes in individual synaptic strengths are integrated and reflected in changes in the way a neuron interconnects with neuronal networks. It is the change in pattern of activities in large networks of neurons that read out as "behavior" and "cognition." Thus, when we seek to modulate synaptic plasticity to slow disease progression, we are concerned with modulating the biochemistry of individual synapses, whereas when we seek to modulate synaptic plasticity to improve cognition in AD, we are concerned with modulating network activity. It remains to be proved whether both of these goals can be accomplished through a single molecular mechanism. There are a myriad of such mechanisms that may be targeted to impact synaptic plasticity. To paraphrase an earlier statement, rigorous testing of multiple mechanism-based hypotheses is precisely what is needed to reach an understanding of the utility of targeting synaptic plasticity in AD. The cyclic nucleotide PDEs may be particularly advantageous to target in this regard. These enzymes are intimately involved in the regulation of cyclic nucleotide signaling, and these signaling cascades are intimately involved in the regulation of synaptic plasticity, as briefly described below.

cAMP and cGMP signaling is ubiquitous in mammals. A wide variety of intercellular communicative, hormonal, and metabolic events trigger the activation of adenylyl and/or guanylyl cyclases to catalyze the formation of cAMP and cGMP from ATP and GTP, respectively. cAMP and cGMP subsequently bind to a variety of effectors including their cognate protein kinases [43], ion channels [44], Epacs [45], and other PDEs [9], resulting in both acute- and long-term changes in cellular function. Both cAMP and cGMP signaling mechanisms are implicated in the regulation of synaptic plasticity at multiple levels [46, 47].

3.1 cAMP

The canonical role of the cAMP/PKA signaling cascade is in the regulation of postsynaptic, protein synthesis-dependent long-term potentiation (L-LTP) [48], widely believed to be an in vitro model of learning and memory [49]. There are considerable data indicating that the cAMP/PKA signaling cascades are also involved in regulation of earlier stages of LTP in the postsynaptic compartment. This includes potentiation of Cam KII induction by PKA-mediated inactivation of protein phosphatases that are responsible for dephosphorylation of Cam KII [50], and PKA phosphorylation of the GluR1 subunit of AMPA receptors to drive insertion of this subunit into the postsynaptic active zone [51] and increase AMPA receptor open channel probability [52]. The cAMP/PKA signaling cascade is also implicated in the regulation of plasticity in the presynaptic compartment. The clearest example is the presynaptic form of LTP characterized at mossy fiber synapses in the dentate gyrus of the hippocampus [53]. Mossy fiber LTP is critically dependent on activation of a calcium/calmodulin-dependent adenylyl cyclase, leading to an increase in presynaptic cAMP and activation of PKA [54] and phosphorylation of the synaptic vesicle-associated protein

Rim1a [55]. This form of PKA-dependent presynaptic plasticity is also observed in cerebellum and at corticothalamic and corticostriatal synapses [53]. Synaptic plasticity also involves adaptive decreases in synaptic strength [56]. Of these, the archetype is NMDA receptor-dependent long-term depression (LTD) in the hippocampus [57], which appears to be critically dependent on the dephosphorylation of PKA substrates. Of particular significance is the selective dephosphorylation of the PKA site Ser845 on GluR1 which decreases the probability of AMPA receptor channel opening and increases AMPA receptor endocytosis [58].

3.2 cGMP

Although less extensively studied, there is a body of evidence implicating cGMP signaling cascades as important pathways for many forms of synaptic plasticity [59]. The canonical role of cGMP in synaptic plasticity is as mediator of the retrograde messenger nitric oxide (NO) at glutamatergic synapses [60–63]. It is also now clear that cGMP signaling cascades participate at several additional levels of regulation that influence hippocampal LTP, including postsynaptic protein synthesis-dependent mechanisms [64]. These distinct presynaptic and postsynaptic functions are perhaps most clearly demonstrated in studies of LTP in visual cortex, where the two guanylyl cycles isoforms are differentially localized to pre- and postsynaptic compartments, and genetic deletions of either isoform have demonstrated separable effects on LTP [65]. Compartmentalization is further indicated by the finding that the source of NO is also an important determinant [66]. Finally, there is evidence indicating a role for cGMP signaling cascades in the depression of synaptic activity [67–69].

4 The Phosphodiesterases

4.1 Enzyme Structure and Function

The PDEs are the family of enzymes that terminate through metabolic inactivation signaling by cAMP and cGMP. Thus, these enzymes are intimately involved in the regulation of cyclic nucleotide signaling throughout the body, including those cyclic nucleotide pathways involved in the regulation of synaptic plasticity. The PDEs are encoded by 21 genes that are functionally separated into 11 families [9, 70]. Further physiological diversity stems from differential mRNA splicing and, to date, more than 60 PDE isoforms have been identified. There is a rapidly expanding body of knowledge about the physiology of these enzymes, from the atomic and structural level to the role in specific signaling processes. This information has both garnered interest in and facilitated drug discovery efforts. Below, we first touch on the current knowledge of the structural features of these enzymes, particularly with regard to drug discovery. We then turn to biological functions and highlight a number of the enzyme families that may be particularly relevant to the treatment of AD.

The PDEs are modular enzymes in which the catalytic domain in the C-terminal portion of the protein is coupled to regulatory elements that reside in the N-terminal region. The 11 PDE families differ most significantly from one another within the unique N-terminal regulatory domains. On the other hand, the C-terminal catalytic domains are highly conserved with respect to specific invariant amino acids, three-dimensional structure, and catalytic mechanism [71]. Nonetheless, subtle differences within the catalytic core impart important family-specific characteristics [72]. To date, essentially all of the pharmaceutical developments around the PDEs have been toward the discovery of catalytic site inhibitors. Structural information from single crystal X-ray crystallography has played an important role in elucidating the important functional differentiating features within the catalytic domains of the 11 gene families that allow for the development of family-specific inhibitors. Indeed, current lead optimization projects without the use of some form of structure-based drug design are becoming practically unthinkable. This area of knowledge is summarized below.

Structures of the catalytic domains of all but two PDE families (PDE6 and PDE11) have been solved. Since the field was last reviewed in 2007 [73], two new PDE families have been added to the list of solved structures, namely PDE8 in its unliganded form as well as in complex with IBMX [74] and PDE10A with various ligands [75, 76]. Characteristics of all PDE structures solved so far are the following features which are also important for the design of new inhibitors:

- The active site contains a glutamine residue that contributes to the binding of the natural substrate cAMP or cGMP through a dual hydrogen bond. The "Glutamine Switch" mechanism [77] suggests that hydrogen-bonding residues surrounding the glutamine serve to either lock it in a fixed conformation (cAMP or cGMP selective PDEs) or allow it to change conformation (PDE1, 2, 3, 10 and 11). Although very elegant in its simplicity, the glutamine switch hypothesis remains somewhat controversial [78, 79]. The glutamine is also nearly invariably involved in hydrogen bonding to PDE inhibitors, although not necessarily through two hydrogen bonds (see [80] and references cited therein).
- A phenylalanine, situated just below the plane of the bound substrate/inhibitor, participates in the substrate binding by π–π interactions. This hydrophobic region, usually referred to as the "Clamp" region, explains why many PDEs appear to have a preference for flat and π-electron-rich inhibitors of the sildenafil type [81].
- The metal ions in the active site may also be targeted for inhibitor binding; however, this approach is not usually addressed by design elements in PDE inhibitors intended for central nervous system (CNS) indications. Specifically,

a good ligand for the metal ion is by its very nature rather polar, thereby adding to the overall polar surface area of the inhibitor to such a degree that transport across the blood–brain barrier becomes exceedingly difficult.

4.2 Compartmentalization of PDE Signaling

The desire for inhibitors selective for different PDE families (and individual isozymes, see below) stems from the fact that cyclic nucleotide signaling is highly compartmentalized within individual cells [82, 83]. Thus, PDE isozymes have distinct signaling roles in individual cell types and there appears to be little or no overlap in function. Compartmentalization is the result of physical localization of signaling pathways to discreet areas of cells and, further, the physical association of the different components of a signal cascade mediated by adaptor and scaffolding proteins. Thus, a scaffold may bring a cyclase, an effector kinase, and a specific PDE isoform together with a cell surface receptor to affect a very localized signaling event. Compartmentalization of PDE-regulated signaling has been most clearly elucidated for the PDE4 family [84]. For example, physical compartmentalization allows only PDE4B to regulate Toll-like receptor signaling in mouse peritoneal macrophages despite the fact that these cells also express PDE4A and PDE4D [85].

The complexity and compartmentalization of PDE-regulated signaling are particularly evident in the CNS [86] and are crucial to the analysis of the different PDEs that may be targeted to impact synaptic plasticity. As noted above, cAMP and cGMP signaling cascades are implicated in the regulation of plasticity in numerous temporally and spatially distinct compartments. It is reasonable to conjecture that different PDE families and isoforms service these distinct signaling compartments, at the level of the individual synapse, neuronal subtype, and brain region. Thus, the challenge is twofold. The first is to determine the role of individual PDE isoforms in the regulation of different aspects of synaptic plasticity at the synaptic and sub-synaptic levels. The second is to relate the effects of manipulating the PDEs at the synaptic level to the impact that may have on neuronal circuits and networks. Fortunately, the localization of the different PDE isoforms throughout the brain continues to be investigated, and the pharmacological tools needed to accomplish these types of analyses are becoming available to allow investigation of function. With regard to AD, the potential targets include PDE2A, PDE4A, 4B, and 4D, PDE5A, PDE7A and 7B, PDE 8B, and PDE9A [86]. We review the current state of knowledge regarding these enzymes with regard to localization, the availability of pharmacological inhibitors and the knowledge to date on the effects of these inhibitors on behavior that may be relevant to AD therapy. We start with PDE4, the most highly pursued drug target among the PDEs, followed by PDE8B and PDE2A as additional cAMP signaling-specific targets. We then turn to PDE5, the most commercially successful PDE target, and finish with PDE2A and PDE9 as the new targets generating the most interest.

4.3 PDE4

PDE4 is the largest and most complex of the PDE gene families and is the major cAMP-regulating enzyme in the body [84]. The PDE4s are encoded by four genes, PDE4A–D, with PDE4A, B, and D expressed appreciably in the CNS. Furthermore, mRNA transcribed from each gene is subjected to alternative N-terminal splicing to yield three major variants. The long variants contain two conserved N-terminal domains, UCR1 and UCR2, with a conserved PKA phosphorylation site at the N-terminal end of UCR1. Phosphorylation at the PKA site stimulates activity and is a key element in the regulation of these variants. The short variants are truncated and lack UCR1 and the PKA site, whereas the supershort variant is further truncated to lack both UCR1 and the N-terminal portion of UCR2. Accounting for genes and splice variants, over 20 PDE4 isoforms have been identified. Thus, the PDE4s provide a rich repertoire for fine tuning cAMP signaling.

PDE4 has been heavily pursued as a therapeutic target. Initial interest in the 1980s was as a CNS target, stemmed from the finding that rolipram, the prototypical PDE4 inhibitor, had clinical antidepressant activity. However, it is the potential to treat inflammatory airway disease that has sustained the most interest. Recently, CNS interest has re-emerged in the potential for PDE4 inhibitors to treat cognitive dysfunction [87], particularly in AD. This latter interest derives from the seminal finding that PDE4 is a key element in the cAMP/PKA signaling cascade involved in protein synthesis-dependent L-LTP in hippocampus and that rolipram potentiates L-LTP in hippocampal slice preparations [88].

The hippocampus has a broad range of functions, but is particularly implicated in the formation of long-term memories. L-LTP putatively represents the molecular mechanism that supports this function [48]. Thus, the robust finding that PDE4 inhibition augments LTP in hippocampus implies that PDE4 inhibitors should facilitate long-term memory formation in vivo. There is ample experimental support for this hypothesis. Administration of rolipram robustly improves the performance of both rodents and nonhuman primates in various long-term memory tasks, under conditions where performance is disrupted by a variety of pharmacological or other manipulations [87, 89]. Rolipram is competitive at the cAMP-binding site of PDE4. The potency at the high affinity site that predominates in brain is approximately 2 nM. The dose of rolipram most often reported as efficacious in rodent cognition assays is 0.1 mg/kg, which yields an estimated free brain concentration of 2–3 nM (unpublished observation calculated from data in the literature).

Studies with rolipram also suggest that PDE4 inhibition may specifically reverse deficits in synaptic function caused by Aβ [90]. Direct application of Aβ to hippocampal slices or in vivo impairs LTP in some systems [91–93]. LTP deficits are also observed in slices prepared from transgenic mice that overexpress Aβ [90]. Significantly, acute rolipram administration to transgenic mice reduced deficits in LTP in slices prepared from these mice, and this beneficial effect was maintained for at least 2 months beyond the end of treatment [90, 93].

Unfortunately, despite more than 30 years of pharmaceutical research, no PDE4 inhibitor has been approved for any indication. The primary obstacle has been severe side effects, notably emesis, nausea, and vasculitis, at exposures that are within the range where therapeutic benefit may begin to be realized. This obstacle led to abandonment of rolipram for the treatment of depression. These side effects of PDE4 inhibitors have also, to date, prevented a thorough exploration of the dose range for efficacy for inflammatory diseases such as asthma and chronic obstructive pulmonary disease (COPD) [94]. It remains to be determined whether there is a sufficient therapeutic index for PDE4 inhibition for the treatment of cognitive dysfunction in AD. The available preclinical data suggest that this may be a challenge. As stated above, the estimated level of PDE4 inhibition associated with improved cognition in rodent models, extracted from the data with rolipram mentioned above, indicates that a significant fractional inhibition may be required. However, it is not possible directly determine the TI in rodents, since rats and mice lack an emetic response. Furthermore, the ferret model of emesis has proved not to be predictive of the emetic potential of PDE4 inhibitors in humans [95]. There are limited data on the effects of rolipram on cognition in nonhuman primates, where it is possible to gage a therapeutic index. Rutten et al. reported positive effects of rolipram in an object retrieval paradigm in cynomolgus monkeys with maximal efficacy at 0.03 mg/kg; however, the next highest dose of 0.1 mg/kg was not tolerated due to emesis [96]. In an earlier study, Ramos et al. found no effect of rolipram in a delayed match to position paradigm in rhesus monkeys at doses up to the maximum tolerated dose of 0.01 mg/kg; a dose of 0.05 mg/kg was not tolerated [97]. Thus, if there is efficacy, the therapeutic index for the treatment of cognitive dysfunction may be similarly low as for the treatment of inflammatory airway disease. However, given the severity of the AD and the fact that the neurodegenerative process may have altered the sensitivity to potential therapeutic and/or adverse effects of PDE4 inhibitors, it is still of interest to investigate PDE4 as a target for the treatment of AD. In fact, Merck & Co. recently completed a Phase II proof of concept study in patients with AD with the PDE4 inhibitor, MK0925, although no results have been published and the compound is no longer listed in the company pipeline. Thus, alternative strategies are worth considering as we await this clinical feedback, as discussed below.

To date, the vast majority of research into the procognitive potential of PDE4 inhibition has relied on the use of inhibitors such as rolipram that are competitive at the catalytic site. These compounds inhibit each of the PDE4 gene families, consistent with the very high homology in the cAMP-binding pocket. There is also no evidence to suggest that such compounds distinguish among the major N-terminal splice variants. An approach for overcoming the narrow therapeutic index of such pan-PDE4 inhibitors may be the development of compounds that interact with specific PDE4 subtypes to capture therapeutic effects while avoiding those subtype(s) that mediate emesis and nausea. The question is, which subtype to target?

Cherry and Davis [98] mapped by immunohistochemistry the distribution of PDE4A, B, and D in mouse brain to many regions relevant to higher order cognitive functions. All three isoforms are expressed in neocortex, albeit with distinctive laminar distributions. PDE4D is most highly expressed in the hippocampus proper, and genetic deletion of PDE4D also potentiates LTP to subthreshold stimuli in hippocampal slices [99], although this was accompanied by poorer performance of the animals in behavioral tasks that measured cognition. PDE4B is most highly expressed in the striatal complex and the dentate gyrus. Nonetheless, in hippocampus, PDE4B expression and subcellular localization respond to the induction of LTP, suggesting a specific role for the 4B isozymes in this form of plasticity [100]. This finding takes on added significance in light of the fact that PDE4B disruption [101] and genetic variation [102, 103] are associated with neuropsychiatric disease.

Targeting specific PDE4 isozymes must also take into consideration particular isozymes that may be involved in the side effects associated with pan-PDE4 inhibition. Based on studies with PDE4 knockout mice in an innovative behavioral approach, Robichard et al. have put forward the hypothesis that PDE4D is specifically involved in the emetic response [104]. Significantly, an inhibitor with ~100-fold selectivity for PDE4D over other family members has been identified and found to cause emesis in early clinical studies in humans [105]. Taken together, these data suggest that inhibitors selective for PDE4A/B over PDE4D may be of particular interest for the treatment of cognitive dysfunction while obviating tolerability issues. The challenge now is to identify compounds with sufficient PDE4B selectivity with which to test this hypothesis.

Compounds with significant PDE4D selectivity have been identified [105]; however, it is unclear how this selectivity is achieved and, therefore, how to utilize the structure–activity relationships around these selective compounds to generate compounds that are selective for other PDE4 isozymes. Recently, Asahi Kasei Pharma (**1**, Fig. 1) and GSK (**2**, Fig. 1) have independently reported on two series of compounds with selectivity for PDE4B over PDE4D [106, 107]. Importantly, the GSK group is beginning to determine the molecular requirements that accompany this selectivity.

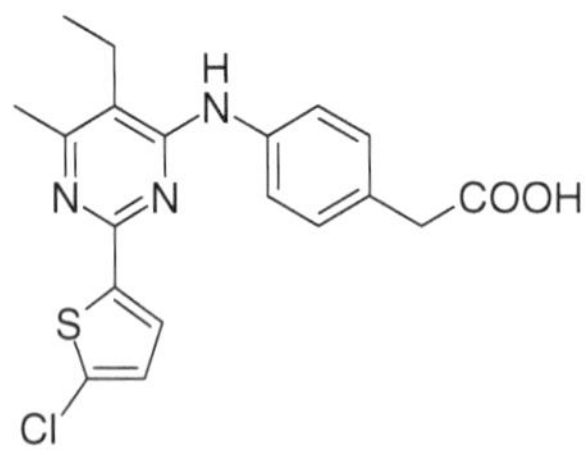

H
N
N
N
COOH
S
Cl
1 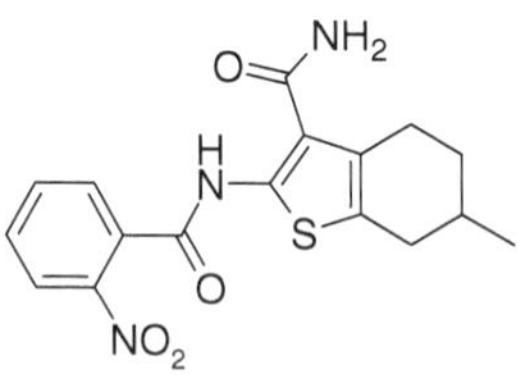
NH_2
O
H
N
S
O
NO_2
2

Fig. 1 PDE4 inhibitiors selective for the PDE4B isoform

Another very interesting advance is the recent disclosure by deCODE Genetics in a series of patents of a new class PDE4 inhibitors that are noncompetitive with respect to the cAMP-binding site [US Patents 12275152, 12275164, and 12275165]. This suggests that deCODE has identified a binding site on PDE4 outside of the substrate-binding pocket through which PDE4 enzymatic activity can be modulated. Although the deCODE compounds are selective for PDE4D over PDE4B, it is possible that knowledge about the PDE4D selectivity mechanism may allow for the development of other classes of compounds selective for the other PDE4 isozymes.

Finally, an area of PDE4 medicinal chemistry that has not yet been explored is the possibility of developing compounds selective for long versus short splice variants. Such an undertaking would be greatly facilitated by crystal structures of PDE4 that include regions of the protein beyond the catalytic domain.

4.4 PDE7 and PDE8B

While PDE4 is the major cAMP metabolizing enzyme, there are two other PDEs that are also selective for cAMP, PDE7, and PDE8. Thus, it is reasonable to investigate whether these PDE families may also play a role in regulating one or more of the many cAMP signaling pathways involved in synaptic plasticity. The physiology and pharmacology of these two enzymes are beginning to be investigated in depth, as reviewed below.

The PDE7 family is composed of two members, PDE7A and PDE7B, which demonstrate high affinity for cAMP but that are insensitive to rolipram [108, 109]. Unlike most other PDEs, PDE7 does not contain defined N-terminal regulatory domains although a consensus site for PKA phosphorylation does exist. While the protein expression profile of PDE7A and 7B is largely unknown, the mRNA levels for both isoforms reveal abundant expression in the CNS. PDE7A mRNA is expressed in the olfactory bulb and tubercle, hippocampus (dentate granule cells), and brain stem nuclei, while the highest level of PDE7B mRNA is localized to the cerebellum, striatum, dentate gyrus, and thalamic nuclei. Moreover, in humans there are three known splice variants for PDE7A that contain unique N- and C-terminal mRNA modifications that likely influence intracellular localization as well as interactions with other proteins. Promoter variants have also been reported for PDE7A, offering additional subtleties with respect to cAMP-responsiveness. PDE7A1 appears to encode a protein that contains peptide sequences in the N-terminal region that directly inhibit PKA catalytic activity [110]. Thus, this splice variant of PDE7A may regulate PKA activity in two ways, through regulation of cAMP levels and through a direct interaction with the PKA catalytic subunit.

The highest level of interest in PDE7 remains as a target to treat inflammatory disease, whereas interest in neurological diseases is just beginning to develop. There is a growing patent and medicinal chemistry literature developing around inhibitors that will serve as useful tools to explore these areas [111]. Omeros Corporation has disclosed in a patent application that in the MPTP mouse model of Parkinson's disease, PDE7 inhibitors restore stride length to prelesioned level when administered alone and also potentiate the activity of L-DOPA [112]. Thus, these apparently potent, brain-penetrant PDE7 inhibitors can serve as much needed tools to investigate the role of PDE7 in brain.

The PDE8 family is encoded by two genes, PDE8A and PDE8B, located on chromosomes 15 and 5, respectively [113]. In vitro, the catalytic properties for both PDE8A and 8B isoforms have been assessed and demonstrate very high affinity (40–60 nM) and specificity for cAMP. Both of the PDE8 mRNAs code for putative N-terminal regulatory elements within their protein structure, although their exact function is still unknown. Each of the putative regulatory domains found in PDE8 (the "REC" domain and "PAS" domain) are unique to this particular PDE family and share homology with highly conserved regulatory domains found in bacteria and mammalian several proteins. In lower organisms, the REC domain has been characterized as a sequence responsible for receiving signals from a particular sensor protein. As yet, it is unclear whether the REC domain in PDE8 plays a similar role in mammals. The PAS domain has been identified in several proteins involved with regulation of circadian rhythms as well as to be a potential site for ligand binding that may influence protein interactions. Alternative splicing of PDE8A results in several isoforms that lack the PAS domain. In addition to alternative start sites within the PDE8 promoter, additional variants are produced from modifications of primary transcripts (for instance, PDE8A2 is a splice variant from PDE8A1).

PDE8A mRNA has been localized to several tissues in the periphery, while the expression of PDE8B is highest in brain, thyroid, and testes. In addition, the PDE8B1 variant appears to be expressed only in the brain, while an equivalent level of expression of the PDE8B3 variant has been reported to occur in brain and thyroid.

The understanding of the role of both PDE7 and PDE8 in the CNS has been hampered by the lack of selective pharmacological tools and the lack of neuronal phenotypes in knockout animals. Nonetheless, there is some information to suggest a specific interest in these two enzymes for the treatment of AD. Recently, the levels of PDE7A, PDE7B, PDE8A, and PDE8B were investigated using specific oligonucleotide probes and in situ hybridization to postmortem brain samples from control and AD patients. Both PDE7 isoforms and PDE8B mRNA were found to be widely distributed in human brain, while PDE8A was not detected. In AD brain samples, the level of PDE7A mRNA was positively correlated with disease stage such that PDE7A mRNA levels decreased in the dentate gyrus with advancing disease progression (Braak stage III–VI). The levels of PDE7B mRNA remained unchanged. PDE8B mRNA levels were the highest in the pyramidal cell layer with advanced AD (Braak stage III–VI) and were also positively correlated with increasing age. This suggests a compensatory relationship between age and cAMP signaling that is enhanced with AD progression.

4.5 PDE5A

PDE5 inhibitors, such as sildenafil **3**, Vardenafil **4**, and Tadalafil **5** (Fig. 2), for the treatment of male erectile dysfunction are the first commercially successful "blockbusters" to arise from pharmaceutical development around the PDE superfamily. This success has generated tremendous interest in PDE5 as a therapeutic target for other disorders [114] and has provided the excellent pharmacological tools that so greatly facilitate such investigation. PDE5 is cGMP-specific. Given the extensive literature on the role of NO/cGMP signaling in synaptic plasticity, there has been considerable interest in PDE5 inhibitors to treat cognitive disorders. However, the effect of PDE5A inhibitors in preclinical models of cognition is enigmatic.

PDE5A inhibitors are robustly active in rodent assays of novel object recognition ([115, 116] and unpublished observation). PDE5A inhibition also attenuates spatial learning impairment in the 14-unit T-maze induced by cholinergic blockade, inhibition of nitric oxide synthase, or in aged rats [117]. Rutten et al. have recently reported that the PDE5 inhibitor sildenafil (Viagra) improves object retrieval performance in nonhuman primates [96]. However, sildenafil failed to effect cognitive deficits in humans suffering from schizophrenia [118]. PDE5 inhibitors also robustly facilitate functional recovery of sensorimotor function after stroke in the rat [119–122]. In these studies, PDE5 inhibitors were administered days after the stroke and had no effect on the infarct volume. Thus, it is argued that the effect of the compounds on sensorimotor recovery is through facilitating the ability of the brain to reorganize after damage; that is, through an effect on plasticity. Recently, Puzzo et al. reported effects of PDE5 inhibition that may be directly relevant to AD. This group found dramatic improvements caused by the PDE5 inhibitor sildenafil on hippocampal LTP measured in vitro in slices and on performance in cognitive tasks in a mouse model of AD, the APP/PS1 mice [123]. These effects were accompanied by an upregulation of CREB phosphorylation and a reduction in the levels of Aβ.

The data reviewed above, from various laboratories and in various model systems, indicate a potentially significant beneficial effect of PDE5 inhibition on brain function in general and synaptic plasticity in particular. The enigma stems from the fact that the expression of PDE5A in forebrain neuronal populations relevant to these effects is very limited. In rat forebrain, PDE5A mRNA was found only in isolated, phenotypically unidentified neurons in one report [124]

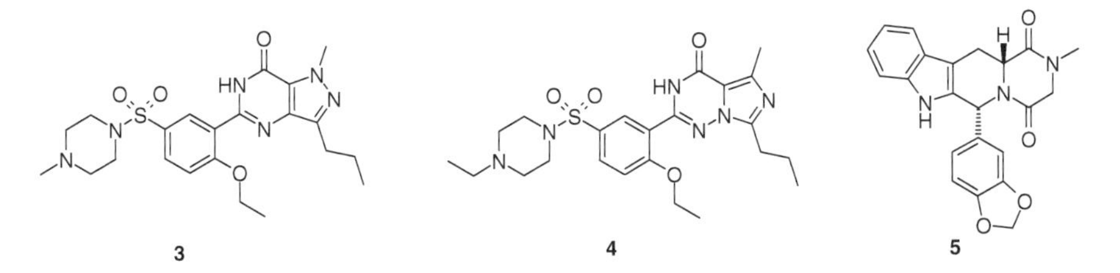

Fig. 2 PDE5A inhibitors approved for clinical uses

and was found not at all in another [125]. In addition, PDE5 protein was not detected [125] or only rarely detected [119] in rat forebrain in studies in which two different antibodies were used. PDE5A mRNA was also not detected in postmortem samples of forebrain from patients suffering from AD [126]. In contrast, PDE5 message and protein are robustly expressed cerebellar Purkinje neurons, some brain stem neurons, and spinal cord [119, 125, 127], as well as the cerebrovasculature [119]. However, it is difficult to reconcile the distribution of the enzyme in these latter neuronal populations with the various effects of the PDE5A inhibitors on brain function that have been reported. Thus, although PDE5A inhibitors are clinically available and are very well tolerated, a better understanding of the mechanisms underlying the effects on brain is warranted to provide meaningful clinical context.

4.6 PDE9A

Of the newly emerging PDE targets, the most interest is being generated around PDE9A and PDE2A, both of which regulate cGMP signaling in the brain. These are reviewed in the final two sections.

PDE9A is a high affinity, cGMP-specific enzyme that is expressed widely throughout the brain, albeit at apparently low levels [124, 126, 128]. PDE9A is the only isoform of this family but exhibits a complex pattern of gene transcripts yielding a total of 20 human splice variants [129, 130]. All splice variants use the same transcriptional start, but generate unique changes in the 5′ region of the mRNA, possibly allowing tissue-specific expression patterns [130, 131]. Functional changes mediated by these variations remain unclear as both the C-terminal catalytic domain and the main part of the N-terminal domain remain unaltered. The primary structure of PDE9A does not contain recognized regulatory domains, such as GAF domains, and the C-terminal homology compared to other PDEs is low, resulting in insensitivity of the enzyme to most known PDE inhibitors [132, 133]. Nonetheless, PDE9A is thought to be key player in regulating cGMP levels as it has the lowest K_m among the PDEs for this nucleotide [132, 133].

Only little is known about the protein expression and localization pattern of PDE9A. Two variants have been examined to date, PDE9A1, which was found in the nucleus, and PDE9A5, which is located in the cytoplasm [131]. A recent immunohistochemical analysis of PDE9A in the trigeminal ganglion confirmed neuronal localization of the protein in the cytoplasm [134]. Significantly more information is available on the expression of PDE9A mRNA, which has been detected in many tissues, reaching peak levels in kidney, brain, spleen, gastrointestinal tissues, and prostate [129, 130, 132]. In the brain, PDE9A mRNA is widely but very moderately expressed [124, 128]. It reaches peak levels in cerebellar Purkinje cells and is furthermore easily detectable in olfactory bulb, hippocampus, and cortical layer V [124]. Here, expression is considered primarily neuronal, but signals have been detected in astrocytes and Schwann cells as well [124, 134]. In human postmortem brain tissue of healthy elderly people and Alzheimer patients,

PDE9A mRNA was detected in cortex, hippocampus, and cerebellum in a pattern comparable to the rodent [126]. No differences in expression were observed in the Alzheimer patients.

Considerable interest in this enzyme was engendered following characterization of BAY 73-6691 (see **10**, Fig. 5, below), the first PDE9A-specific inhibitor [135]. This compound selectively inhibits human PDE9A with an in vitro IC_{50} of 55 nM and a minimum 25-fold window to other PDEs. In a broad pharmacological assessment, BAY 73-6691 enhanced early LTP after weak tetanic stimulation in hippocampal slices prepared from young adult Wistar rats and old, but not young, Fischer 344 X Brown Norway (FBNF1) rats [136]. Significantly, BAY 73-6691 enhanced acquisition, consolidation, and retention of long-term memory in a number of preclinical behavioral paradigms, including a social recognition task, a scopolamine-disrupted passive avoidance task, and a MK-801-induced short-term memory deficit in a T-maze alternation task [136]. Subsequently, it was reported that LTP is enhanced in hippocampal slices prepared from PDE9A knockout mice, and that this effect is mimicked by a PDE9A inhibitor in slices prepared from the rat hippocampus [137]. These latter inhibitors robustly facilitated object recognition memory in both mice and rats and increased baseline cGMP levels in hippocampus, cortex, and striatum [137, 138]. These observations further underline the central role of PDE9A in regulating cGMP levels in the CNS. Taken together, these data suggest that PDE9A inhibition may provide AD patients with some therapeutic benefit. Based on this data, Pfizer Inc. has advanced a PDE9A inhibitor into clinical development for AD. This compound enhanced cGMP levels in the CSF of healthy volunteers, providing proof-of-mechanism to the concept of PDE9A inhibition in humans [139].

In summary, initial doubt around the potential of PDE9A as an effective target for the symptomatic treatment of dementia predicated on the overall modest expression pattern of the gene in the CNS has been superseded by the positive data achieved with selective PDE9A inhibitors outlined above. Two central biological questions still remain to be answered: (1) the nature of a PDE9A sensitive cGMP pool and (2) the subcellular localization pattern of PDE9A in terms of temporal and spatial resolution. Nonetheless, the available data fuel several extensive medicinal chemistry efforts that are summarized below.

Sequence analysis and X-ray crystallographic evidence reveal a number of fundamental differences between PDE9A and other PDEs, but from a chemogenomic perspective the low affinity of PDE9A to IBMX is a clear indicator that PDE9A inhibitors must fulfill other structural requirements than inhibitors of most other PDE isoforms [140]. Full-length PDE9A is inhibited by IBMX with an IC_{50} value of around 230 μM which is significantly lower than for all other PDEs except PDE8 [141]. Nevertheless, a crystal structure of IBMX bound to the PDE9A2 catalytic domain has been obtained by crystallizing the protein with a large excess of IBMX. The X-ray crystal structure reveals a single hydrogen bond between the xanthine N-7 of IBMX and the glutamine 453 of PDE9A, rather than the double hydrogen bond usually observed in complexes of IBMX with other PDEs. A subsequent study of PDE9A crystallized with its natural ligand at low temperature has provided important information about the catalytic mechanism [78].

Although initially elusive, the search for selective PDE9A inhibitors has yielded a number of interesting compounds from various classes. It appears that PDE9A has a very pronounced preference for compounds displaying variations of the purinone scaffold, i.e., flat, aromatic heterobicyclic compounds capable of forming the characteristic double hydrogen bond to the active site glutamine as observed in structures of many other PDE inhibitors such as sildenafil and vardenafil [142]. These structural characteristics are also recognizable in the chemical classes that have resulted from the four major discovery efforts disclosed so far; these chemical classes are discussed below.

Pfizer. Pfizer appears to have been involved in at least two distinct discovery programs centered on PDE9A pharmacology, namely programs in the indications diabetes and cardiovascular disease, as well as neurology. While the peripheral and central indications may have differing requirements of the inhibitor in terms of selectivity profile, pharmacokinetics, and organ distribution, it is interesting to see both programs in comparison.

The starting point for Pfizer's first published PDE9A projects [US20040220186] was compound **6** (Fig. 3), which had been identified by screening a library of PDE inhibitors from previous projects on other PDE isoforms [143]. The compound **6** is a potent inhibitor of PDE9A (IC_{50} = 10 nM) but essentially nonselective with activity on PDE1A-C and PDE5A in the same range and is notably similar to sildenafil. Structural optimization over several iterations led to selective compounds such as **7** (Fig. 3), a 41 nM PDE9 inhibitor with a selectivity factor of 30 or better toward PDE1A-C and PDE5A. The compound **7** is active in vivo in mice (glucose lowering) after oral dosing of 100 mg/kg and above. Other compounds with promising in vitro profiles had no in vivo effect, probably due to poor absorption as a result of relatively high polar surface areas.

Other compounds with carboxylic acid substituents were identified as very potent and reasonably selective, but such compounds almost certainly have very low CNS exposure. Thus, Pfizer's second and currently most advanced PDE9A program aimed at identifying a PDE9A inhibitor for the treatment of cognitive deficits in AD and other neuropsychiatric disorders, and so sought compounds with properties that improve brain penetration. Although relatively little has been disclosed about the in vitro and in vivo profile so far, it is clear that Pfizer's scientists have identified a very potent compound class: numerous compounds in the patent application have IC_{50} values in the single-digit nanomolar region or even below [WO2008139293]. Pfizer has completed Phase I with a PDE9A inhibitor from this

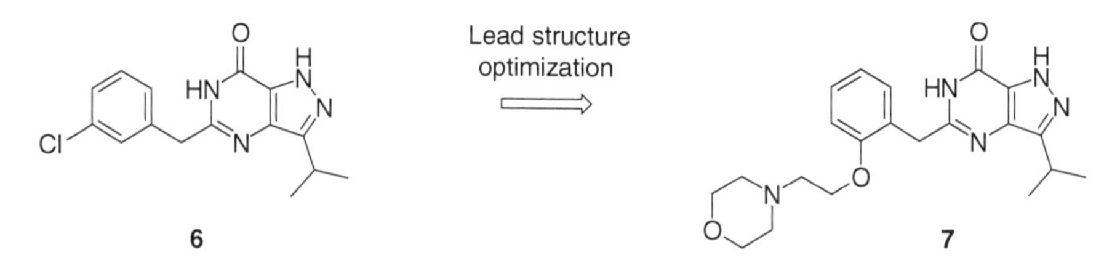

Fig. 3 Lead optimization of PDE9A inhibtors disclosed by Pfizer

series [139], but the structure of the compound has not been disclosed. A couple of examples from the patent application are shown below (**8** and **9**, Fig. 4). Some of the most potent compounds are characterized by a high polar surface, and so the clinical candidate likely to be a compound in which a good in vitro activity and good overall PK and pharmaceutical properties are unified in one molecule.

Bayer. Bayer, the first company to publish detailed pharmacological data for a selective PDE9A inhibitor [135], has also been involved in several compound classes and indications. BAY 736691 (**10**, Fig. 5) [WO2004099211, WO 2004099210, WO 2004018474] belongs to the class of pyrazolopyrimidinones, but several published patent applications describe a second chemical class, the cyanopyrimidinones [WO2004113306, WO2005068436, WO2006125554] exemplified by compound **11** (Fig. 5). Bayer has never disclosed the structure of a clinical candidate from this series, but it appears that there is a high degree of similarity between the SAR in the two series.

ASKA. ASKA Pharmaceutical Co. Ltd of Tokyo has been involved in PDE9A discovery projects for years; the first patent application was filed in 2006 and until now a total of four patent applications on PDE9A inhibitors have been made public [WO2006135080, WO2008018306, WO2008072778, and WO2008072779]. Two distinct compound classes have been disclosed: the first is represented by **12a** and **12b** (Fig. 6) and a class of heterotricyclic compounds (**13**, Fig. 6).

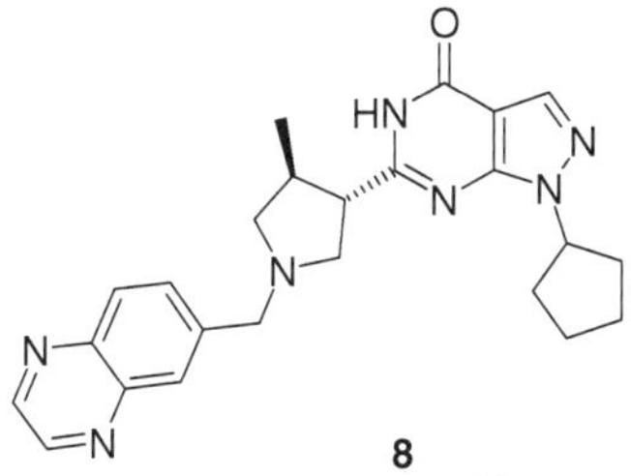

8
IC_{50} = 7 nM

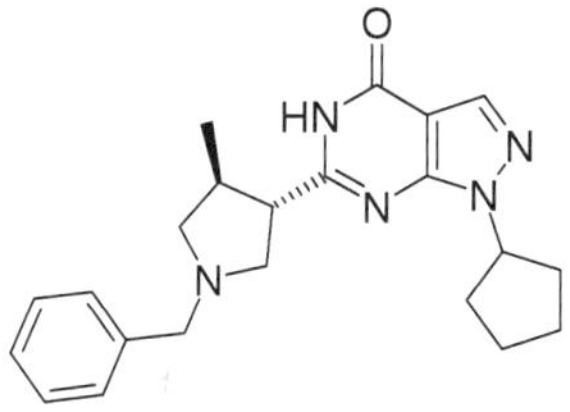

9
IC_{50} = 9 nM

Fig. 4 Examples of PDE9A inhibitors optimized to improve brain penetration

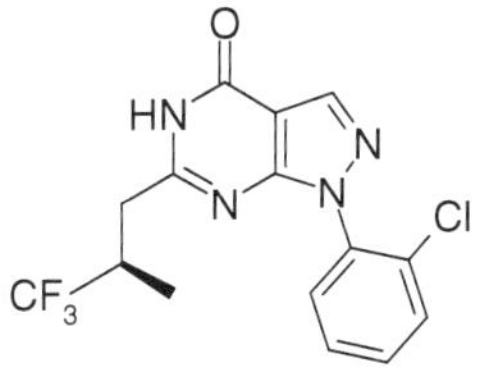

10
BAY 736691
IC_{50} = 55 nM

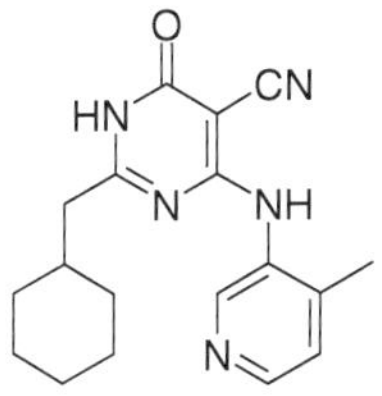

11
IC_{50} = 52 nM

Fig. 5 Pyrazolo- and cyano-pyrimidinone PDE9A inhibitors disclosed by Bayer

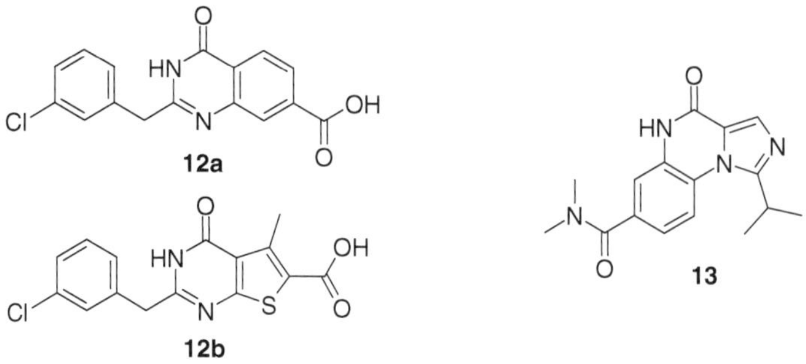

Fig. 6 PDE9A inhibitors disclosed by ASKA. The carboxylic acid group of 12a and 12b is replaced in 13 to improve brain penetration

Although various indications have been claimed for both classes (including CNS indications such as AD and general neuropathy), the carboxylic acid makes any high CNS exposure rather unlikely, and it would appear that these compounds are targeted for peripheral indications such as prostate disease, incontinence, or pulmonary hypertension, although no in vivo data have been published to support these claims. The tricyclic systems, on the other hand, seem more promising in that respect, but the general SAR appears to overlap with that of the Bayer and Pfizer programs, so there is reason to believe that the binding mode of this compound is essentially the same (as is the case for the other ASKA compounds). No structural data have been published so far though. It is unclear whether ASKA is still actively involved in PDE9A-related research and development: there is no mention of the project on the company homepage although the most recent patent application was published in 2008.

Boehringer Ingelheim. The most recent player to enter the increasingly competitive field of PDE9A research is Boehringer Ingelheim with a patent application detailing inhibitors of the pyrazolopyrimidinone type [WO2009068617]. Although no detailed biological data have been disclosed, this focused compound class seems to be quite selective vs. PDE1 and generally rather potent on PDE9A (14, Fig. 7, published with IC50 value as a range as shown).

The similarity to BAY736691 (**10**, Fig. 5) is noticeable, and interestingly one of the original Bayer inventors appears on the Boehringer Ingelheim patent application which seems to indicate that the Boehringer Ingelheim program is based on intellectual property acquired from Bayer. One would expect to see more patent applications from this source in the future.

4.7 PDE2A

PDE2A belongs to the dual substrate PDEs hydrolyzing both cAMP and cGMP [144]. PDE2A is a single gene family with three known splice variants (PDE2A1-3) that differ with respect to their N-terminus [145–147]. It is unclear whether all

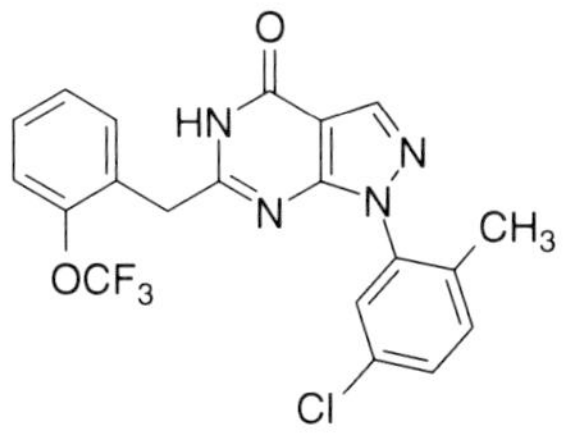

14
IC_{50} between 10-500 nM
Selectivity vs PDE1: 271-fold

Fig. 7 PDE9A inhibitor disclosed by Boehringer Ingelheim

splice variants are shared across species. PDE2A2 and PDE2A3 are the predominant splice variants expressed in the brain where they are associated with membranes [148]. This localization is partially due to N-terminal palmitoylation [149], but was recently shown for PDE2A3 to be mainly mediated through N-terminal acetylation [148]. PDE2A1 is found in soluble fractions and lacks the most N-terminal region present in PDE2A2/3.

PDE2A is unique in that it is allosterically activated by physiological concentrations of cGMP binding to the N-terminal GAF domains, which triggers the degradation of cAMP [146]. This positive cooperativity constitutes a mechanism of crosstalk between distinct cAMP and cGMP-regulated signaling pathways. It also indicates that, although PDE2A remains silent under baseline conditions, it is selectively activated upon neuronal stimulation that causes an increase in cGMP. In primary cultures of forebrain neurons, PDE2A preferentially metabolized cGMP [150], suggesting that in the CNS this enzyme may also serve as an inhibitory feedback regulator of cGMP signaling. As for all other PDEs, it is of central interest to reveal the subcellular localization of PDE2A to understand the impact of this selective activation on cellular function. Several studies show PDE2A expression in various tissues reaching highest levels in the CNS and particularly the limbic system [145, 151, 152]. A recent immunoreactivity study by Stephenson and colleagues substantiates earlier findings on PDE2A expression in the neuronal dendrites and axons, suggesting compartmentalization of the enzyme directly at the input and output region of neurons. Interestingly, a fine punctuate pattern in neurites is pronounced in areas known to be involved in learning and memory formation and affected in AD pathology, like the hippocampus, striatum, and cortex. Here, neuropil localization is accompanied by a lack of PDE2A immunoreactivity in cell bodies. In further studies, PDE2A was detected in membrane rafts [149] and synaptosomal membranes [148], substantiating evidence for a localization at the immediate site of synaptic contacts and thus in a suitable position to hydrolyze the second messengers cAMP and cGMP immediately at the synapse. Inhibition of PDE2A therefore appears attractive as it might offer a selective prolongation of cAMP and cGMP levels directly related to synaptic activation. In fact, one of the highest levels of PDE2A expression in brain appears to be the mossy

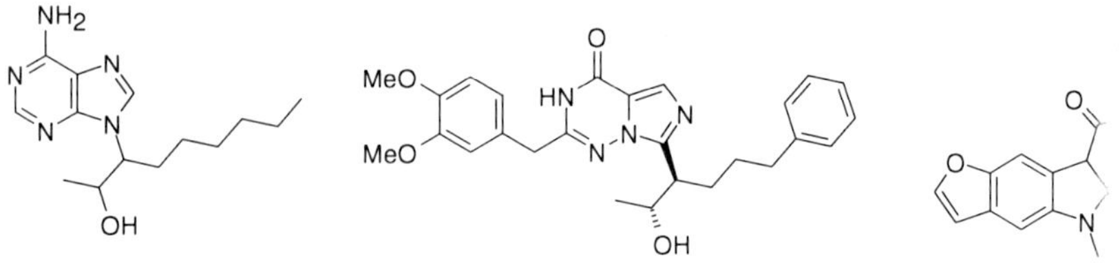

Fig. 8 Prototype PDE2A inhibitors

fibers emanating from the hippocampal dentate granule cells and receivi
from the entorhinal cortex, one of the first brain regions showing morph
signs of pathology in AD [152]. This raises the intriguing possibility that P
involved in regulating presynaptic forms of synaptic plasticity. Perhaps, P
one of the mediators of retrograde NO signaling in the presynaptic termin
through regulating cGMP directly or by regulating cAMP levels in res
cGMP binding to the GAF domain. Interestingly, in a few brain regions
medial habenula and neuronal subsets in the cortex, substantia nigra pars c
or raphe nuclei show somatic staining. This heterogeneous localization
within different neuronal populations indicates divergent roles of PDE2A i
ent cell populations. The CNS expression pattern of PDE2A is prese
mammals, including humans [126], and remains unaltered in postmortem b
Alzheimer patients [117].

Recently, a highly potent and selective PDE2A inhibitor, BAY 6
(**16**, Fig. 8), with an IC_{50} for human recombinant PDE2A of 4.7 nM h
shown to enhance LTP at the CA3/CA1 synapse in hippocampal slices
Systemic administration of BAY 60-7550 to rodents has also been sh
attenuate natural forgetting in young rats and improve age-related imp
on old rats in behavioral tasks addressing episodic short- and long-term m
in rats [116, 153, 154]. The compound also reverses working memory def
mice induced by a time decay or acute treatment with the NMDA receptor a
nist MK-801 [153]. The various temporal stages of memory consolidation, re
from working to short-term and long-term memory, have been suggested
differentially regulated by cAMP and cGMP in either pre- or postsynaptic term
[89]. It was therefore speculated that interference with a dual substrate PDE t
localized both at the pre- and postsynaptic site should have a broad impa
different temporal stages of memory processing. The promnemonic ef
achieved with BAY 60-7550 are in line with this hypothesis [89]. It shoul
noted that BAY 60-7550 penetrates into the CNS very poorly (authors' pers
observations); thus, generalization regarding the effect of PDE2A inhibition
cognitive function awaits confirmatory studies with other compounds. Moreove
PDE2A constitutive knockout mouse line are not available for behavioral studies
genetic deletion of PDE2A is reported to be embryonically lethal.

Based on the CNS expression pattern, positive cooperative kinetics between cAMP and cGMP, and synaptic association of PDE2A, the enzyme is believed to be a very attractive target to support signaling pathways involved in synaptic plasticity and learning and memory. However, to date a clear link from PDE2A to AD is missing. It should also be noted that PDE2A is widely expressed in peripheral tissues as well, including heart, liver, lung, and kidney, where PDE2A inhibitors have various functional effects [155]. With the identification of more brain-penetrating PDE2A inhibitors, it will therefore be important to identify pharmacological windows between centrally mediated effects on cognition and those in the other organs. Toward this end, PDE2 inhibitors have been pursued by a number of research groups for various indications. So far, discovery of potent and selective PDE2 inhibitors with good CNS exposure has proven to be a real challenge; progress is reviewed below.

The main tool compounds available for mechanistic research at present are EHNA **15** (sub-micromolar inhibitor of PDE2A, the first selective inhibitor of PDE2A described in the literature [156]), BAY 607550 **16** (depending on the construct a nanomolar to sub-nanomolar inhibitor of PDE2A, structurally related to EHNA [153]), and the chemically distinct oxindole **17** (double-digit nanomolar inhibitor of PDE2A with good selectivity [105]) as shown in Fig. 8.

All three compounds are unlikely to advance beyond the tool compound level; EHNA is not potent enough to qualify as a development candidate, whereas BAY 607550 has rather poor pharmacokinetic properties and the oxindole has negligible CNS penetration. Still, all three have been immensely useful as mechanistic probes of the PDE2A enzyme and for studying non-CNS pharmacology models.

Bayer. Bayer has been pursuing the structural class around BAY 607550 as documented by various patent applications for CNS and cardiovascular indications [WO2008043461, WO02068423, WO00250078, WO00209713, WO00012504, and WO09840384] although apparently without identifying a clinical candidate.

Pfizer. Pfizer has pursued the oxindole class as well as a class of azaquinazolines [WO2005061497], but again the current development stage remains unclear if this project is uncertain.

Altana Pharma. Altana Pharma (now a part of Nycomed) has been addressing PDE2A through two distinct chemical classes: a BAY 607550-like class **18** [WO2005021037, WO-2004089953] using the EHNA-scaffold and another class of triazolophthalazines **19** [WO2006024640, WO2006072612, WO2006072615] (Fig. 9). No data have been disclosed for individual compounds, but some are reported to inhibit PDE2A in the low nanomolar region. It appears that COPD and inflammation have been the relevant indications for these compounds rather than CNS indications, although the general physicochemical profile might also be compatible with CNS exposure. It is unclear whether Nycomed is actively developing either of these classes of PDE2A inhibitors.

Cell Pathways. The PDE2A inhibitor research program at Cell Pathways has been largely based on substituted indenes (compound **20**) of the type shown in Fig. 10 [EP01749824, US06465494, WO02067936]. Information about pharmacological properties are scarce (the best example reported in the patent literature is a

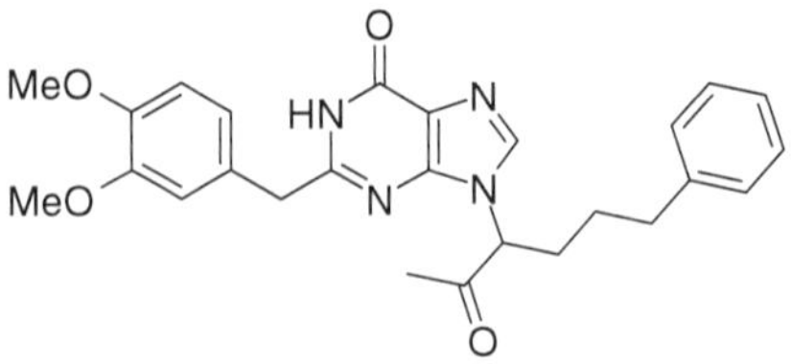

18
Altana Class I

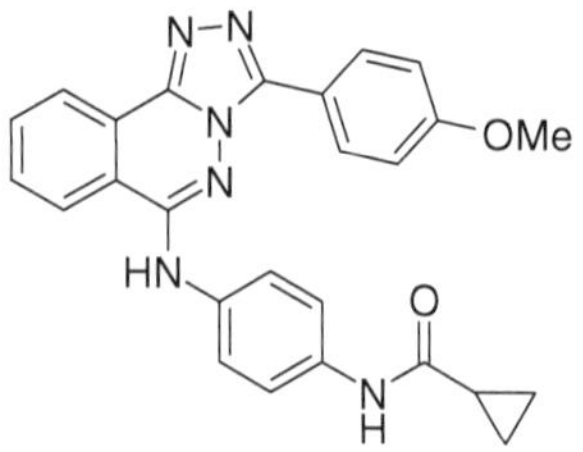

19
Altana Class II

Fig. 9 Representatives of two classes of PDE2A inhibitors disclosed by Altana (Nycomed)

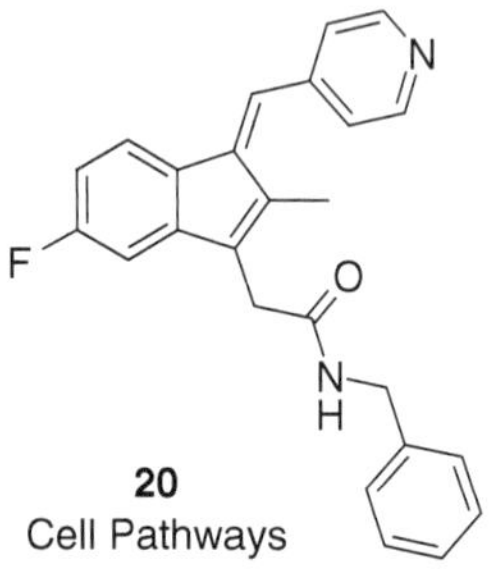

20
Cell Pathways

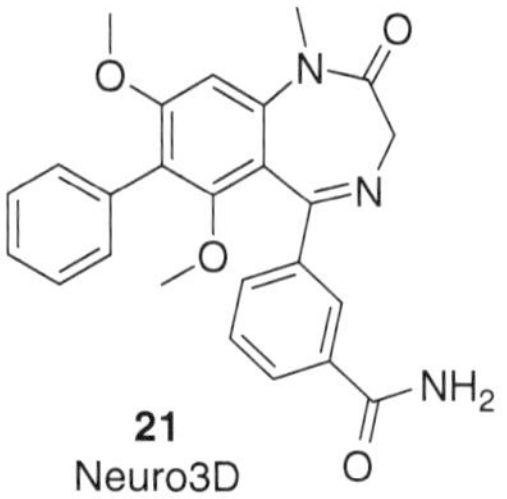

21
Neuro3D

Fig. 10 PDE2A inhibitors disclosed by Cell Pathways and Neuro3D

0.68 μM PDE2A inhibitor), but it seems clear that these compounds are meant for non-CNS indications such as inflammatory bowel disease.

Neuro3D. Finally, Neuro3D (now acquired by Evotec AG) have been involved in PDE2A research with a class of benzodiazepinones **21** (Fig. 10) [EP01548011, WO2004041258] that are reported to be selective although not especially potent inhibitors of PDE2A.

5 Perspective

The suggestion that PDE inhibitors should be explored as a novel approach for the treatment of AD is based on several premises such as follows: (1) AD is principally a disease of synaptic dysfunction, and targeting this dysfunction is a means to impact both the symptoms and the progression of the disease; (2) the cyclic nucleotide signaling cascades offer a molecular entry point for such therapeutic approaches, given the significant roles played in the regulation of synaptic function; and (3) manipulation of PDE activity is a physiologically relevant and pharmaceutically facile way to manipulate cyclic nucleotide signaling. Indeed, these premises form the basis for the development and ongoing clinical trials with both PDE4 and PDE9A inhibitors for the treatment of AD.

While we await important feedback from these clinical trials, there are several points that bear further consideration and investigation. The most important of these is the nature of the synaptic defect that underlies the propensity of an individual to develop AD and that serves as a target for a PDE inhibitor therapy. The "ante" for many therapeutic approaches that target cognition/synaptic dysfunction has been potentiation of LTP at the CA3/CA1 synapse in the hippocampus. This is clearly the case for the PDE inhibitors. There is a wealth of data suggesting that this particular form of synaptic plasticity mediates the long-term memory function in the hippocampus. Given that a deficit in hippocampal-mediated memory function is a hallmark of AD, mechanisms that facilitate hippocampal LTP are certainly reasonable targets to consider for treating those memory deficits. However, even this simple premise must be qualified, given the observation with the PDE4D knockout mice, where increases in hippocampal LTP in slice preparations are associated with deficits in cognitive behavioral tasks. Furthermore, PDE2A, PDE4, PDE5A, and PDE9A inhibitors have all been shown to facilitate hippocampal LTP. As stated above, cAMP and cGMP signaling is intimately involved in the regulation of synaptic function along the entire spatial and temporal continuum of plasticity. Given that PDE function is highly compartmentalized, it is a near certainty that each step in this continuum involves a distinct PDE isozyme. Thus, PDE2A, PDE4, PDE5A, and PDE9A inhibitors may all potentiate hippocampal LTP, but which specifically targets a signaling defect relevant to the synaptic pathology in AD? As a next step to address this issue, it would be very informative to conduct a comparative analysis of the effects of the relevant PDE inhibitors on plasticity at the CA3/CA1 synapse to establish the role and position of each of the cognate enzymes in the complex cascade of events mediating this type of plasticity. These inhibitors then become tools to investigate these specific steps in relevant disease models to determine whether a particular step represents a therapeutically relevant end point. Such an iterative approach should yield interesting insights into the disease and a better focus on the best new therapeutic opportunities.

Finally, the PDEs may be considered more broadly as potential therapeutic targets to treat a range of neuropsychiatric diseases that have as a fundamental pathology synaptic dysfunction. The most obvious examples of these are schizophrenia and autism. However, synaptic dysfunction may also play a principal, though underappreciated, role in neurodegenerative conditions beyond AD. In Parkinson's disease, the loss of dopamine nerve terminals in the striatum appears to precede loss of dopamine neurons in the substantia nigra. The loss of dopamine terminals is also significantly greater than that of cell bodies as the disease progresses. This suggests that therapeutic strategies that preserve dopamine terminals may have a significant impact on both the symptoms of Parkinson's disease and, perhaps, on disease progression. Similar arguments can be made in the treatment of Huntington's disease, where disruption of corticostriatal synapses may be, at least in part, responsible for the loss of cortical BDNF delivery to the vulnerable striatal medium spiny neurons as well as the retrograde transport of BDNF from striatum back to the cortex. Thus, in these cases, maintenance or facilitation of synaptic function through PDE inhibition goes beyond the scope of "cognition enhancement" toward promoting more fundamental brain functions. This is clearly an area for further investigations.

References

1. Gauthier S, Reisberg B, Zaudig M et al (2006) Mild cognitive impairment. Lancet 367: 1262–1270
2. Cummings JL (2004) Alzheimer's disease. N Engl J Med 351:56–67
3. Francis PT, Palmer AM, Snape M et al (1999) The cholinergic hypothesis of Alzheimer's disease: a review of progress. J Neurol Neurosurg Psychiatry 66:137–147
4. Reisberg B, Doody R, Stoffler A et al (2003) Memantine in moderate-to-severe Alzheimer's disease. N Engl J Med 348:1333–1341
5. Chartier-Harlin MC, Crawford F, Houlden H et al (1991) Early-onset Alzheimer's disease caused by mutations at codon 717 of the beta-amyloid precursor protein gene. Nature 353:844–846
6. Murrell J, Farlow M, Ghetti B et al (1991) A mutation in the amyloid precursor protein associated with hereditary Alzheimer's disease. Science 254:97–99
7. Selkoe DJ (2002) Alzheimer's disease is a synaptic failure. Science 298:789–791
8. Roberts GW, Nash M, Ince PG et al (1993) On the origin of Alzheimer's disease: a hypothesis. Neuroreport 4:7–9
9. Conti M, Beavo J (2007) Biochemistry and physiology of cyclic nucleotide phosphodiesterases: essential components in cyclic nucleotide signaling. Annu Rev Biochem 76: 481–511
10. Terry R, Masliah E, Salmon D et al (1991) Physical basis of cognitive alterations in Alzheimer's disease: synapse loss is the major correlate of cognitive impairment. Ann Neurol 30:572–580
11. Scheff S, Price D (2003) Synaptic pathology in Alzheimer's disease: a review of ultrastructural studies. Neurobiol Aging 24:1029–1046
12. Masliah E, Mallory M, Alford M et al (2001) Altered expression of synaptic proteins occurs early during progression of Alzheimer's disease. Neurology 56:127–129
13. Masliah E, Mallory M, Alford M et al (1994) Synaptic and neuritic alterations during the progression of Alzheimer's disease. Neurosci Lett 174:67–72
14. Mirra SS, Heyman A, McKeel D et al (1991) The consortium to establish a registry for Alzheimer's disease (CERAD): Part II. Standardization of the neuropathologic assessment of Alzheimer's disease. Neurology 41:479
15. Braak H, Braak E (1995) Staging of Alzheimer's disease-related neurofibrillary changes. Neurobiol Aging 16:271–278
16. Braak H, Alafuzo I, Arzberger T et al (2006) Staging of Alzheimer disease-associated neurofibrillary pathology using parafin sections and immunocytochemistry. Acta Neuropathol 112:389–404
17. Small DH, Mok SS, Bornstein JC (2001) Alzheimer's disease and Aβ toxicity: from top to bottom. Nat Rev Neurosci 2:595–598
18. Samir K-S, Jessie T, Van Bianca B et al (2006) Mean age-of-onset of familial Alzheimer disease caused by presenilin mutations correlates with both increased Aβ42 and decreased Aβ40. Hum Mutat 27:686–695
19. Hardy J, Higgins G (1992) Alzheimer's disease: the amyloid cascade hypothesis. Science 256:184–185
20. Rowe CC, Ng S, Ackermann U et al (2007) Imaging beta-amyloid burden in aging and dementia. Neurology 68:1718–1725
21. Mufson E, Chen E-Y, Cochran E et al (1999) Entorhinal cortex β-amyloid load in individuals with mild cognitive impairment. Exp Neurol 158:469–490

22. Hyman BT, Van Hoesen GW, Damasio AR et al (1984) Alzheimer's disease: cell-specific pathology isolates the hippocampal formation. Science 225:1168–1170
23. Lue L-F, Kuo Y-M, Roher AE et al (1999) Soluble amyloid beta peptide concentration as a predictor of synaptic change in Alzheimer's disease. Am J Pathol 155:853–862
24. Zheng H, Koo E (2006) The amyloid precursor protein: beyond amyloid. Mol Neurodegener 1:5
25. Tampellini D, Rahman N, Gallo EF et al (2009) Synaptic activity reduces intraneuronal Abeta, promotes app transport to synapses, and protects against Abeta-related synaptic alterations. J Neurosci 29:9704–9713
26. Cirrito J, Yamada K, Finn M et al (2005) Synaptic activity regulates interstitial fluid amyloid-beta levels in vivo. Neuron 48:913–922
27. Calabrese B, Shaked G, Tabarean I et al (2007) Rapid, concurrent alterations in pre- and postsynaptic structure induced by naturally-secreted amyloid-beta protein. Mol Cell Neurosci 35:183–193
28. Shrestha B, Vitolo O, Joshi P et al (2006) Amyloid beta peptide adversely affects spine number and motility in hippocampal neurons. Mol Cell Neurosci 33:274–282
29. Cleary JP, Walsh DM, Hofmeister JJ et al (2005) Natural oligomers of the amyloid-beta protein specifically disrupt cognitive function. Nat Neurosci 8:79–84
30. Arancio O, Puzzo D, Privitera L et al (2008) Amyloid-beta peptide is a positive modulator of synaptic plasticity and memory. Alzheimers Dement 4:T196–T197
31. Giuffrida ML, Caraci F, Pignataro B et al (2009) β-amyloid monomers are neuroprotective. J Neurosci 29:10582–10587
32. Corder E, Saunders A, Strittmatter W et al (1993) Gene dose of apolipoprotein E type 4 allele and the risk of Alzheimer's disease in late onset families. Science 261:921–923
33. Farrer L, Cupples L, Haines J et al (1997) Effects of age, sex, and ethnicity on the association between apolipoprotein E genotype and Alzheimer disease. A meta-analysis. ApoE and Alzheimer disease meta analysis consortium. JAMA 278:1349–1356
34. Corder EH, Saunders AM, Risch NJ et al (1994) Protective effect of apolipoprotein E type 2 allele for late onset Alzheimer disease. Nat Genet 7:180–184
35. Deary IJ, Whiteman MC, Pattie A et al (2002) Ageing: cognitive change and the APOE epsilon 4 allele. Nature 418:932
36. Filippini N, MacIntosh BJ, Hough MG et al (2009) Distinct patterns of brain activity in young carriers of the APOE-epsilon4 allele. Proc Natl Acad Sci USA 106:7209–7214
37. Reiman EM, Chen K, Alexander GE et al (2005) Correlations between apolipoprotein E epsilon4 gene dose and brain-imaging measurements of regional hypometabolism. Proc Natl Acad Sci USA 102:8299–8302
38. Busch RM, Lineweaver TT, Naugle RI et al (2007) Apoe-epsilon4 is associated with reduced memory in long-standing intractable temporal lobe epilepsy. Neurology 68:409–414
39. Palop JJ, Chin J, Roberson ED et al (2007) Aberrant excitatory neuronal activity and compensatory remodeling of inhibitory hippocampal circuits in mouse models of Alzheimer's disease. Neuron 55:697–711
40. Ponomareva NV, Korovaitseva GI, Rogaev EI (2008) EEG alterations in non-demented individuals related to apolipoprotein E genotype and to risk of Alzheimer disease. Neurobiol Aging 29:819–827
41. Ponomareva NV, Selesneva ND, Jarikov GA (2003) EEG alterations in subjects at high familial risk for Alzheimer's disease. Neuropsychobiology 48:152–159
42. Wang C, Wilson W, Moore S et al (2005) Human APOE4-targeted replacement mice display synaptic deficits in the absence of neuropathology. Neurobiol Dis 18:390–398
43. Francis S, Corbin J (1999) Cyclic nucleotide-dependent protein kinases: intracellular receptors for cAMP and cGMP action. Crit Rev Clin Lab Sci 36:275–328
44. Kaupp UB, Seifert R (2002) Cyclic nucleotide-gated ion channels. Physiol Rev 82: 769–824

45. Bos JL (2006) Epac proteins: multi-purpose cAMP targets. Trends Biochem Sci 31: 680–686
46. Malenka RC, Bear MF (2004) LTP and Ltd: an embarrassment of riches. Neuron 44:5–21
47. Citri A, Malenka RC (2008) Synaptic plasticity: multiple forms, functions, and mechanisms. Neuropsychopharmacol 33:18–41
48. Kandel E (2001) The molecular biology of memory storage: a dialogue between genes and synapses. Science 294:1030–1038
49. Abel T, Nguyen P, Barad M et al (1997) Genetic demonstration of a role for PKA in the late phase of LTP and in hippocampus-based long-term memory. Cell 88:615–626
50. Lisman J (1989) A mechanism for the Hebb and the anti-Hebb processes underlying learning and memory. Proc Natl Acad Sci USA 86:9574–9578
51. Esteban J, Shi S, Wilson C et al (2003) PKA phosphorylation of AMPA receptor subunits controls synaptic trafficking underlying plasticity. Nat Neurosci 6:136–143
52. Banke T, Bowie D, Lee H et al (2000) Control of GluR1 AMPA receptor function by cAMP dependent protein kinase. J Neurosci 20:89–102
53. Lysetskiy M, Földy C, Soltesz I (2005) Long- and short-term plasticity at mossy fiber synapses on mossy cells in the rat dentate gyrus. Hippocampus 15:691–696
54. Weisskopf M, Castillo P, Zalutsky R et al (1994) Mediation of hippocampal mossy fiber long-term potentiation by cyclic AMP. Science 265:1878–1882
55. Nicoll R, Schmitz D (2005) Synaptic plasticity at hippocampal mossy fibre synapses. Nat Rev Neurosci 6:863–876
56. Kemp A, Manahan-Vaughan D (2007) Hippocampal long-term depression: master or minion in declarative memory processes? Trends Neurosci 30:111–118
57. Bear MF, Abraham WC (1996) Long-term depression in hippocampus. Annu Rev Neurosci 19:437–462
58. Lee H, Barbarosie M, Kameyama K et al (2000) Regulation of distinct ampa receptor phosphorylation sites during bidirectional synaptic plasticity. Nature 405:955–959
59. Kleppisch T, Feil R (2009) Cgmp signalling in the mammalian brain: role in synaptic plasticity and behaviour. In: Schmidt H, Stasch J-P, Hofmann F (eds) CGMP: generators, effectors and therapeutic implications. Springer, Berlin, pp 549–579
60. Hawkins R, Son H, Arancio O (1998) Nitric oxide as a retrograde messenger during long-term potentiation in hippocampus. Prog Brain Res 118:155–172
61. Arancio O, Kandel E, Hawkins R (1995) Activity-dependent long-term enhancement of transmitter release by presynaptic 3', 5'-cyclic GMP in cultured hippocampal neurons. Nature 376:74–80
62. Arancio O, Kiebler M, Lee C et al (1996) Nitric oxide acts directly in the presynaptic neuron to produce long-term potentiation in cultured hippocampal neurons. Cell 87:1025–1035
63. Son H, Lu Y-F, Zhuo M et al (1998) The specific role of cGMP in hippocampal LTP. Learn Mem 5:231–245
64. Lu Y-F, Kandel ER, Hawkins RD (1999) Nitric oxide signaling contributes to late-phase LTP and CREB phosphorylation in the hippocampus. J Neurosci 19:10250–10261
65. Haghikia A, Mergia E, Friebe A et al (2007) Long-term potentiation in the visual cortex requires both nitric oxide receptor guanylyl cyclases. J Neurosci 27:818–823
66. Hopper RA, Garthwaite J (2006) Tonic and phasic nitric oxide signals in hippocampal long-term potentiation. J Neurosci 26:11513–11521
67. Zhuo M, Kandel E, Hawkins R (1994) Nitric oxide and cGMP can produce either synaptic depression or potentiation depending on the frequency of presynaptic stimulation in hippocampus. Neuroreport 5:1033–1036
68. Reyes-Harde M, Potter BVL, Galione A et al (1999) Induction of hippocampal LTD requires nitric-oxide-stimulated PKG activity and ca2+ release from cyclic ADP-ribose-sensitive stores. J Neurophysiol 82:1569–1576
69. Wei J, Jin X, Cohen E et al (2002) cGMP-induced presynaptic depression and postsynaptic facilitation at glutamatergic synapses in visual cortex. Brain Res 927:42–54

70. Lugnier C (2006) Cyclic nucleotide phosphodiesterase (PDE) superfamily: a new target for the development of specific therapeutic agents. Pharmacol Ther 109:366–398
71. Ke H, Wang H (2006) Structure, catalytic mechanism, and inhibitor selectivity of cyclic nucleotide phosphodiesterases. In: Beavo JA, Francis SH, Houslay MD (eds) Cyclic nucleotide phosphodiesterases in health and disease. CRC, Boca Raton, FL
72. Scapin G, Patel SB, Chung C et al (2004) Crystal structure of human phosphodiesterase 3b: atomic basis for substrate and inhibitor specificity. Biochemistry 43:6091–6100
73. Ke H, Wang H (2007) Crystal structures of phosphodiesterases and implications on substrate specificity and inhibitor selectivity. Curr Top Med Chem 7:391–403
74. Wang H, Yan Z, Yang S et al (2008) Kinetic and structural studies of phosphodiesterase-8a and implication on the inhibitor selectivity. Biochemistry 47:12760–12768
75. Verhoest PR, Chapin DS, Corman M et al (2009) Discovery of a novel class of phosphodiesterase 10a inhibitors and identification of clinical candidate 2-[4-(1-methyl-4-pyridin-4-yl-1 h-pyrazol-3-yl)-phenoxymethyl]-quinoline (pf-2545920) for the treatment of schizophrenia. J Med Chem 52:7946–7949
76. Wang H, Liu Y, Hou J et al (2007) Structural insight into substrate specificity of phosphodiesterase 10. Proc Natl Acad Sci 104:5782–5787
77. Zhang KYJ, Card GL, Suzuki Y et al (2004) A glutamine switch mechanism for nucleotide selectivity by phosphodiesterases. Mol Cell 15:279–286
78. Liu S, Mansour MN, Dillman KS et al (2008) Structural basis for the catalytic mechanism of human phosphodiesterase 9. Proc Natl Acad Sci 105:13309–13314
79. Zoraghi R, Corbin JD, Francis SH (2006) Phosphodiesterase-5 Gln817 is critical for cGMP, vardenafil, or sildenafil affinity: its orientation impacts cGMP but not cAMP affinity. J Biol Chem 281:5553–5558
80. Sandner P, Svenstrup N, Tinel H et al (2008) Phosphodiesterase 5 inhibitors and erectile dysfunction. Expert Opin Ther Pat 18:21–33
81. Sung BJ, Hwang KY, Jeon YH et al (2003) Structure of the catalytic domain of human phosphodiesterase 5 with bound drug molecules. Nature 425:98–102
82. Smith FD, Scott JD (2006) Anchored cAMP signaling: onward and upward – a short history of compartmentalized cAMP signal transduction. Eur J Cell Biol 85:585–592
83. Baillie G, Scott J, Houslay M (2005) Compartmentalisation of phosphodiesterases and protein kinase a: opposites attract. FEBS Lett 579:3264–3270
84. Houslay MD, Adams DR (2003) PDE4 cAMP phosphodiesterasesmodular enzymes that orchestrate signalling cross-talk, desensitization and compartmentalization. Biochem J 370:1–18
85. Jin SLC, Lan L, Zoudilova M et al (2005) Specific role of phosphodiesterase 4b in lipopolysaccharide-induced signaling in mouse macrophages. J Immunol 175:1523–1531
86. Menniti FS, Faraci WS, Schmidt CJ (2006) Phosphodiesterases in the CNS: targets for drug development. Nat Rev Drug Discov 5:660–670
87. Rose G, Hopper A, De Vivo M et al (2005) Phosphodiesterase inhibitors for cognitive enhancement. Curr Pharm Des 11:3329–3334
88. Barad M, Bourtchouladze R, Winder DG et al (1998) Rolipram, a type IV-specific phosphodiesterase inhibitor, facilitates the establishment of long-lasting long-term potentiation and improves memory. Proc Natl Acad Sci USA 95:15020–15025
89. Reneerkens O, Rutten K, Steinbusch H et al (2009) Selective phosphodiesterase inhibitors: a promising target for cognition enhancement. Psychopharmacol 202:419–443
90. Gong B, Vitolo O, Trinchese F et al (2004) Persistent improvement in synaptic and cognitive functions in an Alzheimer mouse model after rolipram treatment. J Clin Invest 114:1624–1634
91. Cullen W, Suh Y, Anwyl R et al (1997) Block of LTP in rat hippocampus in vivo by beta-amyloid precursor protein fragments. Neuroreport 8:3213–3217
92. Itoh A (1999) Impairments of long-term potentiation in hippocampal slices of beta-amyloid infused rats. Eur J Pharmacol 382:167–175

93. Vitolo O (2002) Amyloid beta-peptide inhibition of the PKA/CREB pathway and long-term potentiation: reversibility by drugs that enhance cAMP signaling. Proc Natl Acad Sci USA 99:13217–13221
94. Giembycz MA (2009) Can the anti-inflammatory potential of PDE4 inhibitors be realized: guarded optimism or wishful thinking? Br J Pharmacol 155:228–290
95. Sturton G, Fitzgerald M (2002) Phosphodiesterase 4 inhibitors for the treatment of COPD. Chest 121:192S–196S
96. Rutten K, Basile JL, Prickaerts J et al (2008) Selective PDE inhibitors rolipram and sildenafil improve object retrieval performance in adult cynomolgus macaques. Psychopharmacology (Berl) 196:643–648
97. Ramos BP, Birnbaum SG, Lindenmayer I et al (2003) Dysregulation of protein kinase a signaling in the aged prefrontal cortex: new strategy for treating age-related cognitive decline. Neuron 40:835–845
98. Cherry JA, Davis RL (1999) Cyclic AMP phosphodiesterases are localized in regions of the mouse brain associated with reinforcement, movement, and affect. J Comp Neurol 407:287–301
99. Rutten K, Misner D, Works M et al (2008) Enhanced long-term potentiation and impaired learning in phosphodiesterase 4d-knockout (PDE4d) mice. Eur J Neurosci 28:625–632
100. Ahmed T, Frey J (2005) Phosphodiesterase 4b (PDE4b) and cAMP-level regulation within different tissue fractions of rat hippocampal slices during long-term potentiation in vitro. Brain Res 1041:212–222
101. Millar JK, Pickard BS, Mackie S et al (2005) DISC1 and PDE4B are interacting genetic factors in schizophrenia that regulate cAMP signaling. Science 310:1187–1191
102. Fatemi S, King D, Reutiman T et al (2008) PDE4B polymorphisms and decreased PDE4B expression are associated with schizophrenia. Schizophr Res 101:36–49
103. Numata S, Ueno S, Iga J et al (2008) Positive association of the PDE4B (phosphodiesterase 4B) gene with schizophrenia in the japanese population. J Psychiatr Res 43:7–12
104. Robichaud A, Stamatiou P, Jin S et al (2002) Deletion of phosphodiesterase 4D in mice shortens alpha(2)-adrenoceptor-mediated anesthesia, a behavioral correlate of emesis. J Clin Invest 110:1045–1052
105. Chambers R, Abrams K, Castleberry T et al (2006) A new chemical tool for exploring the role of the PDE4D isozyme in leukocyte function. Bioorg Med Chem Lett 16:718–721
106. Kranz M, Wall M, Evans B et al (2009) Identification of PDE4B over 4D subtype-selective inhibitors revealing an unprecedented binding mode. Bioorg Med Chem 17:5336–5341
107. Naganuma K, Omura A, Maekawara N et al (2009) Discovery of selective PDE4B inhibitors. Bioorg Med Chem Lett 19:3174–3176
108. Giembycz MA, Smith SJ (2006) Phosphodiesterase 7 (PDE7) as a therapeutic target. Drugs Future 31:207–229
109. Michaeli T (2006) Pde7. In: Beavo JA, Francis SH, Houslay MD (eds) Cyclic nucleotide phosphodiesterases in health and disease. CRC, Boca Raton, FL, pp 195–204
110. Han P, Sonati P, Rubin C et al (2006) Pde7a1, a camp-specific phosphodiesterase, inhibits camp-dependent protein kinase by a direct interaction with c. J Biol Chem 281:15050–15057
111. Gil C, Campillo NE, Perez DI et al (2008) PDE7 inhibitors as new drugs for neurological and inflammatory disorders. Expert Opin Ther Pat 18:1127–1139
112. Bergmann JE, Cutshall NS, Demopulos GA et al (2008) Use of PDE7 inhibitors for the treatment of movement disorders, US 20080260643
113. Vasta V (2006) cAMP-phosphodiesterase 8 family. In: Beavo JA, Francis SH, Houslay MD (eds) Cyclic nucleotide phosphodiesterases in health and disease. CRC, Boca Raton, FL
114. Sandner P, Hutter J, Tinel H et al (2007) PDE5 inhibitors beyond erectile dysfunction. Int J Impot Res 19:533–543
115. Prickaerts J, Sik A, van Staveren W et al (2004) Phosphodiesterase type 5 inhibition improves early memory consolidation of object information. Neurochem Int 45:915–928

116. Rutten K, Prickaerts J, Hendrix M et al (2007) Time-dependent involvement of cAMP and cGMP in consolidation of object memory: studies using selective phosphodiesterase type 2, 4 and 5 inhibitors. Eur J Pharmacol 558:107–112
117. Devan B, Duffy K, Bowker J et al (2005) Phosphodiesterase type 5 (PDE5) inhibition and cognitive enhancement. Drugs Future 30:725
118. Goff D, Cather C, Freudenreich O et al (2009) A placebo-controlled study of sildenafil effects on cognition in schizophrenia. Psychopharmacol 202:411–417
119. Menniti F, Ren J, Coskran T et al (2009) PDE5A inhibitors improve functional recovery after stroke in rats: optimized dosing regimen and implications for mechanism. J Pharmacol Exp Ther 331:1–9
120. Zhang L, Zhang RL, Wang Y et al (2005) Functional recovery in aged and young rats after embolic stroke: treatment with a phosphodiesterase type 5 inhibitor. Stroke 36:847–852
121. Zhang L, Zhang Z, Zhang RL et al (2006) Tadalafil, a long-acting type 5 phosphodiesterase isoenzyme inhibitor, improves neurological functional recovery in a rat model of embolic stroke. Brain Res 1118:192–198
122. Zhang R, Wang Y, Zhang L et al (2002) Sildenafil (viagra) induces neurogenesis and promotes functional recovery after stroke in rats. Stroke 33:2675–2680
123. Puzzo D, Staniszewski A, Deng SX et al (2009) Phosphodiesterase 5 inhibition improves synaptic function, memory, and amyloid-beta load in an Alzheimer's disease mouse model. J Neurosci 29:8075–8086
124. Van Staveren W, Steinbusch H, Markerink-Van Ittersum M et al (2003) Mrna expression patterns of the cGMP-hydrolyzing phosphodiesterases types 2, 5, and 9 during development of the rat brain. J Comp Neurol 467:566–580
125. Kotera J, Fujishige K, Omori K (2000) Immunohistochemical localization of cGMP-binding cGMP-specific phosphodiesterase (PDE5) in rat tissues. J Histochem Cytochem 48:685–694
126. Reyes-Irisarri E, Markerink-Van Ittersum M, Mengod G et al (2007) Expression of the cGMP-specific phosphodiesterases 2 and 9 in normal and Alzheimer's disease human brains. Eur J Neurosci 25:3332–3338
127. Kruse LS, Sandholdt NTH, Gammeltoft S et al (2006) Phosphodiesterase 3 and 5 and cyclic nucleotide-gated ion channel expression in rat trigeminovascular system. Neurosci Lett 404:202–207
128. Andreeva S, Dikkes P, Epstein P et al (2001) Expression of cGMP-specific phosphodiesterase 9A mRNA in the rat brain. J Neurosci 21:9068–9076
129. Guipponi M, Scott HS, Kudoh J et al (1998) Identification and characterization of a novel cyclic nucleotide phosphodiesterase gene (PDE9A) that maps to 21q22.3: alternative splicing of mRNA transcripts, genomic structure and sequence. Hum Genet 103:386–392
130. Rentero C, Monfort A, Puigdomenech P (2003) Identification and distribution of different mRNA variants produced by differential splicing in the human phosphodiesterase 9A gene. Biochem Biophys Res Commun 301:686–692
131. Wang P, Wu P, Egan RW et al (2003) Identification and characterization of a new human type 9 cGMP-specific phosphodiesterase splice variant (PDE9A5). Differential tissue distribution and subcellular localization of PDE9A variants. Gene 314:15–27
132. Fisher DA, Smith JF, Pillar JS et al (1998) Isolation and characterization of PDE9A, a novel human cGMP-specific phosphodiesterase. J Biol Chem 273:15559–15564
133. Soderling SH, Bayuga SJ, Beavo JA (1998) Identification and characterization of a novel family of cyclic nucleotide phosphodiesterases. J Biol Chem 273:15553–15558
134. Kruse LS, Moller M, Tibaek M et al (2009) PDE9A, PDE10A, and PDE11A expression in rat trigeminovascular pain signalling system. Brain Res 1281:25–34
135. Wunder F, Tersteegen A, Rebmann A et al (2005) Characterization of the first potent and selective PDE9 inhibitor using a cGMP reporter cell line. Mol Pharmacol 68:1775–1781
136. van der Staay F, Rutten K, Bärfacker L et al (2008) The novel selective PDE9 inhibitor bay 73-6691 improves learning and memory in rodents. Neuropharmacol 55:908–918

137. Menniti FS, Kleiman R, Schmidt C (2008) PDE9A-mediated regulation of cGMP: impact on synaptic plasticity. Schizophr Res 102:38–39
138. Schmidt CJ, Harms JF, Tingley FD et al (2009) PDE9A-mediated regulation of cGMP: developing a biomarker for a novel therapy for Alzheimer's disease. Alzheimers Dement 5: P331
139. Nicholas T, Evans R, Styren S et al (2009) Pf-04447943, a novel PDE9A inhibitor, increases cGMP levels in cerebrospinal fluid: translation from non-clinical species to healthy human volunteers. Alzheimers Dement 5:P330–P331
140. Kubinyi H, Müller G (eds) (2004) Chemogenomics in drug discovery: a medicinal chemistry perspective. Wiley-VCH, Weinheim
141. Huai Q, Wang H, Zhang W et al (2004) Crystal structure of phosphodiesterase 9 shows orientation variation of inhibitor 3-isobutyl-1-methylxanthine binding. Proc Natl Acad Sci USA 101:9624–9629
142. Wang H, Ye M, Robinson H et al (2008) Conformational variations of both phosphodiesterase-5 and inhibitors provide the structural basis for the physiological effects of vardenafil and sildenafil. Mol Pharmacol 73:104–110
143. Deninno MP, Andrews M, Bell AS et al (2009) The discovery of potent, selective, and orally bioavailable PDE9 inhibitors as potential hypoglycemic agents. Bioorg Med Chem Lett 19:2537–2541
144. Stroop S, Beavo J (1991) Structure and function studies of the cGMP-stimulated phosphodiesterase. J Biol Chem 266:23802–23809
145. Rosman GJ, Martins TJ, Sonnenburg WK et al (1997) Isolation and characterization of human cDNAs encoding a cGMP-stimulated 3′, 5′-cyclic nucleotide phosphodiesterase. Gene 191:89–95
146. Sonnenburg WK, Mullaney PJ, Beavo JA (1991) Molecular cloning of a cyclic GMP-stimulated cyclic nucleotide phosphodiesterase cDNA. Identification and distribution of isozyme variants. J Biol Chem 266:17655–17661
147. Yang Q, Paskind M, Bolger G et al (1994) A novel cyclic GMP stimulated phosphodiesterase from rat brain. Biochem Biophys Res Commun 205:1850–1858
148. Russwurm C, Zoidl G, Koesling D et al (2009) Dual acylation of PDE2A splice variant 3: targeting to synaptic membranes. J Biol Chem 284:25782–25790
149. Noyama K, Maekawa S (2003) Localization of cyclic nucleotide phosphodiesterase 2 in the brain-derived triton-insoluble low-density fraction (raft). Neurosci Res 45:141–148
150. Suvarna NU, O'Donnell JM (2002) Hydrolysis of N-methyl-D-aspartate receptor-stimulated cAMP and cGMP by PDE4 and PDE2 phosphodiesterases in primary neuronal cultures of rat cerebral cortex and hippocampus. J Pharmacol Exp Ther 302:249–256
151. Sadhu K, Hensley K, Florio VA et al (1999) Differential expression of the cyclic GMP-stimulated phosphodiesterase PDE2A in human venous and capillary endothelial cells. J Histochem Cytochem 47:895–906
152. Stephenson DT, Coskran TM, Wilhelms MB et al (2009) Immunohistochemical localization of PDE2A in multiple mammalian species. J Histochem Cytochem 57:933–949
153. Boess F, Hendrix M, van der Staay F et al (2004) Inhibition of phosphodiesterase 2 increases neuronal cGMP, synaptic plasticity and memory performance. Neuropharmacol 47:1081–1092
154. Domek-Lopacinska K, Strosznajder JB (2008) The effect of selective inhibition of cyclic GMP hydrolyzing phosphodiesterases 2 and 5 on learning and memory processes and nitric oxide synthase activity in brain during aging. Brain Res 1216:68–77
155. Bender A (2006) Calmodulin-stimulated cycli nucleotide phosphodiesterease. In: Beavo JA, Francis SH, Houslay MD (eds) Cyclic nucleotide phosphodiesterases in health and disease. CRC, Boca Raton, FL, pp 35–54
156. Podzuweit T, Nennstiel P, Muller A (1995) Isozyme selective inhibition of cGMP-stimulated cyclic nucleotide phosphodiesterases by erythro-9-(2-hydroxy-3-nonyl) adenine. Cell Signal 7:733–738

Top Med Chem 6: 91–147
DOI: 10.1007/7355_2010_11

Published online: 6 October 2010

Glutamate and Neurodegenerative Disease

Eric Schaeffer and Allen Duplantier

Abstract As the main excitatory neurotransmitter in the mammalian central nervous system, glutamate is critically involved in most aspects of CNS function. Given this critical role, it is not surprising that glutamatergic dysfunction is associated with many CNS disorders. In this chapter, we review the literature that links aberrant glutamate neurotransmission with CNS pathology, with a focus on neurodegenerative diseases. The biology and pharmacology of the various glutamate receptor families are discussed, along with data which links these receptors with neurodegenerative conditions. In addition, we review progress that has been made in developing small molecule modulators of glutamate receptors and transporters, and describe how these compounds have helped us understand the complex pharmacology of glutamate in normal CNS function, as well as their potential for the treatment of neurodegenerative diseases.

Keywords Alzheimer's disease, AMPA, Excitatory amino acid transporter, Glutamate, Huntington's disease, mGluR, Neurodegeneration, NMDA, Parkinson's disease

Contents

E. Schaeffer (✉)
CHDI Foundation, 300 Alexander Park, Suite 100, Princeton, NJ 08540, USA
e-mail: eric_schaeffer@comcast.net

A. Duplantier
Pfizer Global Research and Development, Eastern Point Road, Groton, CT 06340, USA
e-mail: allen.j.duplantier@pfizer.com

1 Introduction

As the major excitatory neurotransmitter in the mammalian central nervous system, glutamate plays a critical role in virtually all aspects of behavior, perception, and cognition. Decades of research have elucidated the molecular mechanisms involved in glutamate synthesis, vesicular trafficking, and synaptic transmission. In addition to its role as a neurotransmitter, glutamate is also an essential amino acid required for protein synthesis, and as such is present at much higher levels in the brain, than are most other neurotransmitters. Glutamate is synthesized from glutamine in the synaptic nerve terminal, where it is translocated into synaptic vesicles through the action of vesicular glutamate transporters (VGLUTs) [1]. When an action potential is fired, glutamate is released into the synaptic cleft, where it binds to and activates a wide variety of different receptors. Following release glutamate is then cleared from the synapse by the action of a family of excitatory amino acid transporters (EAATs), expressed predominantly on astrocytes [2]. Once taken up into the astrocyte, the glutamate is converted to glutamine and is stored in this form until additional levels of glutamate are required by the neuron. This intimate relationship between neurons and astrocytes is unique to this excitatory neurotransmitter, and is thought to be an important mechanism for maintaining its potent excitatory activity in a quiescent state until required for synaptic function.

Glutamate signaling is mediated through several distinct classes of ligand-gated ion channels (ionotropic glutamate receptors) [3, 4] as well as a family of G-protein-coupled receptors (metabotropic glutamate receptors) [5]. There are three families of ionotropic glutamate receptors (1) *N*-methyl D-aspartate (NMDA) receptors, (2) 2-amino-3-(3-hydroxy-5-methylisoxazol-4-yl) propionic acid (AMPA) receptors, and (3) kainate (KA) receptors. Each of these receptor families was originally named for the ligand shown to selectively activate them. These pharmacological classifications are still used today, although with the molecular cloning of the protein subunits that make up these receptors we now have a better appreciation of the complexity and heterogeneity of each of these receptor families [4]. The metabotropic family of glutamate receptors (mGluRs) consists of eight receptor subtypes, and has been divided into three subfamilies or groups (see Sect. 1 for more detail). These receptors are found both pre- and postsynaptically, and are typically involved in modulating synaptic responses to glutamate. Unlike the ionotropic receptors which are generally involved in fast synaptic transmission, the metabotropic receptors tend to be involved in mediating slower responses to synaptic glutamate release, consistent with their G-protein-coupled signaling mechanism [5].

Under normal physiological conditions, glutamate signaling is a tightly regulated process. However, following acute neuronal injury as well as in chronic neurodegenerative conditions, these signaling systems can become dysfunctional due to either excessive or inadequate levels of extracellular glutamate. Prolonged exposure to elevated levels of glutamate may lead to overactivation of excitatory glutamate receptors, which in turn can result in an abnormally high influx of calcium into the cell [6]. This can occur most notably with the NMDA receptor, which as will be discussed below is preferentially permeable to calcium ions, but may also involve certain subtypes of AMPA receptors. Such prolonged increases in intracellular calcium can serve as a toxic trigger for activation of a number of calcium-sensitive intracellular processes including mitochondrial membrane depolarization, caspase activation, production of reactive oxygen species, and eventually cell death. This process by which elevated glutamate levels leads to prolonged receptor activation is referred to as excitotoxicity, and has been shown to play an important role in both acute neuronal injury and chronic neurodegenerative diseases [6]. Although the cause of elevated glutamatergic signaling can vary from increased release, to reduced uptake, to increased receptor number or activity, the resulting prolonged elevation in intracellular calcium and the downstream excitotoxic cascade appears to be a common mechanism involved in diverse CNS degenerative conditions.

Although excessive activation of glutamate receptors can be deleterious, reduced glutamatergic signaling is also associated with various pathological states. This has been particularly well established in schizophrenia, where reduced glutamatergic tone in prefrontal brain regions is thought to be a key underlying neurochemical deficit. Indeed, NMDA antagonists such as ketamine and phencyclidine give rise to psychotic-like symptoms and cognitive deficits in healthy subjects, and can exacerbate symptoms in schizophrenics. Current theories suggest that a prefrontal hypoglutamatergic state is particularly involved in the social withdrawal and cognitive deficits associated with schizophrenia, and treatments which enhance glutamatergic neurotransmission may be useful in treating these symptom domains [7]. Although these approaches are being explored extensively in the area of psychotic disorders, a more general role may exist for direct or indirect glutamate agonists as cognition enhancers, with potential relevance to neurodegenerative disease, where cognitive deficits are well known.

A wide variety of approaches for modulating glutamatergic transmission have been investigated as potential therapies for treating neurodegenerative as well as psychiatric conditions. These may involve modulating pre- or postsynaptic signaling through the use of selective receptor agonists and antagonists, as well as directly modulating the release or reuptake of the neurotransmitter. In this chapter, we have focused on some of the key evidence linking different glutamate receptors and other related targets with neurodegenerative diseases, and review examples of small molecule approaches that target different components of the glutamatergic synapse.

2 NMDA Receptors

2.1 NMDA Receptor Biology

NMDA receptors are composed of four subunits derived from three subunit classes (NR1, NR2A–D, and NR3A and B). A typical NMDA receptor is a heterotetramer consisting of two NR1 and two NR2 subunits, with NR3 sometimes substituting for NR2 [8, 9]. These ligand-gated cation channels flux both Na^+ and Ca^{2+} ions in response to glutamate stimulation with a greater permeability to Ca^{2+} than is seen for other ionotropic glutamate receptors [10, 11]. The degree of cation preference is influenced by the specific subunit composition, with NR2-containing receptors being highly permeable to Ca^{2+} ions [12]. In addition to the agonist glutamate, NMDA receptors also require binding of the coagonist glycine or D-serine in order for the channel to open [13, 14]. A unique property of the NMDA receptor is the fact that the channel pore is typically blocked by Mg^{2+} at resting membrane potentials. When the neuronal cell membrane is depolarized in response to an excitatory stimulus, this Mg^{2+} block is removed, and in the presence of glutamate and glycine the channel will flux Ca^{2+} ions. The term “coincidence detector” has been used to describe the NMDA channel, since it will only open when two events – postsynaptic cell depolarization and glutamate release – occur coincidentally [15]. This property is thought to make the NMDA receptor well suited to play a role in synaptic plasticity, learning and memory, and there is considerable evidence in support of these functions [16].

As described above, prolonged increases in extracellular glutamate, associated with a variety of acute and chronic neurodegenerative conditions, may lead to elevated intracellular calcium due to overstimulation of the NMDA receptor. The rationale for NMDA blockade as an approach for the treatment of stroke, mediated by reduced excitotoxic damage and enhanced cell survival has been well supported by animal model studies. However, despite the strong rationale for this mechanistic approach, clinical experience with NMDA antagonists has been largely negative, due to poor toleration and a general lack of efficacy [17, 18]. There tends to be a narrow window of opportunity after an ischemic event when excitotoxic damage occurs, and for practical reasons it has been difficult to treat stroke patients within this limited time window. Another important consideration is that, although blocking NMDA receptors proximal to a lesion may attenuate cell death associated with excitotoxicity, it will also interfere with the survival-promoting effects of NMDA activation that are a necessary part of the recovery process [17]. While the role of the NMDA receptor in excitotoxicity has been extensively studied, it has more recently been shown that NMDA activation also plays a critical role in normal cellular physiology and can be an important neuroprotective mechanism [19]. Blockade of NMDA receptors more distal to an ischemic lesion will interfere with a prosurvival function leading to apoptosis and further functional impairment [20]. Thus, the response of neurons to NMDA activation follows a bell-shaped curve, with too much, or too little stimulation being detrimental to neuronal health and survival [21].

The NMDA receptor is linked with a large intracellular protein complex via the C-terminal tails of the NR1 and NR2 subunits. This protein complex, which is localized to the intracellular side of the postsynaptic membrane, positions the proteins involved in transducing the NMDA receptor signal in close proximity to where Ca^{2+} enters the cell [22]. The signaling pathways downstream from the NMDA receptor that mediate the prosurvival functions involves activation of protein kinases such as the PI3K/Akt cascade, which may be activated via influx of calcium [23, 24]. NMDA receptor activation has also been shown to result in a calcium-dependant phosphorylation of the transcription factor CREB by multiple protein kinases [25], which has been linked with prosurvival effects both in vitro [26, 27] and in vivo [28]. The body of work linking NMDA activation with antiapoptotic activity and prosurvival signaling is now extensive, and would suggest that while NMDA antagonists may have utility in treating neurodegenerative disease, this will have to be balanced with maintaining normal function to promote endogenous recovery processes.

While modulation of NMDA function has been explored with regard to the treatment of acute neuronal injury, it has also received considerable attention with respect to the etiology and treatment of several chronic neurodegenerative conditions. As the etiologies of diseases such as Alzheimer's disease (AD), Parkinson's disease (PD), and Huntingon's disease (HD) have been studied, it has become apparent that normal excitatory neurotransmission and plasticity is compromised in various brain regions, and that dysregulation of neurotransmitter systems can occur early in the disease process, often before frank neuropathology is detectable.

Alzheimer's disease is classically characterized by two neuropathological hallmarks (1) β-amyloid plaques, containing the amyloid β-peptide (Aβ) and (2) neurofibrillary tangles, consisting mainly of hyperphosphorylated tau protein. While these proteinaceous deposits are clearly present late in the disease process, it is increasingly clear that synaptic pathology occurs before these protein aggregates are present, and it is the deterioration of synaptic function which correlates most directly with symptom progression [29, 30]. Studies looking at the effects of the Aβ peptide on cell function, as well as transgenic mouse models, which overexpress the human amyloid precursor protein (hAPP), have shed light on the relationship between AD pathology and glutamate receptors. In rodent studies, it has been shown that administration of Aβ peptides (Aβ1-42 and Aβ25-35) promotes a cascade of events involving astroglial depolarization, increases in extracellular glutamate and activation of NMDA receptors. Consistent with a critical role for the NMDA receptor in mediating the toxic effects of Aβ is the fact that treatment of rats with the NMDA antagonist dizocilpine, reduced the toxicity observed [31]. Similar studies have shown that the toxic effects of direct injection of Aβ1-40 into the rat hippocampus can be blocked by the weak NMDA antagonist memantine (Compound **9**, Fig. 2) [32]. Interestingly, it has also been shown that more modest activation of NMDA receptors can lead to increased production of Aβ, through a shift from normal processing of hAPP via α-secretase, to pathological processing via β-secretase. This leads to the plausible hypothesis of a positive feed-forward loop, where NMDA activation enhances Aβ production, which in turn leads to further, and potentially toxic NMDA activity [33].

Beyond the excitotoxic role of NMDA receptors, there is also evidence suggesting that earlier in the AD process, Aβ may play a role in downregulation of NMDA receptors through increased endocytosis and decreased surface expression. This would have clear consequences for reduced glutamatergic transmission, which may occur early in the disease. A number of studies have supported this mechanism by showing that incubation of Aβ1-42 peptide with cultured neurons [34, 35] can result in reduced surface expression of NMDA receptors. In addition to reducing receptor expression, Aβ has also been shown to have a direct or indirect modulating effect on NMDA function [35, 36], and NMDA-mediated plasticity [37]. Thus, by modulating NMDA receptor expression and/or function, Aβ may have profound effects on glutamatergic signaling in AD. Based on the above findings, the directionality of NMDA modulation may need to differ depending on the stage of disease, with activators being useful in early disease and blockers being more appropriate later.

Reduced NMDA function in AD, as well as in psychiatric conditions such as schizophrenia, is thought to be associated with cognitive impairment. This has led to an interest in the identification of compounds which function as agonists at this receptor. While direct agonists that bind to the glutamate-binding site (including NMDA itself) are generally excitotoxic, a more subtle approach for NMDA activation has been proposed through activation of the glycine coagonist site [38]. Current approaches include drugs which bind directly to the glycine site, as well as inhibitors of the glycine transporter as a means of elevating the endogenous ligand. These agents have shown promise in animal studies [39], and preliminary clinical data suggest the potential for efficacy in treating the negative and cognitive symptoms of schizophrenia [40].

NMDA receptors are expressed throughout the basal ganglia, and prominently on the medium spiny neurons (MSNs) of the striatum, where they play important roles in regulating motor and cognitive functions. It is thus not surprising that the NMDA receptor has been explored as a therapeutic target in the treatment of conditions that involve degeneration of striatal neurons themselves, as well as populations of dopaminergic neurons that project to the striatum. In PD, the loss of dopaminergic innervation of the striatum results in overactivity of the indirect pathway, and a consequent hyperactivation of the subthalamic nucleus (STN) (the role of glutamatergic neurotransmission and PD has been recently reviewed by Johnson et al. [41]). The hyperactivity of the STN is thought to play a critical role in the etiology of PD motor symptoms, and may also lead to excitotoxicity, and further degeneration of the remaining substantia nigra compacta (SNc) neurons. Because NMDA receptors play an important role in regulating the excitatory transmission of the STN and striatum [42], it has been suggested that blockade of NMDA receptors may be a beneficial therapeutic strategy. Consistent with this idea, it has been demonstrated that treatment with NMDA antagonists can attenuate symptoms in both acute [43–45] and chronic [46–48] models of PD. Further work has shown that direct infusion of NMDA antagonists into various basal ganglia regions provides antiparkinsonian effects in PD animal models, suggesting that reduction of NMDA activity in the indirect pathway and output nuclei of the basal ganglia may provide therapeutic benefit [43, 49]. The acute efficacy observed in these studies is consistent with a symptomatic benefit. However, longer term treatment with NMDA antagonists may have the potential to impact disease progression through attenuation of excitotoxic damage of dopaminergic neurons in the SNc. In addition, NMDA antagonists, through modulation of acetylcholine release from striatal interneurons, may have a beneficial effect on levodopa (L-DOPA)-induced dyskinesias, suggesting a benefit of using these drugs as adjunctive therapy [50–52].

Huntington's disease is an inherited neurodegenerative disorder, caused by an expansion of a poly-glutamine repeat region within the huntingtin (htt) gene [53, 54]. Although many brain regions are affected, the MSNs of the striatum have been a focus of research in this disease, since these neurons appear to be especially sensitive to degeneration and death [55]. While the gene responsible for this disease was identified more than 15 years ago, the mechanisms which link the mutant form of the htt protein (mHtt) with neuronal cell death remain obscure. With that said, there are a number of studies which have linked mHtt with dysregulation of glutamate neurotransmission [56–58]. As in other neurodegenerative diseases, excitotoxicity has been a leading hypothesis for cell death in HD, and mHtt has been suggested to enhance NMDA function and calcium signaling [59, 60]. A physical association has been shown between Htt and the NMDA receptor-PSD95 complex, suggesting a direct role for Htt in modulating NMDA signaling [61, 62]. Other studies have suggested that mHtt may modulate NMDA function through influencing gene expression of the receptor subunits NR1A and NR2B [63]. Studies in HD animal models have supported dysregulation of NMDA function as an etiological factor in the disease, and altered NMDA function has been observed in corticostriatal synapses as well as striatal neurons from HD models [64]. Finally, recent work has shown that extrasynaptic NMDA receptors, which are involved in mediating the neurotoxic effects of NMDA receptor activation, are selectively upregulated in HD mouse models. Interestingly, low doses of the weak NMDA antagonist memantine, shown to selectively block extrasynaptic (but not synaptic) NMDA receptors, are able to ameliorate some of the neuropathological and behavioral manifestations of the disease [65, 66].

Due to the widespread expression of NMDA receptors throughout the brain, and their involvement in a wide range of CNS functions, there continues to be concern about side effects from nonselective NMDA antagonists, ranging from cognitive and motor impairment, to psychotomimetic effects [67]. To partly address these concerns, there has been more recent interest in the potential of subtype selective NMDA antagonists as a path to obtaining some of the beneficial effects of these drugs, with less of the safety and side effect liabilities. The NR2B subunit is highly expressed throughout the basal ganglia, and the availability of NR2B selective antagonists has provided the opportunity to test the potential benefits of subtype selective agents.

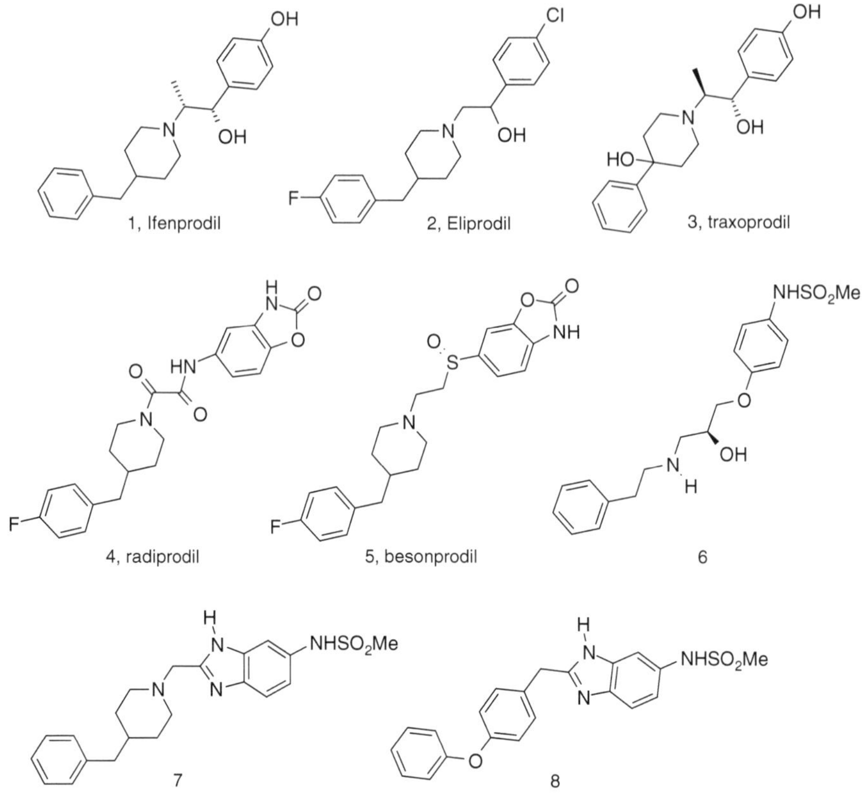

Fig. 1 NR2B selective antagonists

2.2 NMDA Receptor Modulators

As previously stated, the NMDA receptor is composed of two NR1 subunits and two subunits from the group of NR2A-D and/or NR3A-B. The glutamate-binding site exists at the cleft between the NR1 and NR2/3 subunits and the allosteric glycine-binding site is located solely on the NR1 subunit. Compounds that discriminate how they bind among the multiple NMDA-binding sites can have varying effects on the pharmacology and functional outcomes of receptor activation. The concerns over excitotoxicity due to full agonism of the NMDA receptor have been lessened by targeting the allosteric glycine site that is believed to be unsaturated in vivo. The glycine site can also be modulated by increasing the extracellular levels of the endogenous ligands, glycine, and D-serine. This can be accomplished by blocking the reuptake of glycine via inhibition of the glyT-1 transporter or by inhibiting D-amino acid oxidase (DAAO), the enzyme responsible for the metabolic breakdown of D-serine. In addition, regulation of kynurenic acid, a degradation product of tryptophan that is an endogenous competitive antagonist of the strychnine-insensitive glycine-binding site, is yet another avenue for modulating the NMDA receptor. Compounds that positively modulate the NMDA receptor through either direct agonism of the glycine site [68, 69], inhibition of the glycine transporter [68, 70, 71], inhibition of DAAO [72, 73], or reduction of the endogenous levels of kynurenic acid have been targeted for the treatment of schizophrenia and will not be further discussed here.

Regarding negative modulation of the NMDA receptor, high-affinity (noncompetitive) NMDA channel blockers (e.g., dizocilpine and phencyclidine) and competitive NMDA antagonists are known to have serious side effects at minimal effective clinical doses. However, there is an example of a low-affinity channel blocker, (memantine, compound **9**, Fig. 2) that appears to block pathological, but not physiological activation of the NMDA receptor. Memantine is a prototype of a fast off-rate NMDA receptor antagonist that preferentially blocks NMDA receptors in their open state (only after channel opening) during the chronic, low level excitation that may be associated with AD. However, memantine dissociates from the receptor in response to the strong short-lived depolarizations that normally trigger NMDA receptor activity. It should be noted that memantine is not selective at concentrations above 10–50 μM where it can interact with other CNS targets (e.g., serotonin, dopamine uptake, nicotinic acetylcholine receptors) and potentially confound the mechanism of its therapeutic efficacy [74]. Interestingly, Mg^{2+}, an endogenous NMDA receptor channel blocker, may play a critical role in determining memantine's NMDA selectivity profile. At physiological concentrations of Mg^{2+}, memantine is more selective for NR2C and NR2D subunits over NR2A and NR2B subunits [75]. The preclinical data, mechanism of action, and clinical efficacy of memantine have been thoroughly reviewed [74, 76–78]. Antagonists of the glycine site of the NMDA receptor have also been reported [79–81] with the goal of increasing the therapeutic window [82, 83], but this approach has largely been deprioritized in recent years as efforts have shifted towards the pursuit of selective noncompetitive antagonists. Thus far, it has been difficult to obtain selective NR2A, NR2C, or NR2D antagonists that are also bioavailable and brain penetrant [84–86] and most reports have centered on the NR2B subunit.

Medicinal chemistry efforts around the NR2B receptor have predominantly focused on the phenylethanolamine series of NR2B selective noncompetitive antagonists, and this series is represented by **1** (ifenprodil), **2** (eliprodil), **3** (traxoprodil or CP-101606), **4** (radiprodil or RGH-896), and **5** (besonprodil or CI-1041) (Fig. 1) [87, 88]. In preclinical studies, CI-1041, when coadministered with

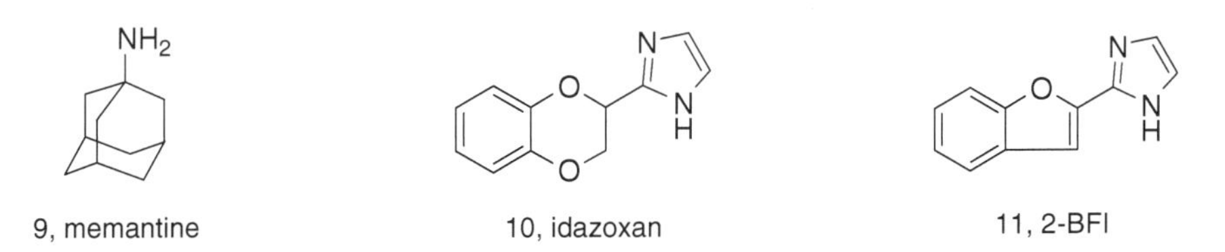

Fig. 2 NMDA receptor modulators

L-DOPA, completely prevented the induction of dyskinesias in three out of four PD model monkeys, suggesting that antagonism of NR2B-containing NMDA receptor subtypes may prevent L-DOPA-induced dyskinesias in PD patients [89]. More recent monkey studies using CI-1041 showed that NR2B receptors play a greater role in L-DOPA-induced dyskinesias than NR2A receptors [90]. In contrast to nonselective NMDA antagonists, the NR2B selective antagonist **6** (propanolamine) ($IC_{50} = 50$ nM), did not cause increased locomotion in rodents, suggesting potential for an improved therapeutic window [91]. In the absence of crystallographic information of the binding site, homology modeling of the NR2B modulatory domain with ifenprodil suggests that the closed conformation of the R1–R2 domain, rather than the open, constitutes the high-affinity binding site [92]. Subsequent computational efforts have led to a pharmacophore model that was used to generate alternative indole substitution off of the piperidine nitrogen of ifenprodil [93]. Analogs incorporating other heterocyclic groups, such as the 5-substituted benzimidazoles **7** ($K_i = 0.99$ nM) and **8** ($K_i = 0.68$ nM) were shown to inhibit NR2B receptors via direct binding to the amino-terminal domain of the NR2B subunit and to effectively protect rat primary cortical neurons against NMDA-induced excitotoxicity [94]. Compounds **7** and **8** significantly reversed neuronal death at concentrations 1.5 and 10-fold their K_i, respectively. Another class of noncompetitive NMDA receptor modulators are the ligands for imidazoline I_2 receptors, such as **10** (Idazoxan) and **11** (2-BFI) (Fig. 2) [95]. Compounds **10** and **11** reversibly blocked Ca^{2+} influx in cortical neurons in a fashion similar to that of memantine. NMDA receptor selectivity was not reported. Nevertheless, these compounds suppressed NMDA receptor-mediated calpain activity as a result of blocking NMDA receptor function, rather than through direct inhibition of calpain.

3 AMPA Receptors

3.1 AMPA and Kainate Receptor Biology

AMPA receptors are primarily responsible for mediating fast excitatory neurotransmission in the CNS. Within this section, we also discuss the kainate (KA) receptors, which although structurally and pharmacologically are closely related to AMPA receptors, are much less well understood with regard to their CNS function. In fact, although the agonists AMPA and KA do distinguish between these receptors, virtually all competitive antagonists that have been discovered show cross-reactivity between these two receptor classes, further supporting their close relationship. Given the difficulty in functionally differentiating the AMPA and KA receptors, more recent classification schemes have moved away from referring to them by their preferred agonist, and instead utilize a molecular nomenclature based on receptor subunit composition.

Molecular cloning led to the identification of four AMPA receptor subunits termed GluR1-4, as well as five KA subunits termed GluR5-7 and KA1-2 [96]. As with

NMDA receptors, the AMPA and KA receptor subunits assemble to form heterotetramers which function as ligand-gated ion channels, with permeability to Na^+, K^+, and Ca^{2+} ions [97–99]. Although subunits will generally coassemble with other members from the same subfamily, there are still many possible combinations of heterotetramers, and this structural diversity can lead to important functional differences. An additional level of heterogeneity exists among the AMPA subunits which are alternatively spliced to yield either a “flip” or “flop” isoform. The presence of these different splice variants confers distinct desensitization properties on the channels [100]. In addition, the GluR2 subunit is particularly important for determining ion selectivity, in that channels containing this subunit show greater permeability to Ca^{2+} [101, 102]. AMPA receptors are broadly expressed throughout the CNS [103]. All subunits are expressed in many brain regions, and different heterotetrameric combinations may be expressed within a single cell. While AMPA receptors are predominantly expressed postsynaptically [104], more recent work has shown that at least in some cases these receptors may be found on presynaptic terminals, and regulate neurotransmitter release [105–107].

AMPA and KA receptors show fast-gating kinetics, desensitize rapidly and mediate rapid glutamatergic neurotransmission [97]. In contrast to the NMDA receptor which requires the coagonist glycine to be activated, AMPA and KA receptors only require the presence of glutamate to induce channel opening. Given the ability of AMPA receptors to flux Ca^{2+} ions, it is not surprising that in the presence of high levels of extracellular glutamate, hyperactivation of these channels may lead to neuronal injury and death. Thus, AMPA receptor antagonists have received attention for the treatment of a variety of both acute and chronic neurodegenerative conditions. Many studies have shown that in models of global and focal ischemia, high levels of synaptic glutamate are released, and AMPA receptors are overstimulated [108, 109]. AMPA antagonists have been demonstrated to have a positive impact on neuronal survival in these experimental paradigms [110, 111]. Epilepsy is another condition involving excessive glutamate release and AMPA receptor activation [112, 113], and AMPA antagonists have been used successfully both as anticonvulsants to control seizures [114], and as neuroprotectants to reduce post seizure neuronal damage [115, 116]. Indeed, some currently marketed anticonvulsant agents have AMPA antagonist activity, which may be in part responsible for their efficacy [114]. Traumatic brain injury is another example of acute CNS damage leading to excessive glutamate release. AMPA antagonists have been shown to be effective when administered either before or after the trauma in animal models of brain injury [111]. While AMPA receptor antagonists show a distinct efficacy profile from NMDA antagonists, these two classes of agents are also distinct in terms of their side effect burden. AMPA antagonists are generally not associated with the psychotomimetic and cognitive liabilities of the NMDA receptor blockers and as such may present fewer hurdles to clinical advancement. With that said, because of the widespread distribution of AMPA receptors, AMPA antagonists are likely to have side effects of their own, some of which may be mitigated by the use of subtype selective agents.

In addition to the involvement of AMPA receptors in mediating the effects of acute neuronal injury, there is also substantial evidence implicating this receptor family in more chronic neurodegenerative conditions. AMPA receptors have been shown to play an important role in well-known forms of synaptic plasticity termed long-term potentiation (LTP) and long-term depression (LTD) [117, 118]. Studies have shown that LTP is associated with increased insertion of AMPA receptors into the postsynaptic membrane, while LTD involves reduced surface expression of AMPA receptors [117–120]. Additional evidence has linked deficits in LTP and LTD with animal models of AD [121], thus supporting the possibility that aberrant AMPA receptor expression or function may play a role in this disease. Several studies have shown decreased AMPA binding sites early in AD, consistent with decreased AMPA receptor expression [122–124]. One mechanism that may link AD pathology with decreased AMPA receptor expression is the induction by the Aβ peptide of the cysteine protease, caspase [125]. The induction of caspase by Aβ, which has been seen in AD brain as well as cultured rat neurons treated with the Aβ peptide, is a key step in Aβ-mediated apoptotic cell death. Induction of this protease has been suggested to lead to AMPA cleavage, which is further supported by the presence of a caspase-3 cleavage site within AMPA subunits. This action of caspase shows some specificity in that levels of NMDA receptors are unaffected by the Aβ peptide, consistent with this receptor not being a caspase substrate. In addition to a putative effect on AMPA receptor degradation, the Aβ peptide may have a direct modulatory effect on these receptors. For example, treatment of cultured hippocampal neurons with Aβ1-42 peptide reduces AMPA-evoked current. Furthermore, this effect is selective, with the Aβ1-42 peptide enhancing, rather than decreasing activity of NMDA receptors [126]. In addition, different forms of the Aβ peptide can have very different effects, with Aβ1-42 reducing AMPA function, while Aβ1-40 potentiates AMPA activity, arguing for a fairly specific interaction between the peptide and AMPA subunits [127]. In addition to a direct modulatory role on AMPA function, data suggest that the Aβ peptide can also influence postsynaptic expression of the AMPA receptor through a variety of possible mechanisms. For example, treatment with Aβ1-42 peptide during the induction of LTP has been shown to reduce autophosphorylation of CamKII and the subsequent phosphorylation of the GluR1 subunit [128]. In addition, in cultured neurons expressing Aβ endogenously (from APP transgenic mice) or when exogenously applied, a reduction in PSD95 can be observed, which in turn is associated with a reduction in AMPA receptor expression at the postsynaptic membrane [129].

The interaction between Aβ and AMPA receptors is complex and may vary with cell type and the specific form of the Aβ peptide. For instance Aβ1-42 toxicity is attenuated in retinal neurons following blockade of AMPA receptors [130], while Aβ25-35 toxicity in cerebellar granule cells is enhanced by AMPA antagonists [131]. While the reasons for these mechanistic differences are not entirely clear, the effects of Aβ on AMPA function may be mediated indirectly via increases in extracellular glutamate, rather than a direct modulatory effect of the peptide on the AMPA receptor itself. Additional support for a role of AMPA receptors in AD comes from hAPP transgenic animals. In cultured neurons from these animals, there is a marked reduction in excitatory postsynaptic currents, and specifically those mediated by AMPA as opposed to NMDA receptors [132]. An additional mechanism by which Aβ has also been suggested to mediate AMPA receptor down-regulation is through enhanced phosphorylation of the GluR2 subunit which stimulates endocytosis [133, 134]. The above evidence indicating reduced AMPA function in AD would suggest that AMPA activators will be beneficial in the treatment of this disease.

Given that a hyperglutamatergic state exists in PD, and the key role of AMPA-mediated neurotransmission in the basal ganglia, AMPA antagonists might be predicted to provide therapeutic benefit in the treatment of this disorder. However, there have been a number of studies showing that AMPA antagonists, when administered alone, are not effective at reversing the motor symptoms in animal models of PD involving acute treatment with cataleptic agents (e.g., D2 receptor antagonists) [135, 136]. Similarly, AMPA blockade does not provide benefit in more chronic models of PD including 6-hydroxydopamine (6-OHDA)-lesioned rodents and 1-methyl-4-phenyl-1,2,3,6-tetrahydropyridine (MPTP)-treated primates [137, 138]. Although AMPA antagonists have not generally shown efficacy on their own, when administered in conjunction with L-DOPA, they are effective in reversing motoric symptoms, suggesting therapeutic potential in patients as an adjunct to L-DOPA treatment [137–139]. Interestingly, AMPA antagonist efficacy has been most clearly shown in models involving toxin-induced degeneration of SNc neurons, which may be more relevant to the human disease than acute models involving D2 receptor antagonists. As discussed above for NMDA antagonists, AMPA blockade may have the potential to reduce degeneration of dopaminergic SNc neurons, but this is not likely to occur in the short-term studies described above. Beyond their potential utility as a symptomatic therapy in PD, AMPA antagonists may have greater utility in treating the motoric side effects associated with long-term treatment with L-DOPA. It has now been shown in both rodent and primate models of PD that L-DOPA-induced dyskinesias may be attenuated by AMPA receptor blockade [140–142]. In addition, an AMPA positive allosteric modulator, which increases AMPA function, was reported to exacerbate these dyskinesias, lending further support to the key role of AMPA receptors in mediating the deleterious side effects of L-DOPA treatment [140].

As described above for the NMDA receptor, reduced glutamatergic signaling through the AMPA receptor has also been associated with cognitive deficits, and thus the identification of agents capable of increasing AMPA activity is an area of active pharmaceutical research. Positive allosteric modulators of the AMPA receptor have shown efficacy in a variety of animal models of cognitive impairment, and early clinical studies have suggested that this class of agents is well tolerated in humans [143, 144]. Although clinical efficacy of these compounds remains to be established, a number of trials are underway to further explore the potential of these agents in a variety of diseases involving cognitive deficits, including AD, schizophrenia, mild cognitive impairment, depression, and fragile X syndrome [145].

3.2 AMPA Receptor Modulators

Compounds that positively modulate the AMPA receptor have been reviewed [146, 147]. The first reports of AMPA receptor modulators, appearing over 20 years ago, were plant lectins (e.g., concanavalin A) that inhibited rapid non-NMDA receptor desensitization [148]. Shortly after, the nootropic drug **12** (aniracetam) (Fig. 3), which was reported to effect learning and memory, was found to selectively

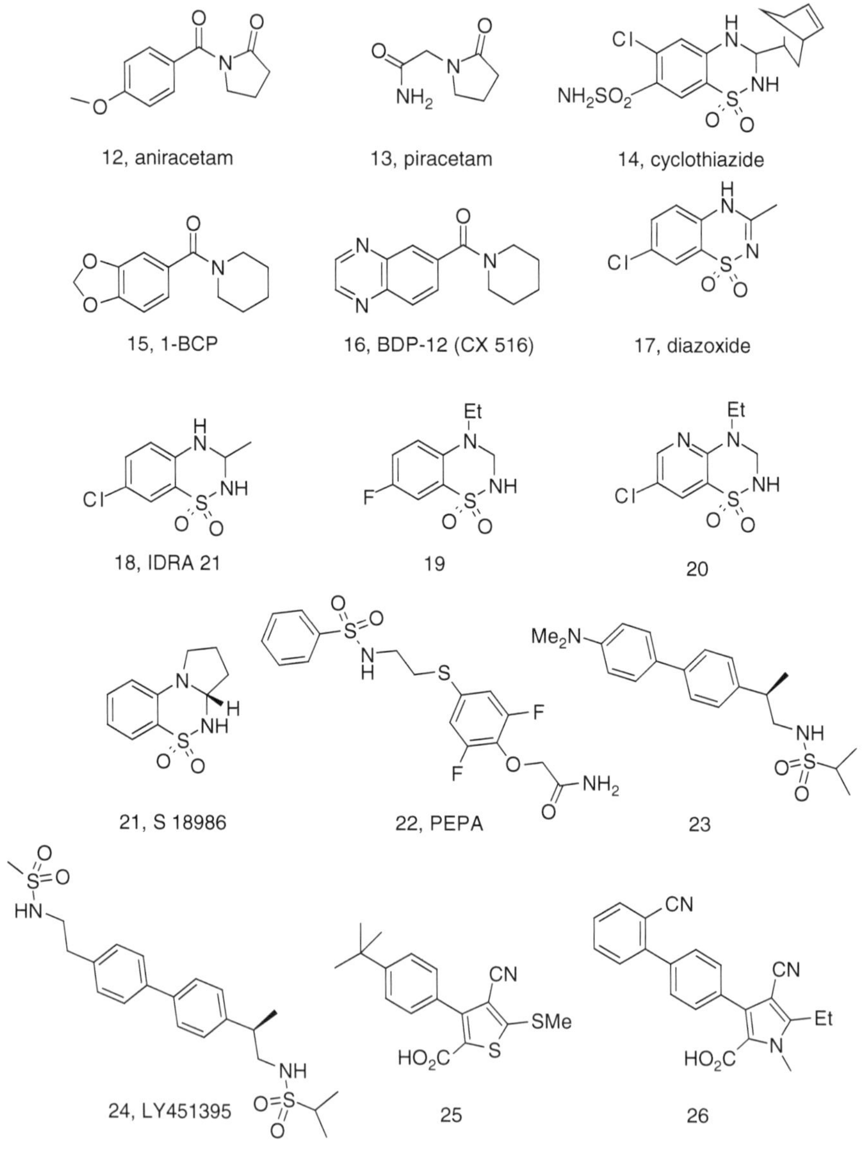

12, aniracetam

13, piracetam

14, cyclothiazide

15, 1-BCP

16, BDP-12 (CX 516)

17, diazoxide

18, IDRA 21

19

20

21, S 18986

22, PEPA

23

24, LY451395

25

26

Fig. 3 Structures of positive AMPA receptor modulators

potentiate the responses mediated by the quisqualate (later renamed AMPA) receptor channel and provided evidence for reversible non-NMDA allosteric potentiation [149]. Recently, the binding mode of aniracetam and other nootropic agents such as **13** (piracetam) at the AMPA allosteric binding site was reported [150]. A more potent AMPA modulator, **14** (cyclothiazide), caused glutamate to induce long bursts of channel openings and greatly increased the number of repeated openings at 10 μM [151]. These early modulators were useful tools for evaluating the AMPA receptor in vitro, but were less useful in vivo due to poor pharmacokinetic (PK) properties and limited brain exposure (aniracetam is rapidly metabolized to anisoyl γ-aminobutyric acid and cyclothiazide does not cross the blood–brain barrier). Efforts to improve brain penetration and metabolic stability led to the discovery of **15** (1-BCP) [152] and **16** (BDP-12 or CX 516) [153], both of which were shown to improve performance in memory task experiments. The in vivo effects of these compounds in rodents as well as an initial human study with CX516 showing improved memory in aged individuals have been reviewed [143]. Recent updates to the clinical assessment of CX 516 reported that it failed to improve delayed verbal recall in a group of subjects with mild cognitive impairment, and did not improve cognition in schizophrenic patients when added to the antipsychotic drugs clozapine, olanzapine, or risperidone [154]. However, CX 516 is a low potency agent ($EC_{50} > 1$ mM) with a short human half-life ($T_{1/2} = 1$ h); hence, it is unlikely that even the selected high dose (900 mg, three times daily) provided adequate exposure. Low potency thiazides such as **17** (diazoxide) require 100–500 μM concentrations to stimulate AMPA receptors, which are levels also known to bind to potassium channels [155]. Saturation of the C=N bond of **17** led to a threefold increase in potency (compound **18**, IDRA 21). Subsequent structure–activity relationships (SAR) developed around IDRA 21 led to the observation that ethyl substitution at the 5′-position gave a 30-fold improvement in affinity ($EC_{50} =$ 22 μM), and the *N*-ethyl analog, **19**, gave a positive effect in an object recognition test in rats demonstrating cognition enhancing potential [156, 157]. In efforts to improve metabolic stability, fluorination of this ethyl group by a single fluorine atom did not affect potency, but led to unexpected toxicity [158], whereas addition of multiple fluorine atoms led to decreased potency. A pyridyl analog of **19** (**20**) was also active in the object recognition test as well as a rat social recognition test [0.3 mg/kg, intraperitoneally (i.p.)] [159]. Tying the *N*-ethyl group onto the thiadiazine ring (**21**, S 18986) selectively improved aged mouse performance in a test of long-term/declarative memory flexibility and exerted a beneficial effect in a short-term/working memory test [160]. In efforts to understand the binding mode of the thiazide series, a crystal structure of the allosteric binding site of GluR2 bound to a set of thiazide derivatives was recently reported and revealed that these compounds maintain a hydrogen bond with the Ser754 hydroxyl, supporting a partial selectivity for the flip variant of the AMPA receptor [161]. The interaction of the NH hydrogen bond donor in the 4-position of cyclothiazide appears to be a major determinant of the receptor desensitization kinetics [162].

Further SAR development around these early modulators led to novel sulfonamides with improved potency. Compound **22** (PEPA), a flop-preferring allosteric modulator of AMPA receptor desensitization, was reported to be greater than 100-fold more potent than aniracetam [163]. Furthermore, the biaryl sulfonamide **23** was shown to be greater than 100-fold more potent than cyclothiazide [164]. A binding mode at the dimer interface of the GluR2/4 receptor for the biarylsulfonamide class of compounds has been proposed based on docking, analysis of hydrogen bonding patterns, and calculated energies [165]. The bis-sulfonamide **24** (LY451395) was evaluated in a phase II AD clinical trial, but unfortunately did not show a significant improvement on the AD Assessment Scale – Cognitive Subscale [166]. However, toxicological issues prevented clinical evaluation at the maximum tolerated dose, and it is unclear whether sufficient exposure was achieved to test the hypothesis. Alternative chemical lead matter unrelated to the thiazides and sulfonamides was discovered from a high-throughput screen (HTS) using human cloned homomeric AMPA receptors, and led to the discovery of thiophene **25** [167], and its pyrole analog **26** [168].

In 2001, Nikam and Kornberg reviewed the SAR and proposed pharmacophore models for AMPA receptor antagonists and negative allosteric modulators (NAMs) [169]. From a structural point of view, AMPA antagonists were divided into three classes. (a) The first class are closely related analogs of AMPA and kainic acid that are amino acids with generally poor physicochemical properties, making bioavailability and brain penetration difficult to achieve. (b) The decahydroisoquinolines (e.g., **27**; Fig. 4) are also amino acids with poor physical properties. For the interested reader, molecular modeling information is available that suggests where the carboxylic acid, amino group, and tetrazole bind at the receptor for this series [170]. (c) The quinoxaline-2,3-diones, represented by **28** (CNQX), **29** (NBQX) and **30** (PNQX), generally have selectivity issues versus the NMDA-associated glycine-binding site, but SAR exists to differentiate AMPA antagonism [169]. It is noteworthy that the pyrazoloquinazolone **31** has shown high AMPA affinity (K_i = 100 nM) and is greater than 1,000-fold selective over the glycine site of the NMDA receptor [171].

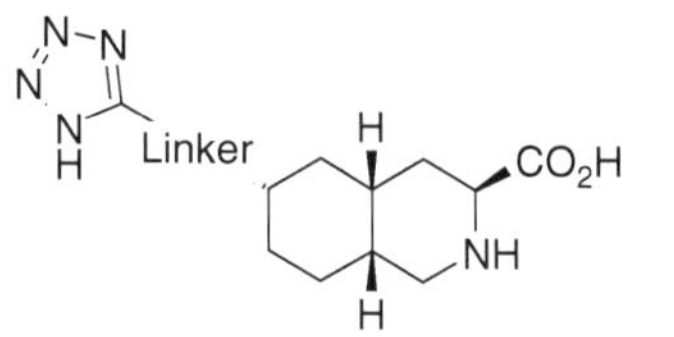

27

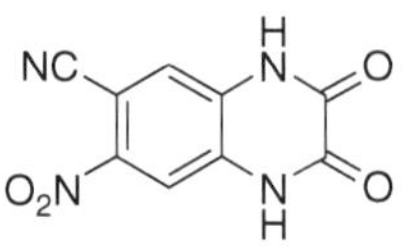

28, CNQX

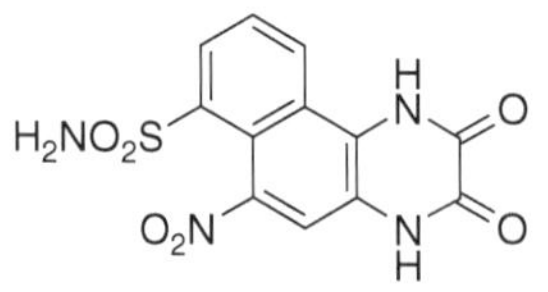

29, NBQX

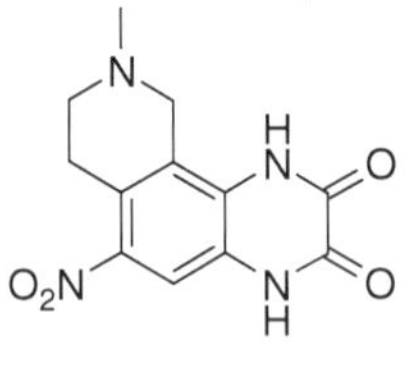

30, PNQX

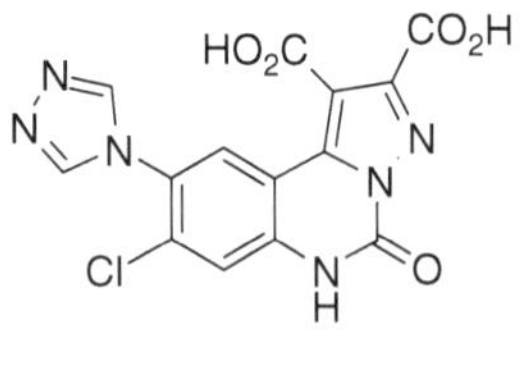

31

Fig. 4 AMPA receptor antagonists

Noncompetitive AMPA receptor antagonists have the theoretical advantage of counteracting excitotoxicity even at high glutamate concentrations, and show less adverse side effects compared to competitive antagonists. Radioligand binding assays [172] have aided in the identification of allosteric modulators of the AMPA receptor. Reported NAMs have been derived from 2,3-benzodiazepines, starting from **32** (GYKI 52466; $IC_{50} = 2.7\ \mu M$; Fig. 5) [169]. Several closely related analogs such as **33** (CFM-2) and **34** had similar potency, and **34** noncompetitively inhibited AMPA receptor-mediated toxicity in primary mouse hippocampal cultures ($IC_{50} = 1.6\ \mu M$) and blocked kainate-induced calcium influx in rat cerebellar granule cells ($IC_{50} = 6.4\ \mu M$) [173]. Chiral analogs, **35** (LY300164, talampanel) and **36** (LY30370), were also active in vivo [174]. Moreover, talampanel significantly reduced seizures in humans [175], and the more potent LY30370 was shown to be a powerful neuroprotective agent in a model of AMPA receptor-mediated excitotoxicity [174]. Condensing the seven-membered fused ring of LY30370 to a 1,2-dihydrophthalazine ring system exemplified by **37** (SYM 2207) gave similar AMPA potency ($IC_{50} = 1.8\ \mu M$) [176]. Replacement of the dioxolane moiety of

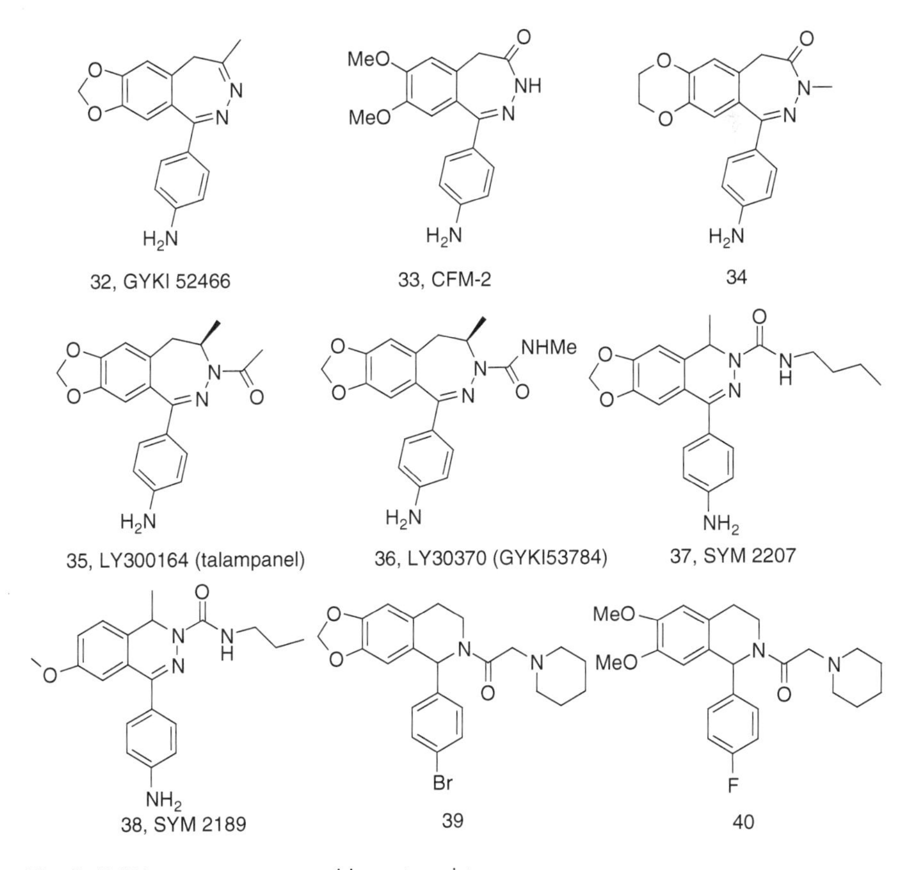

Fig. 5 AMPA receptor noncompetitive antagonists

SYM 2207 with methoxy **38** (SYM 2189) was also equipotent, but had the added benefit of reduced sedative side effects [177]. Other analogs such as the tetrahydroquinolines **39** and **40** had potencies comparable to that of talampanel [178, 179]. For the interested reader, the EC_{50}s for anticonvulsant activity of a large set of 2,3-benzodiazepines, annealated 2,3-benzodiazepines, 1,2-phthalazines, and tetrahydroisoquinolines were tabulated and subjected to a QSAR analysis [180]. Also, much has been learned in recent years about how compounds bind to the AMPA receptor binding domain. The structures of the binding domain (S1S2) of the GluR3 (flip) AMPA receptor subunit bound to glutamate and AMPA and the GluR2 (flop) subunit bound to glutamate were determined by X-ray crystallography [181].

4 Metabotropic Glutamate Receptors

4.1 Metabotropic Glutamate Receptor Biology

The metabotropic glutamate receptors (mGluRs) are a family of G-protein-coupled receptors, which modulate excitatory and inhibitory neurotransmission, both pre- and postsynaptically. This class of receptors consists of eight members, which are grouped based on sequence homology, ligand binding, and G-protein coupling specificity. The group I receptors (mGluR1 and mGluR5) signal through Gq and activate the phospholipase C pathway, leading to increases in intracellular calcium and activation of protein kinase C [182, 183]. These receptors are preferentially localized postsynaptically in neurons. They are also expressed in glial cells, although their function here is less well studied [184]. Group II receptors (mGluR2 and mGluR3) couple to Gi/Go, signal through the inhibition of adenylate cyclase, and are typically localized presynaptically [185]. Group III receptors (mGluR4, 6, 7, and 8) also couple with Gi/Go and signal through inhibition of adenylate cyclase. Like group II, these receptors are also generally located presynaptically, where they modulate ion channel activity and neurotransmitter release [186, 187].

Generally the mGluRs play a neuromodulatory role, serving to either reduce or potentiate synaptic transmission. For this reason, mGluRs have received a great deal of attention as drug targets in the treatment of neurodegenerative diseases. In AD, some of the key modulatory functions of mGluRs may be disrupted, potentially as a direct consequence of Aβ peptide expression. In cultured prefrontal cortical neurons, activation of group II mGluRs will potentiate NMDA function, and mGluR5 agonists will increase GABA transmission. In both of these cases, mGluR signaling is known to occur via the enzyme PKC, and following application of the Aβ peptide, the potentiating effects of these receptors is abolished. It has been suggested that this Aβ-mediated inhibition is likely to occur through blockade of PKC activation [36]. These effects on mGluR signaling are potentially quite significant, since GABA and NMDA neurotransmission are both critically involved in the normal cognitive function of prefrontal neurons [188, 189], and this mechanism may partly explain how Aβ can contribute to cognitive decline in AD. Additional evidence comes from studies showing that group I mGluR signaling is impaired in prefrontal brain regions in AD and deteriorates as the disease progresses [190]. Although postsynaptic signaling appears to be impaired in AD, other studies have shown that a presynaptic dysfunction also may exist. It has been shown that introduction of the Aβ peptide into hippocampal slice preparations can lead to a strong downregulation of the synaptic vesicle protein synaptophysin [191]. Furthermore, in both animal models and postmortem AD brains, downregulation of key signaling enzymes like calcineurin [192] and decreased expression of synaptophysin are consistent with presynaptic dysfunction [126, 193, 194]. In addition, synaptic vesicle recycling and transmitter release appear to be negatively impacted by Aβ produced by mouse neurons overexpressing hAPP [132]. Taken together, the above data suggest that in AD there are fundamental deficits in presynaptic structure and function, and that mGluR signaling may also be impaired. In light of these observations, approaches to enhance nerve terminal function through modulation of mGluRs may represent a viable therapeutic strategy.

In the case of the group I receptors, activation is generally associated with increased neuronal excitability, and antagonists of mGluR1 and mGluR5 will typically attenuate neuronal activity [195, 196]. Since group I agonists potentiate glutamatergic signaling, it is not surprising that they can enhance the potency of NMDA-induced cell toxicity [197, 198]. Conversely, as might be predicted, blockade of group I receptors tends to be neuroprotective, and antagonists of these receptors have been shown to reduce neuronal death in vitro in response to a variety of toxic stimuli [199, 200]. Antagonists of mGluR1 have also been shown to be neuroprotective in in vivo models of acute neuronal damage including traumatic brain injury and cerebral ischemia [199, 201, 202]. Treatment with an mGluR1 antagonist has been shown to improve recovery in a spinal cord injury model, although this was not seen with blockade of mGluR5, showing that in some cases the two group I receptors serve different functions [203]. In fact, in contrast to the neuroprotective role seen with mGluR1 antagonists, it is mGluR5 agonists that have been shown to protect cultured neurons from apoptotic stimuli [204, 205]. mGluR5 is also expressed in glial cells and seems to play key roles in regulating their activity. For instance, stimulation of mGluR5 results in PKC activation and an attenuation of microglial activation and inflammation [206]. In a similar fashion, mGluR5 agonists have been shown to reduce cell death in astrocyte–neuron cocultures [207], as well as excitotoxicity in oligodendrocytes [208].

Group I mGluRs are widely expressed throughout the basal ganglia, where they play a modulatory function by counteracting the effects of dopamine and potentiating NMDA activity in striatal neurons [209–211]. Thus, antagonists at these receptors would be expected to attenuate the hyperactivity characteristic of the PD striatum, and to have antiparkinsonian effects. Consistent with this, the mGluR5 NAM MPEP (compound 79, Fig. 10) has shown efficacy in treating the akinesia observed in PD animal models including haloperidol- and 6-OHDA-treated rats [212–214]. Interestingly, mGluR5 has been shown to work synergistically with other receptors with which it is coexpressed. For instance, at submaximal doses,

MPEP coadministered with an A2a antagonist will reverse the motor symptoms in a 6-OHDA-treated rat [215, 216]. It has been shown that mGluR5 and A2a are coexpressed in D2-containing MSNs, and it has more recently been demonstrated that they can physically associate, providing a mechanistic explanation for the interaction of drugs modulating these two targets [217–220]. In a similar fashion, negative modulation of mGluR5 by MPEP can potentiate the effects of NMDA receptor blockade. In the 6-OHDA model, motor function was improved by cotreatment with MPEP and a low dose of the nonselective NMDA antagonist MK801 [221]. By lowering the minimum effective dose of MK801, MPEP may allow for the beneficial effects of NMDA blockade, while avoiding some of the undesirable side effects. Additional studies have sought to further localize the sites at which mGluR5 receptors may modulate basal ganglia function. Direct infusion of mGluR5 agonists into the striatum causes activation of neurons in the indirect pathway and a reduction in motor function [222, 223]. Furthermore, antagonism of mGluR5 with MPEP reduces activity at the striatopallidal synapse [224], and group I agonists excite STN and SNr neurons in brain slices treated with haloperidol. These results are consistent with mGluR5 blockade reducing hyperactivity of neurons in the indirect pathway, and provide a rationale for their potential efficacy in treating motor symptoms in PD patients [225]. Interestingly, mGluR5 antagonists may also have beneficial effects in treating the nonmotor symptoms of PD, including cognitive and psychiatric symptoms. MPEP demonstrated efficacy in reversing a cognitive impairment in a 6-OHDA-lesioned mouse model [226, 227]. In addition, there have been a number of studies in which efficacy has been demonstrated for MPEP in animal models of depression and anxiety [227, 228]. These results indicate that mGluR5 antagonists may have the potential to treat a range of motor and nonmotor symptoms in this disease.

Studies have shown that mGluR5 receptors are upregulated in MPTP-treated primates following chronic treatment with L-DOPA, suggesting that activation of mGluR5 may also be involved in side effects associated with this treatment [229]. This suggests the exciting possibility that mGluR5 antagonism may also be used to treat L-DOPA-induced dyskinesias. In support of this, treatment of 6-OHDA-lesioned rats, following chronic L-DOPA, with an mGluR5 antagonist, attenuated the L-DOPA-induced side effects [230, 231]. These data suggest that in addition to the potential benefits of mGuR5 antagonists in treating PD symptoms, they may be an effective adjunct when coadministered with L-DOPA therapy.

MGluR5 receptors have also been investigated as potential therapeutic targets in HD. Given the hyperglutamatergic state believed to play a role in this disease, blockade of postsynaptic mGluR5 receptors may provide neuroprotection against excitotoxic injury. This was explicitly tested in the R6/2 model of HD, which expresses the N-terminal polyQ region of the Htt protein. This is an aggressive disease model, with motoric symptoms apparent by 2 months, and death typically occurring before 4 months of age. In R6/2 mice treated with MPEP, an improvement in motor coordination, as measured by rotorod performance, was evident. In addition, MPEP-treated animals survived approximately 2 weeks longer than vehicle-treated controls [232].

The group II and III receptors are predominantly presynaptic and serve to regulate the release of neurotransmitters at the synaptic terminal, and are particularly important for the regulation of glutamate release [233]. Agonists that stimulate group II and III mGluRs have been the focus of recent research, and such compounds have been demonstrated to be neuroprotective in a number of in vitro and in vivo paradigms.

The group II mGluRs have also been extensively studied in the context of basal ganglia function and appear to be involved in regulating some of the key circuits that are dysfunctional in PD. It has been shown that activation of presynaptic group II receptors reduces excitatory transmission at the STN–SNr synapse, suggesting that agonists at these receptors may provide symptomatic benefit [234, 235]. In support of this, mGluR2 agonists have been shown to improve motor function in haloperidol- [234] and reserpine-treated rats [236]. However, somewhat surprisingly, activation of mGluR2 receptors does not show a similar benefit in more chronic PD models [237]. This raises some questions about whether activation of mGluR2 receptors will be effective in the disease state, which is presumed to be more closely related to chronic, rather than acute models of dopamine depletion. However, although the responsiveness of mGluR2 receptors to activation at the STN–SNr synapse may be attenuated in the PD brain, these receptors are expressed in other basal ganglia circuits where their regulation may differ. Consistent with this possibility, the efficacy of an mGluR2 agonist at reducing transmission at the corticostriatal synapse is actually increased in rats treated with 6-OHDA, and this effect is lost following L-DOPA treatment [238]. This result argues that in contrast to the STN–SNr synapse, mGluR2 receptor responsiveness to agonism may actually be enhanced at cortical terminals within the dopamine-depleted striatum.

In a similar fashion to what has been described for group II receptors, treatment of rat brain slices with the group III selective agonist, L-AP4, is associated with reduced activity at both the striatopallidal and the STN–SNr synapse [239–241]. The ability of group III receptor activation to reduce indirect pathway activity is also observed in vivo, with L-AP4 treatment of both acute and chronic PD animal models, leading to an improvement in motor function [240, 242]. Further support for the site of action of group III agonists comes from studies showing that direct infusion of L-AP4 into the pallidum attenuates motor symptoms in reserpine-treated rodents [242]. The above effects of L-AP4 were lost in a mGluR4 KO mouse, strongly implicating this receptor subtype in mediating the effects of this drug [240].

4.2 *Metabotropic Glutamate Receptor Modulators*

Both agonists and antagonists of groups I, II, and III mGluRs were thoroughly reviewed by Schoepp et al. in 1999 [243], and in the same year Pin et al. reviewed the structural features of the mGluR-binding site along with pharmacophore models of the mGluRs [244]. Competitive mGluR ligands have historically been amino acid analogs derived from glutamate, where either (a) the conformation was fixed through mono-or bicyclic ring structures, (b) the linker between the two acid moieties was varied, (c) substituents were inserted into the glutamic acid structure, or (d) one of the acid groups were replaced with a bioisostere. These analogs have generally displayed poor selectivity between the mGluRs and have shown poor CNS exposure. The poor selectivity can be attributed to the high degree of sequence homology in the agonist-binding site between mGluRs, and especially among those receptors within the same group. Amino di-acids (or acid isosteres), present in nearly all of the mGluR competitive agonists and antagonists, are generally poor substrates for penetration across the blood–brain barrier (unless transporter assisted, but designing this into a molecule is not very well understood).

Allosteric modulators bind to a site on the receptor other than the glutamate-binding site, which greatly increases the possibilities for identifying subtype selective agents. In addition, they are typically not capable of activating a receptor on their own and will only potentiate the effects of a direct agonist like glutamate. For this reason, they tend to potentiate normal or physiological receptor function, and thus might be expected to have a better safety profile. The discovery of allosteric modulators of the mGluRs has provided lead chemical structures without the amino di-acid functionality and with physicochemical properties suitable for brain penetration. This section will touch upon some of the recent advances of amino acid derivatives, but the main focus will be on negative and positive allosteric modulators of the mGluRs. It should be noted that much of the current literature on mGluR modulators has been focused on psychosis, anxiety, and pain. The inclusion of in vivo animal model data pertaining to these diseases is intended as a means for the medicinal chemist to ascertain compound exposure, brain penetration, and an in vivo pharmacological response from the corresponding mechanism.

4.2.1 mGluR1

The early noncompetitive modulators of mGluR1 have been reviewed [245]. The first reported mGluR1 NAMs were the oxime ethyl ester **41** [(+/−)-CPCCOEt; hmGluR1b, IC_{50} = 1.5 μM] and its phenyl amide analog **42** (PHCCC) (Fig. 6), shown to inhibit receptor signaling without affecting glutamate binding [246]. The interaction site of **41** on mGluR1 was initially discovered, using chimeric human mGluR1 (hmGluR1) and hmGluR5 receptors and site-directed mutagenesis, to be located in the transmembrane (TM) domain of hmGluR1b. Subsequent studies with the more active (−)-CPCCOEt isomer used molecular modeling based on the α-carbon template of the TM helices of bovine rhodopsin to suggest a more precise binding mode [247]. Compound **43** (BAY36-7620), structurally dissimilar to **41**, and a much more potent rat mGluR1 (rmGluR1) NAM (IC_{50} = 160 nM), was shown to be neuroprotective in a rat acute subdural hematoma model [40–50% efficacy at 0.01 and 0.03 mg/kg/h, intravenously (i.v.)] and protected against pentylenetetrazole-induced convulsions in the mouse (MED = 10 mg/kg, i.v.) [248, 249]. Perhaps due to its high lipophilicity, generally leading to poor

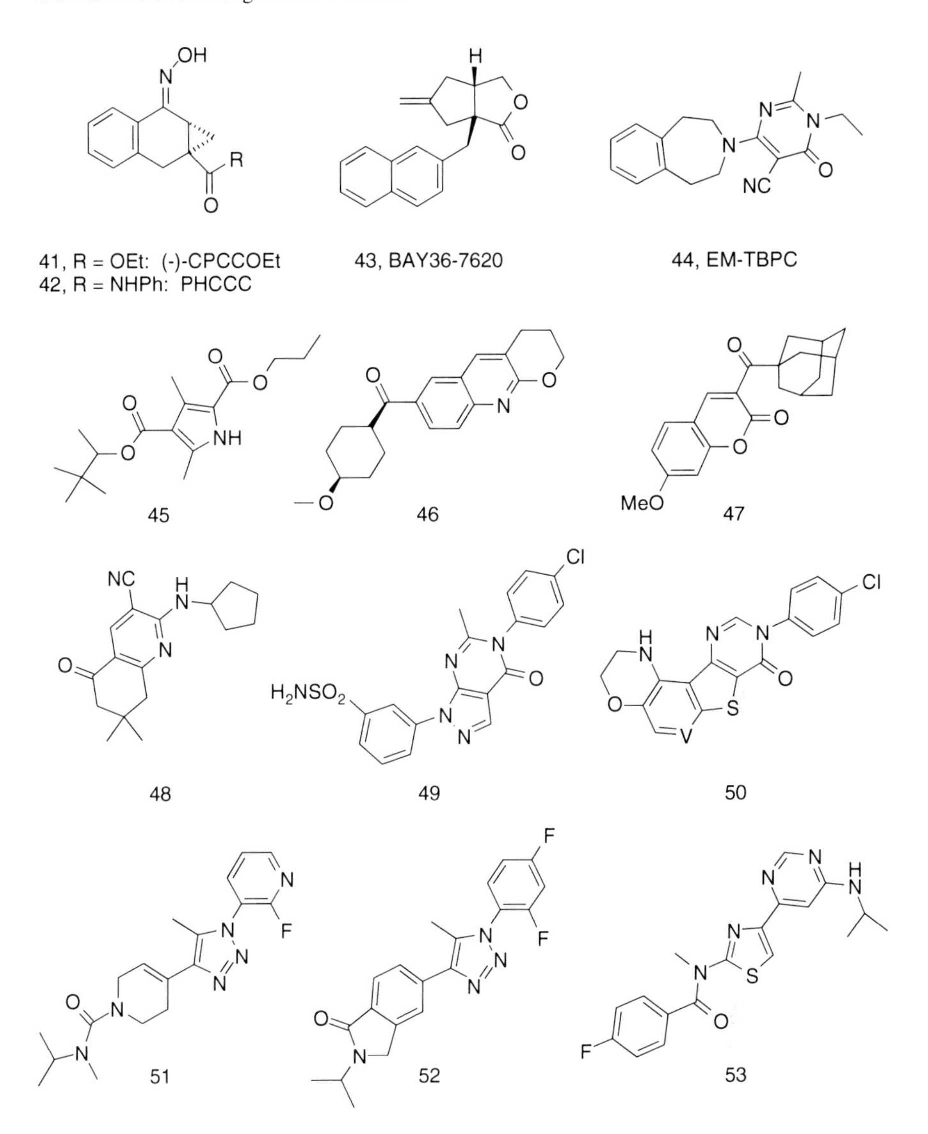

Fig. 6 Noncompetitive antagonists of mGluR1

pharmacokinetics, BAY36-7620 displayed low receptor occupancy in the rat brain when dosed at 10 mg/kg, subcutaneously (s.c.) [250]. Noteworthy with respect to achieving adequate CNS exposure for the mGluR1 NAMs is the potential differences between the rat and human allosteric binding sites of the receptor. Compound **44** (EM-TBPC), yet another structurally unique mGluR1 NAM, was shown to be potent against rmGluR1 (EC_{50} = 130 nM) but weak against hmGluR1. Site-directed mutagenesis has located the key amino acid residues of mGluR1 that differentiate the allosteric binding sites between the two species [251].

In efforts to obtain novel chemical matter, Micheli et al. noted the 3–5 bond distance spacing between the two carboxylic acid groups within reported mGluR1 antagonists, and used this information to do a similarity search of their compound collection to provide pyrrole **45** (rmGluR1a, $IC_{50} = 15.8$ nM) [252]. Pyrrole **45** was orally active in both the early and late phases of the formalin test in mice [$ED_{50} = 0.3$ mg/kg, orally (p.o.)]. Other efforts to identify novel chemical matter led to quinoline **46** (JNJ16259685; rmGluR1 $IC_{50} = 3$ nM, hmGluR1 $IC_{50} = 0.55$ nM), which was shown to have high receptor occupancy in rat brain when dosed s.c., but low oral bioavailability (1%) in rats precluded oral administration [253]. The SAR of this quinoline series was further evaluated via a 3D-QSAR model from a comparative molecular field analysis of 45 analogs [254]. With the binding mode becoming better understood, many researchers have turned to in silico modeling to guide SAR development. For example, using a set of known mGluR1 NAMs, a pharmacophore hypothesis was proposed and subsequent virtual screening led to the adamantyl coumarine **47** (rmGluR1 $IC_{50} = 60$ nM) [255] and the hydroquinolinone **48** (rmGluR1 $IC_{50} = 78$ nM; rmGluR5 $IC_{50} = 49$ μM) [256].

Efforts to identify potent mGluR1 NAMs with desirable PK properties led to the discovery of pyrazolopyrimidinone **49** with moderate hmGluR1 activity ($IC_{50} = 127$ nM), but with high oral bioavailability (100%) and a moderate half-life ($T_{1/2} = 1.5$ h) [257]. Brain exposure was not provided, but one may speculate that the sulfonamide NH_2 might prohibit **49** from crossing the blood–brain barrier. Pyrimidinone **50** (hmGluR1 $IC_{50} = 2.9$ nM) was active in a rat pain model ($ED_{50} = 5.1$ mg/kg, p.o.) [258], and a 3D-QSAR analysis of this triazofluorenone series has been reported [259]. Another compound derived from a HTS is the aryltriazole **51**, which was considered to be a balanced lead based on potency (hmGluR1 $IC_{50} = 5.8$ nM, mouse mGluR1 $IC_{50} = 3.1$ nM), lipophilicity ($\log D = 2.1$), solubility (>170 μM), and metabolic stability, and was active in a mouse pain model at 30 mg/kg, p.o. [260]. However, **51** had a short rat half-life, high clearance, and low oral bioavailability. Subsequent SAR development to improve the PK led to isoindolinone **52** (hmGluR1 $IC_{50} = 4.3$ nM, rmGluR1 $IC_{50} = 3.6$ nM) with improved rat oral bioavailability (46%), half-life ($T_{1/2} = 0.7$ h), and clearance (CLp = 20 ml/min/kg) [261]. Oral administration at 1 mg/kg provided total brain exposures of 0.45 nmol/g and resulted in an antipsychotic-like effect in a rat prepulse inhibition (PPI) assay. A series of aryl thiazoles was also derived from the above HTS effort, and SAR development led to compound **53** with a similar pharmacological and pharmacokinetic profile as **52** – active in a PPI disruption model (MED 1.0 mg/kg, p.o.) and a mouse hyperlocomotion model (MED = 0.3 mg/kg, p.o.) [262].

4.2.2 mGluR2 and 3

The best characterized mGluR2/3 agonist is **54** (LY354740; hmGluR2 $K_i = 75$ nM, hmGluR3 $K_i = 93$ nM), a bicyclic conformationally constrained analog of glutamic acid (Fig. 7). Early reports of LY354740 showed similar efficacy to diazepam in multiple anxiety models, but without the undesirable side effects associated with

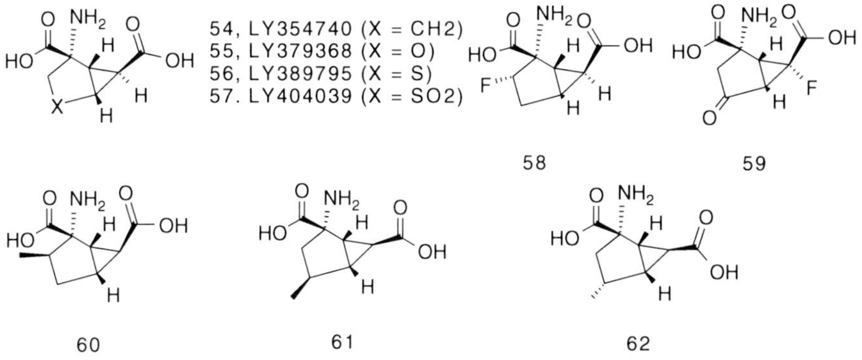

54, LY354740 (X = CH2)
55, LY379368 (X = O)
56, LY389795 (X = S)
57. LY404039 (X = SO2)

58 59

60 61 62

Fig. 7 mGluR2/3 agonists

diazepam (e.g., sedation, deficits in neuromuscular coordination, interaction with CNS depressants, memory impairment) [263, 264]. LY354740 and the closely related more potent analogs, **55** (LY379368; hmGluR2 K_i = 14 nM, hmGluR3 K_i = 5.8 nM) and **56** (LY389795; hmGluR2 K_i = 41 nM, hmGluR3 K_i = 5 nM), are believed to cross the blood–brain barrier and block seizures induced by group I mGluR activation in mice [265]. In addition, these analogs were shown to have analgesic effects in a variety of pain models in the rat, but the animals built up a tolerance to the effect after 4 days of once-daily dosing [266]. Oxidation of the sulfur atom in LY389795 to its corresponding sulfone led to **57** (LY404039) with potency similar to LY354740 (hmGluR2 K_i = 149 nM, hmGluR3 K_i = 92 nM) [267]. In vitro, LY404039 suppressed electrically evoked excitatory activity in rat striatal slices and serotonin-induced L-glutamate release in rat prefrontal cortex, suggesting that it modulates glutamatergic activity in the limbic and forebrain regions of the brain [268]. LY404039 was also active in vivo, blocking PCP-evoked ambulations in rat [267]. In humans, there is evidence that mGluR2/3 agonists might play a role in treating working memory impairment related to deficits in NMDA receptor function, as LY354740 (100 and 400 mg) produced a significant dose-related improvement in working memory (19 healthy subjects) during ketamine infusion [269]. Since LY354740 was shown to have low systemic availability due to poor intestinal permeability, an N-linked alanyl prodrug (LY544344) was developed that dramatically improved the bioavailability in rats and dogs [270]. Subsequently, LY2140023, a prodrug of LY404039, was reported to be active in a phase IIb clinical trial against the positive and negative symptoms of schizophrenia [271]. Other efforts to improve the bioavailability of LY354740 led to the closely related fluorinated analog, **58**, which was shown to have similar mGluR2/3 binding potencies, but improved oral activity in PCP-induced hyperactivity (ED_{50} = 5.1 mg/kg) and head-weaving behavioral (ED_{50} =0.26 mg/kg) models [272]. Another fluorinated analog, **59**, was shown to be a very potent mGluR2/3 agonist (K_i's = 0.57 and 2.1 nM, respectively). Dosed orally, **59** was extremely potent in the aforementioned PCP-induced hyperactivity (ED_{50} =0.30 mg/kg) and head-weaving (ED_{50} = 0.090 mg/kg) models. Placement of a methyl group around the 3- and 4-positions of LY354740 reduced the mGluR2/3 binding affinities 2–13-fold (compounds **60, 61** and **62**), but interestingly, **60** was found to have mGluR2/3 antagonist properties, **61** was a full agonist at both receptors, and **62** was an mGluR2 agonist/mGluR3 antagonist [273]. Selectivity for only mGluR2 or mGluR3 agonist activity has been difficult to achieve, and this was the first report of an mGluR2 agonist devoid of mGluR3 agonist activity.

An alternative strategy for obtaining chemical lead matter selective for mGluR2 has been through targeting an mGluR2 allosteric binding site. Early chemistry efforts in this area were reviewed by Rudd and McCauley [274], and most recently by Fraley, who also included a review of the patent activity around mGluR2 modulators [275]. For a more detailed review of the chemistry in this area, the reader is directed to these reviews. The first reported example of a selective mGluR2 PAM was the sulfonamide **63** (LY487379), discovered through an HTS screen and found to potentiate glutamate agonism, shifting its potency by twofold (EC_{50} = 270 nM) (Fig. 8) [276]. Compound **63** did not potentiate a chimeric mGluR2/1 receptor (prepared by fusing the glutamate site containing the amino-terminal region of the mGluR2 receptor to the transmembrane domain of the mGlu1 receptor), demonstrating that it did not bind to the agonist-binding site, but rather to the transmembrane region of mGluR2. Compound **63** demonstrated activity in a rodent model of anxiety (3 mg/kg, i.p.), which could be blocked with an mGluR2/3 antagonist confirming the selectivity of this compound. A structurally distinct chemical lead, **64** (LY487379; EC_{50} = 1,700 nM, 52% potentiation), was also derived from an HTS hit and was found to have improved potency compared to **63** in an hmGlu2 GTPγS functional assay (EC_{50} = 93 nM, 128% potentiation) [277]. A similar analog, **65**, although not brain penetrable, was shown to inhibit both ketamine-evoked norepinephrine release and hyperactivity in rats when dosed intracerebroventricularly (i.c.v.) [278]. The carboxylic acid analog **66** (EC_{50} = 33 nM) showed activity at 32 mg/kg, i.p., in a variety of antipsychotic and anxiolytic models in the mouse [279]. In efforts to improve brain penetration, the acidic moiety within compounds **64–66** was removed to provide analogs such as pyridine **67** with modest potency (EC_{50} = 340 nM, 33% potentiation). Pyridine **67** had low oral bioavailability (3%), but i.p. administration (20 mg/kg) provided a brain/plasma ratio of 1.2 and total brain exposure equal to 330 nM [280]. Although the total brain concentrations were near the mGluR2 EC_{50} value, the unbound brain concentration was not reported making it difficult to interpret the relevance of this exposure. Subsequent efforts around this series have been focused on improving potency and brain penetration [281–284].

High-throughput screening by other groups has led to the aza-benzimidazole **68** (rmGluR2, EC_{50} = 64 nM) [285], benzimidazole **69** (mGluR2, pEC_{50} = 6.9) [286], and the cyclic carbamate **70** (rmGluR2, EC_{50} = 30 nM) [287]. In the rat, **68** had good oral bioavailability (79%) and low clearance (28 ml/min/kg); **69** had moderate bioavailability (22%) and moderate clearance (32 ml/min/kg), as well as a brain–blood ratio of 1.3 with brain C_{max} of 32 ng/g; and **70** had good oral bioavailability (64%), but high clearance (102 ml/min/kg). Compound **70** attenuated

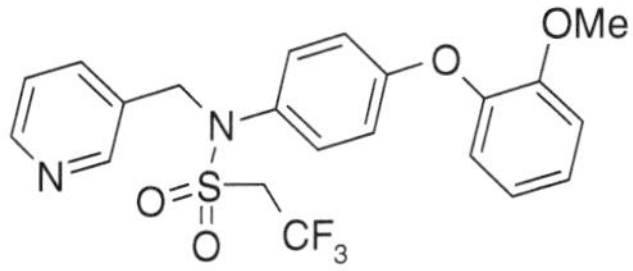

63, LY487379 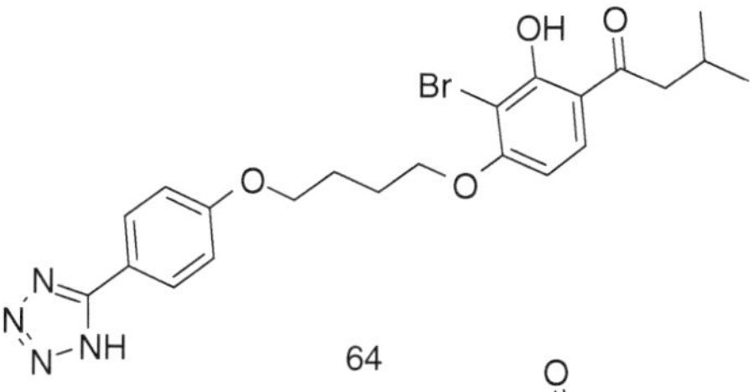 64

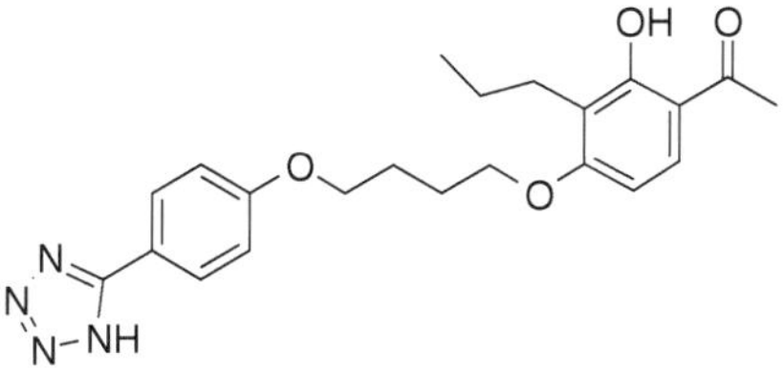

65 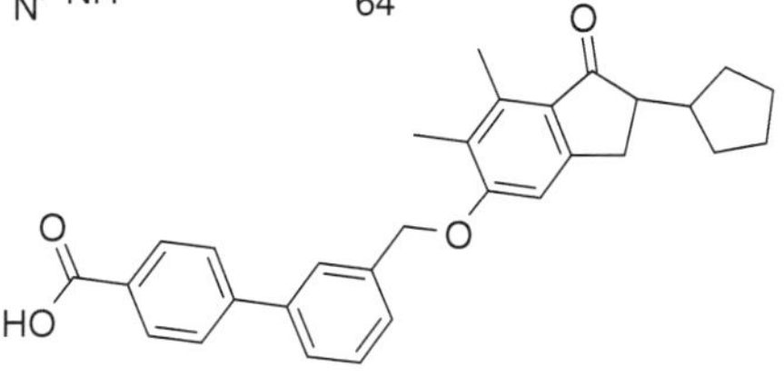 66

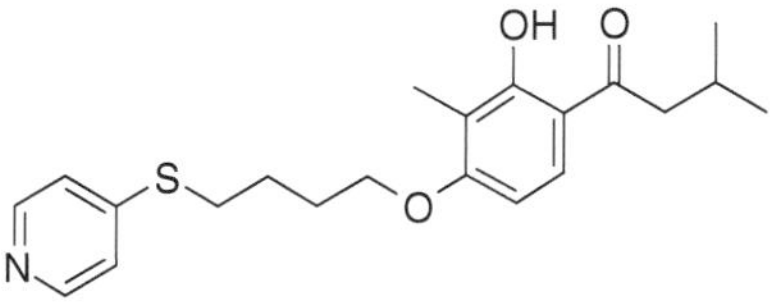

67 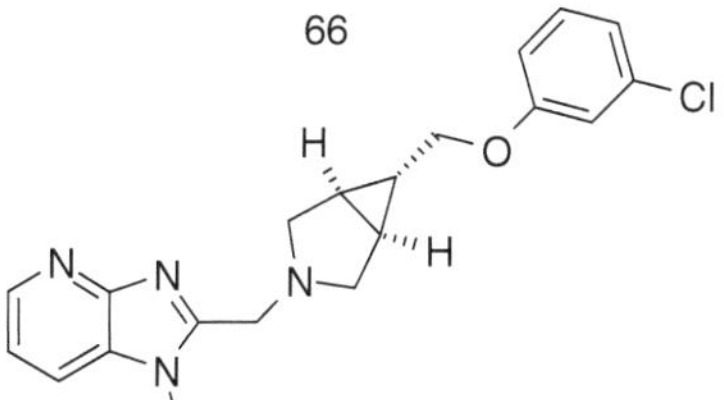 68

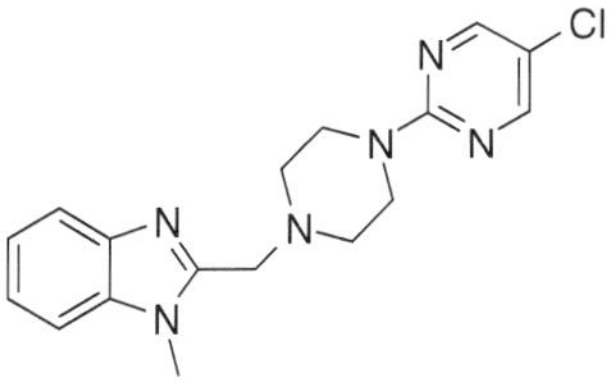

69 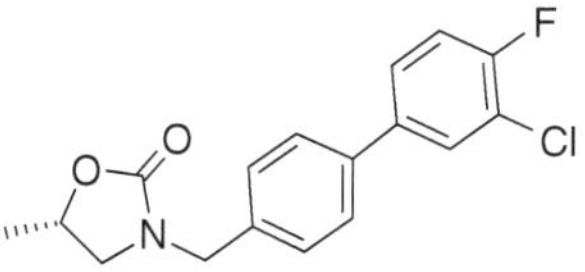 70

Fig. 8 mGluR2 positive allosteric modulators

methamphetamine-induced locomotor activity in mice (MED = 10 mg/kg, s.c.) with a free brain exposure of 34 nM, similar to its mGluR2 EC_{50} value.

Although many of the examples of mGluR2 PAMs in the literature have been centered on anxiety and psychosis models, given that this target is intimately tied to the modulation of glutamate neurotransmission, we expect mGluR2 modulators to also have broad application in the treatment of neurodegenerative diseases.

4.2.3 mGluR4

Positive allosteric modulators of mGluR4 and their role in PD have recently been reviewed [288, 289], but compared to the enormous efforts around mGluRs 1, 2, and 5, relatively little has been reported about modulators of mGluR4. The mGluR1 partial antagonist, **71** (PHCCC), was the first reported robust mGluR4 PAM (EC_{50}

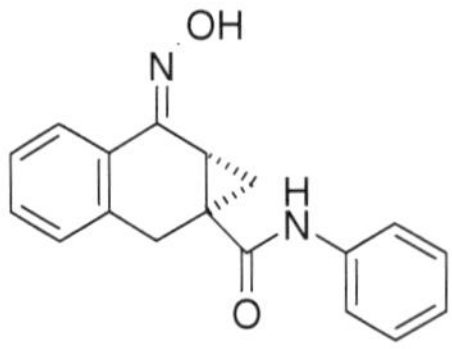

Fig. 9 mGluR4 positive allosteric modulators

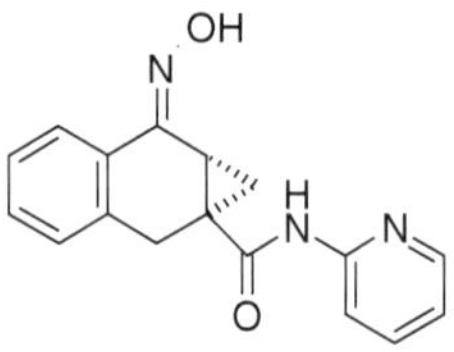

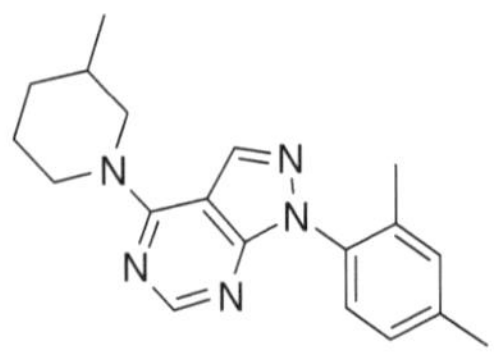

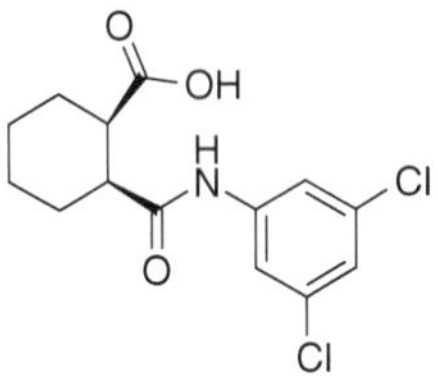

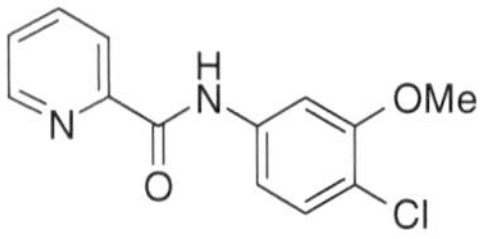

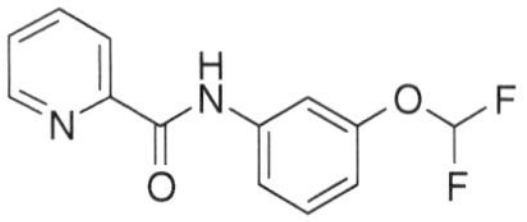

= 4.1 μM, 5.5-fold leftward shift of the glutamate dose–response curve) (Fig. 9) [289, 290]. SAR development around the structure of **71** led to the pyridyl analog **72** with improved mGluR4 potency [EC_{50} = 380 nM, 121% of the maximal glutamate response (Glu max)] and selectivity versus other mGluRs [291]. A structurally diverse lead compound, **73** (EC_{50} = 4.6 μM, 12–27-fold shift), had the advantage of having no H-bond donors, providing a higher probability of getting across the blood–brain barrier [292]. Another mGluR4 PAM, *cis*-cyclohexane carboxylic acid **74** (EC_{50} = 0.74 μM, 127% Glu max), although not brain penetrable, was shown to decrease haloperidol-induced catalepsy and reserpine-induced akinesia in rats when administered i.c.v. [293, 294]. Continued efforts to improve brain availability led to the pyridine amides **75** (hmGluR4 EC_{50} = 240 nM, 182% Glu max) and **76** (hmGluR4 EC_{50} = 340 nM, 235% Glu max) [295]. Although **75** and **76** had poor in vitro PK parameters, when dosed 10 mg/kg, i.p. in the rat they provided ~6 μM total brain concentrations of drug (fraction unbound in brain was not reported). Future directions for mGluR4 modulators are focused on improving potency, selectivity, PK, and brain penetration.

4.2.4 mGluR5

Since orthosteric antagonists of mGluR5 with desirable drug-like properties have been difficult to achieve, many researchers have turned to allosteric modulation of the receptor. Through the use of high-capacity functional assays based on recombinantly expressed mGluR subtypes, the first subtype selective mGluR5 antagonists (**77**, SIB-1757 and **78**, SIB-1893) were identified (Fig. 10) [296]. Subsequent SAR development around these leads led to the discovery of the diaryl alkyne, **79**

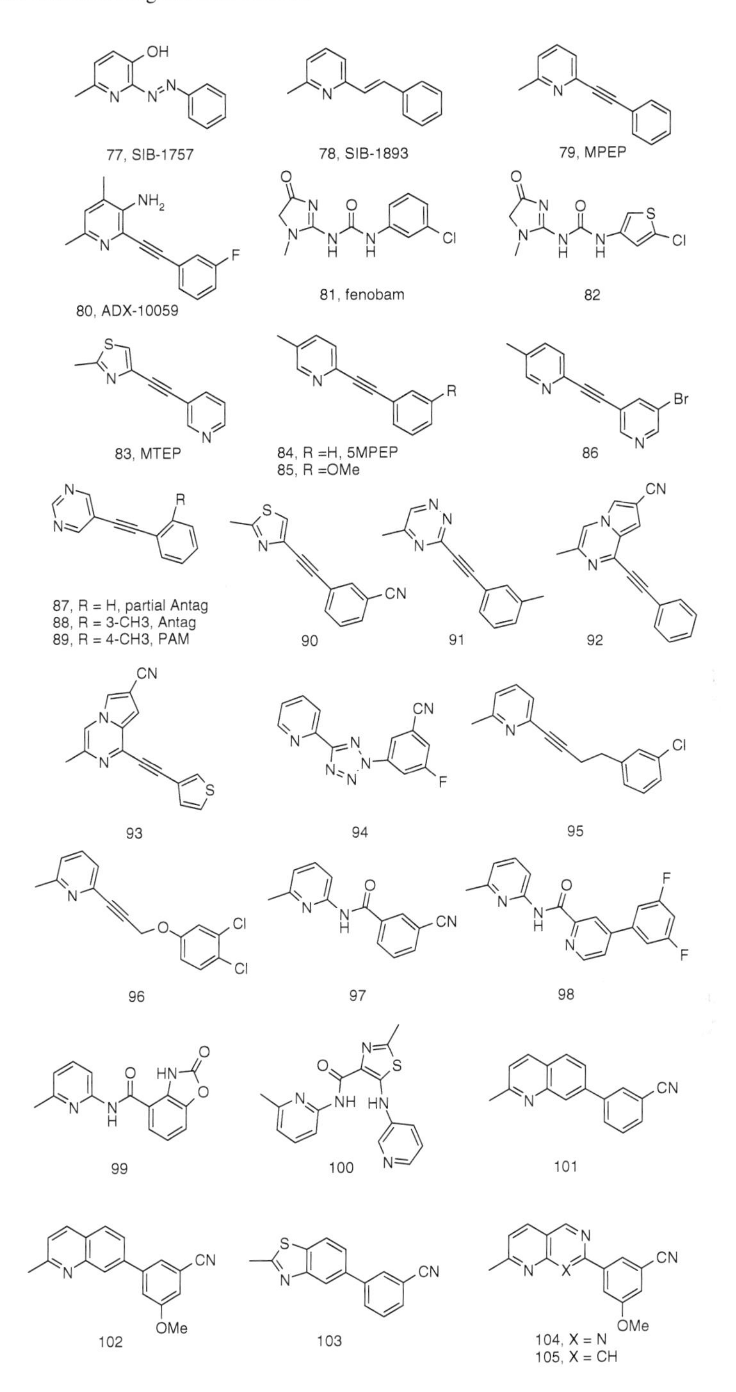

Fig. 10 Negative allosteric modulators of mGluR5

(MPEP), a potent (hmGluR5a $IC_{50} = 36$ nM) subtype selective mGluR5 antagonist with good brain penetration [297]. As previously stated, MPEP has been a useful tool compound for exploration of the in vivo effects associated with mGluR5 blockade. Several mGluR5 NAMs have entered into clinical trials. Compound **80** (ADX-10059) is an analog of MPEP that has been in phase II trials for several indications. This compound was not effective in a 50-patient trial for reduction of moderately severe dental anxiety, but did show positive results in patients with gastroesophageal reflux disease and in patients with migraine. However, the efficacious CNS exposure of ADX-10059 used to treat these diseases was not determined and safety issues forced the discontinuation of the development of ADX-10059 for chronic indications [298]. A structurally distinct mGluR5 NAM, **81** (fenobam), was discovered in the 1970s as a non-benzodiazepine with in vivo anxiolytic activity [299], and is currently under clinical development for fragile X syndrome. Subsequent medicinal chemistry efforts around the structure of fenobam seeking to improve potency have not been successful [300]. Even subtle changes such as replacement of the phenyl group with 3-thienyl (**82**, $IC_{50} = 434$ nM) led to a tenfold decrease in functional activity [301]. In efforts to relate CNS exposure with in vivo efficacy for mGluR5 NAMs, MPEP and some of its closely related analogs [e.g., **83** (MTEP)] were successfully radiolabeled (^{11}C or ^{18}F) and used for PET imaging in monkeys [302]. More recently, the radio tracer ^{11}C-ABP-688 was evaluated in rats, mice, and humans and showed high levels of uptake in the hippocampus, striatum, and cortex [303–305]. ^{11}C-ABP-688 is suggested to be a suitable PET ligand for imaging mGluR5 distribution in humans. Future use of mGluR5 PET ligands in receptor occupancy studies will help to guide dose selection for clinical proof of concept studies.

The enormous amount of SAR development around the structure of MPEP, including rational drug design to replace the potentially toxic alkyne moiety, as well as the discovery of non-MPEP-based mGluR5 NAMs have been extensively reviewed [306–309]. Noteworthy are the findings that small changes to the structure of mGluR5 modulators can have significant effects on the pharmacology. For example, **84** 5MPEP, the 5-methylpyridyl analog of MPEP, was shown to be a neutral allosteric ligand as it binds to the MPEP-binding site, but does not lead to a functional response [310]. Other closely related 5-methylpyridyl analogs (**85** and **86**) were reported to be partial antagonists of mGluR5. It was also discovered that small substituent changes around the phenyl group on the 5-phenylethynylpyrimidine scaffold yielded compounds with distinct profiles ranging from partial antagonists (**87**, R = H, $IC_{50} = 480$ nM), to full antagonists (**88**, R = 3-methyl, $IC_{50} =$ 7 nM) to PAMs (**89**, $EC_{50} = 3.3$ μM) [311, 312]. Tool compounds with diverse mGluR5 pharmacology may prove to be useful as it is unclear whether full or partial mGluR5 antagonists are necessary to achieve clinical efficacy with an adequate therapeutic window.

New compounds derived from MPEP and MTEP are numerous, and the SAR directions can be generally divided into three strategies: (a) variation of the aryl groups and their substituents, (b) replacement of the alkyne linker, and (c) fusion of the alkyne linker to one of the two aryl groups. Much of this work has been reported in medicinal chemistry communications, and highlights of those reports are summarized here. A significant advancement in SAR was realized when the 3-cyano analog of MTEP (**90**) was found to improve functional activity by 490-fold [313]. Substitution at the 3- or 4-position of the phenyl ring of MPEP with an additional aryl group did not offer potency advantages and substantially increased molecular weight. Replacement of the methylpyridine group of MPEP with methyltriazines (**91**, $IC_{50} = 2.3$ nM) [314] and methyl pyrolopyrazines (**92**, $IC_{50} = 1.3$ nM; **93**, $IC_{50} = 0.40$ nM) led to novel analogs with potent functional activity [315]. Replacement of the alkyne moiety of MPEP with tetrazole led to **94** (mGlu5 Ca^{2+} flux $IC_{50} = 4$ nM) with good rat brain penetration (total brain conc. = 2.4 μM, 3 mg/kg, i.p.) and excellent receptor occupancy ($ED_{50} = 1.3$ mg/kg, i.p.) [316]. Elongating the alkyne linker of MPEP led to 1-butyne **95** ($IC_{50} = 5$ nM) and 3-oxypropyne **96** ($IC_{50} = 15$ nM) [317]. Replacement of the alkyne with an amide linker led to compounds **97** ($IC_{50} = 5.3$ μM), **98** ($IC_{50} = 160$ nM), and **99** ($IC_{50} = 60$ nM), the latter being designed with the assistance of structural overlays with fenobam [318, 319]. Compound **100** ($IC_{50} = 25$ nM) with a unique aminothiazole amide linker replacement was reported to be active in a Vogel model (MED = 3 mg/kg, p.o.), but due to very high clearance it was speculated that the in vivo activity could be linked to a reactive metabolite [320]. Tying the linker into the pyridine ring was performed concomitantly by two different groups and led to quinolines **101** ($K_i = 110$ nM) and **102** ($IC_{50} = 0.8$ nM), benzothiazole **103** ($K_i = 2{,}100$ nM), pyridopyrimidine **104** ($IC_{50} = 1.2$ nM), and naphthyridine **105** ($IC_{50} = 6.2$ nM) [321–324]. Many of the above analogs of MPEP and MTEP were shown to have activity in animal models of anxiety, gastroesophageal reflux, and pain. Furthermore, the medicinal chemistry goals around the follow-up of MPEP and MTEP are being realized: potency has been improved, mGluR subtype selectivity retained, desirable PK and brain penetration achieved, and the potentially toxic alkynyl group (via metabolic activation) has been shown to be replaceable.

Structurally novel chemical leads have also been identified through the use of HTS screening. The tri-aryl analogs **106** and **107** were derived from an HTS hit and were found to be potent mGluR5 NAMs (Ca^{2+} flux $IC_{50} = 41$ and 23 nM, respectively) (Fig. 11) [325]. These compounds had high oral bioavailability and showed good rat PK, but were not active in a rat model of anxiety, likely due to poor receptor occupancy (33–52%, 10 mg/kg, i.p.). Compounds **108** ($K_i = 1$ nM) and **109** (FLIPR $IC_{50} = 80$ nM), although derived from an HTS hit, have structural similarities to MPEP [326, 327]. Three structurally unique mGluR5 NAMs (**110**, **111**, and **112**) were recently discovered [328]. SAR development around **112** was interesting, in which small changes to the substituents around the fused bicycle changed the pharmacology from a full antagonist to an mGluR5 PAM (**113**, $EC_{50} = 7.6$ μM, 73% Glu max). Finally, compound **114** ($IC_{50} = 32$ nM) was derived from an HTS hit and was found to have a good PK profile in rats and to have robust anxiolytic-like effects in several animal models of fear and anxiety [329]. It will be interesting to see how these new chemical leads influence the direction of medicinal chemistry design in the mGluR5 area.

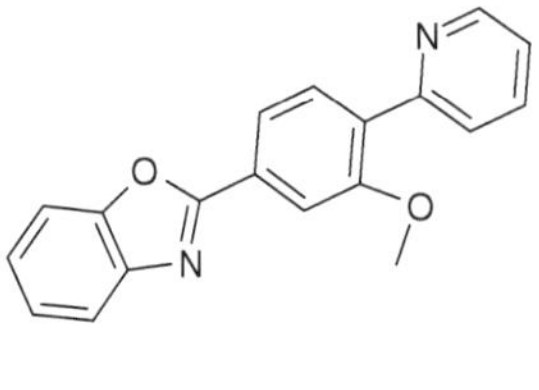

106

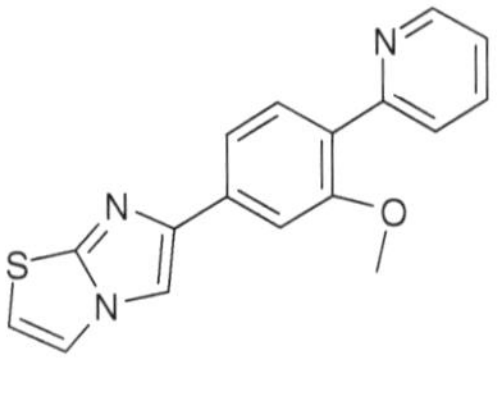

107

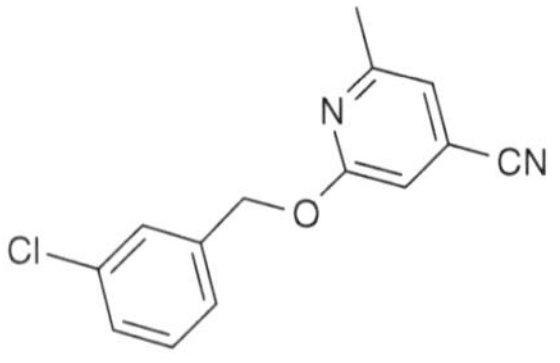

108

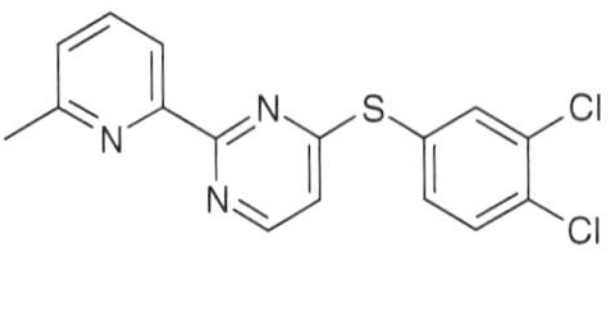

109

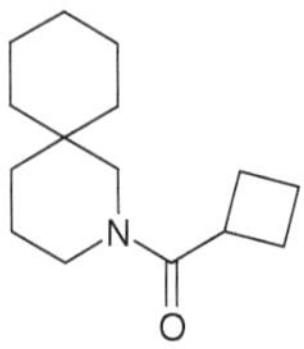

110

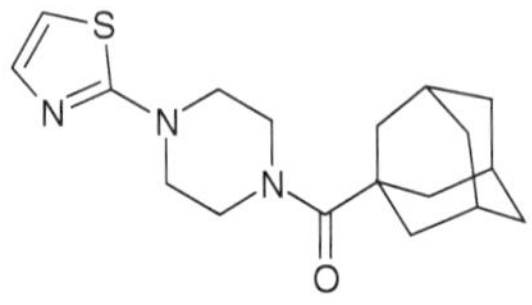

111

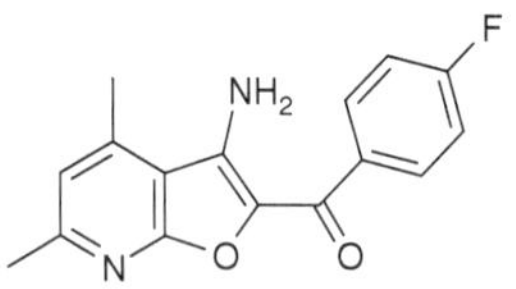

112

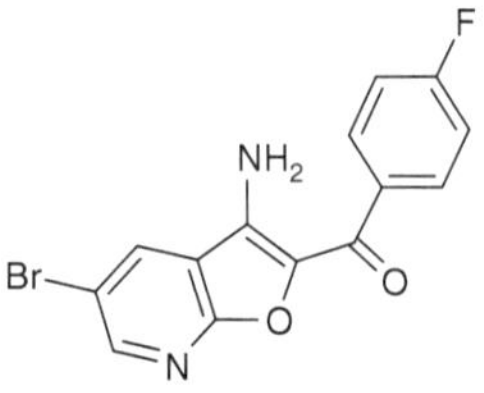

113 (mGluR5 PAM)

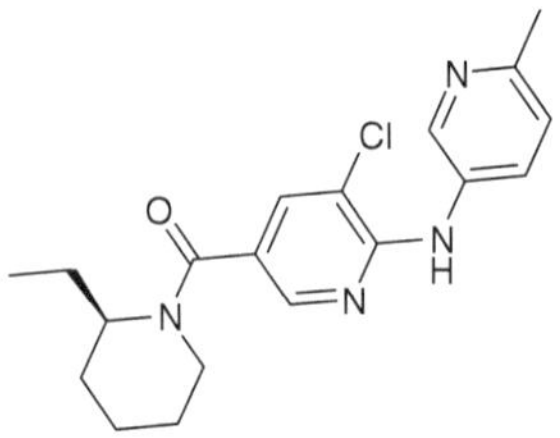

114

Fig. 11 Negative allosteric modulators (non-MPEP-based) of mGluR5

Less attention has been given to positive allosteric modulators of the mGluR5 receptor which have also been reviewed [330]. As discussed above, mGluR5 activators may enhance NMDA function. Thus, based on the NMDA hypofunction hypothesis of schizophrenia, mGluR5 PAMs may prove to play a role in treating cognitive deficits in this disorder. One of the first mGluR5 PAM leads was the benzaldazines, represented by **115** (Fig. 12). These compounds were found to have low micromolar mGluR5 PAM activity, and similar to the MPEP series of mGluR5 NAMs, it was discovered that small changes to the substituents around either of the phenyl rings could alter the allosteric modulator pharmacology from positive to negative to neutral. Noteworthy is the potential chemical reactivity (and potential toxicity) of the azine functionality rendering it less attractive as a lead structure. Efforts to find alternative lead matter has led to the discovery of the diphenyl pyrazole **116** (hmGluR5 $EC_{50} = 10$ nM, fourfold potentiation; rmGluR5 $EC_{50} = 20$ nM, 4.3-fold potentiation) [331]. Compound **116** reversed amphetamine-induced disruption of PPI in a dose-responsive manner (3, 10, and 30 mg/kg, s.c.). However, when this compound was evaluated in a rat cognition model, an inverted U-shaped dose curve was observed, with lower doses improving recognition and higher doses having no effect [332]. Both **115** and **116** have been shown to bind at the same allosteric binding site as MPEP. Another mGluR5 PAM, **117**

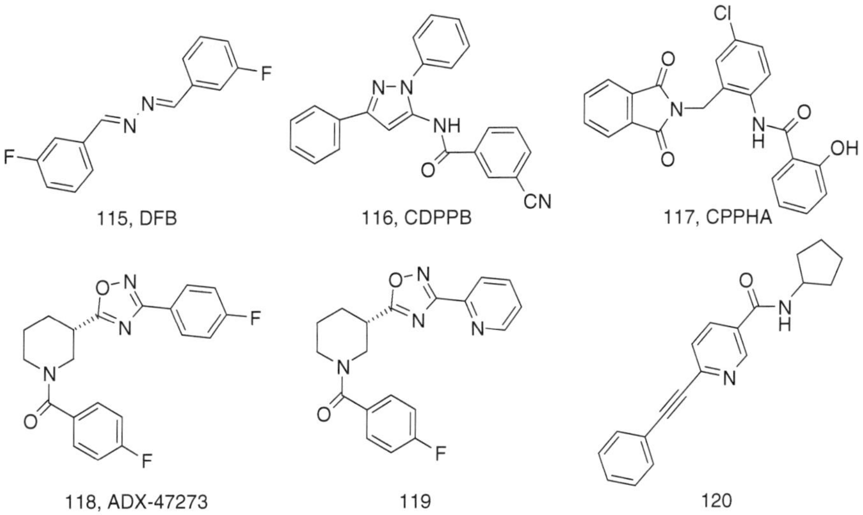

115, DFB
116, CDPPB
117, CPPHA
118, ADX-47273
119
120

Fig. 12 Positive allosteric modulators of mGluR5

(hmGluR5 $EC_{50} = 250$ nM, 7.1-fold potentiation), was identified which does not bind at the MPEP site. However, SAR around **117** has thus far not led to improved activity [333]. Other efforts have led to the identification of **118** ($EC_{50} = 168$ nM, 107% Glu max) and subsequent pyridyl analogs (e.g., **119**, $EC_{50} = 348$ nM) with selective mGluR5 PAM activity [295]. Finally, di-aryl alkyne **120** ($EC_{50} = 30$ nM) was found to be selective versus mGluR1 (not evaluated against other mGluRs) and to be brain penetrable in rodents [334].

4.2.5 mGluR 6, 7, and 8

Agonists and antagonists for group III mGluRs have generally been derived from glutamic acid, where the conformation has been restricted, the linker between the two acid moieties has been varied, substituents inserted, and/or one of the carboxylic acids has been replaced with an acid bioisostere. These efforts have been reviewed and will not be discussed here [335]. Subtype selectivity has been challenging with direct agonists and antagonists, and so like the group I and II mGluRs, allosteric modulation of the group III mGluRs has been an area of active pursuit. For mGluR7, an allosteric modulator has been identified (**121**, Fig. 13) that directly activates receptor signaling through an allosteric site in the transmembrane domain [336]. Compound **121** was shown to be orally active and brain penetrable, elevating the plasma stress hormones corticosterone and corticotropin in an mGluR7-dependent fashion, based on a comparison of activity in mGluR7 +/+ versus −/− mice (1 and 6 mg/kg, p.o.). Compound **121** also reversed haloperidol-induced catalepsy in rats, reduced apomorphine-induced rotations in unilateral 6-

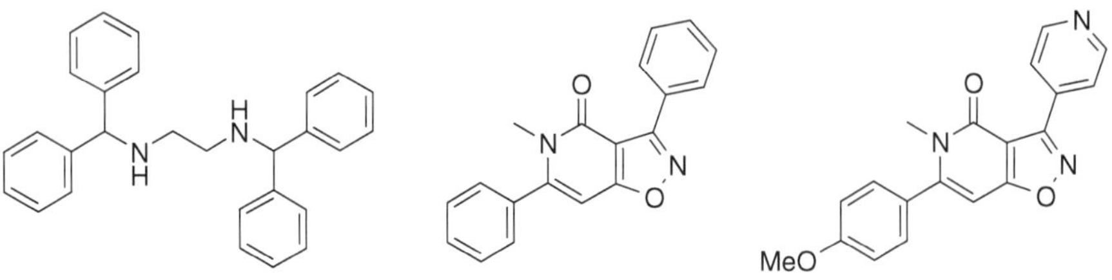

121, AMN082 122, MDIP 123, MMPIP

Fig. 13 mGluR7 modulators

OHDA-lesioned rats, and reversed the increased reaction time to respond to a cue in bilateral 6-OHDA-lesioned rats, suggesting mGluR7 agonism may be a useful approach in treating PD [337]. HTS screening efforts around mGluR7 have led to the identification of **122**, the first subtype selective mGluR7 NAM [338, 339]. SAR development around **122** seeking to lower logP and improve physicochemical properties led to **123** (mGluR7 IC50 = 26 nM) with good oral bioavailability, low clearance, and good total brain availability. These tools should prove useful in further defining the role that mGluR7 plays in the CNS.

5 Glutamate Transporters

5.1 Glutamate Transporter Biology and the Role of Astrocytes

Glutamate is synthesized from glutamine in the presynaptic nerve terminal by the enzyme glutaminase. Once formed, glutamate is loaded into synaptic vesicles by the action of the VGLUT. After glutamate is released from the neuron, its action is terminated by removal from the synapse by the action of another membrane protein known as the EAAT. These two families of transporters are critical for normal glutamatergic neurotransmission, but as is often the case, their dysfunction can be associated with significant CNS pathology.

Molecular characterization has revealed that the VGLUT family consists of three isoforms, VGLUT1, 2, and 3. VGLUT1 and 2 are found on synaptic vesicles within terminals of glutamatergic neurons, while VGLUT3 is expressed predominantly in neuronal soma and dendrites as well as in astrocytes. VGLUT3 is expressed on glutamatergic as well as nonglutamatergic neurons, indicating that this isoform functions in cells that release other transmitters in addition to glutamate [1, 340–342]. VGLUT1 is the most highly expressed isoform within this family, and is responsible for the majority of activities in the CNS. The transport of glutamate across the synaptic vesicle membrane is driven by an electrochemical gradient, which is created by a vesicular ATPase activity. This gradient in turn provides the energy to transport the neurotransmitter across the vesicle membrane.

VGLUT1 and 2 are specific markers of glutamatergic terminals, and as such are useful indicators of the status of these neurons in the healthy or diseased brain. Decreased expression of VGLUT1 has been described in the AD brain and correlated with decreased cognitive function [343]. In PD, alterations in VGLUT levels have also been described, with increases in some regions and decreases in others [344]. It remains unclear to what extent these changes in VGLUT expression are part of the underlying pathology of these diseases, as opposed to simply serving as a marker of glutamatergic neuronal degeneration and death, which may ultimately have a variety of underlying causes.

The second class of glutamate transporters is the EAATs. These transporters are responsible for removing glutamate from the synaptic cleft following release, and thus play a critical role in regulating the extent and duration of the excitatory signal. Indeed, it can be readily appreciated that reduced expression or function of these transporters will result in prolongation of elevated synaptic glutamate levels, with potentially deleterious consequences. Molecular cloning of this transporter has led to the identification of five family members (EAAT1–5), with EAAT1 and 2 expressed in astrocytes, and EAAT3 and 4 expressed in neurons [340, 345, 346]. EAAT5 appears to have a restricted expression and is found only in the retina [347]. EAATs are composed of three identical subunits, and the energy required for transport is generated by a Na^{+}/K^{+} ATPase that is physically associated with the transporter protein complex [348]. Interestingly, some of the EAATs (particularly EAAT4 and 5) can function as glutamate-gated chloride channels and are able to generate a chloride current in the absence of any glutamate transport [349]. The astrocytic EAATs, and especially EAAT2, are primarily responsible for removing synaptic glutamate following vesicular exocytosis from the nerve terminal [346]. EAAT2 is thus critically involved in regulating the duration of the glutamate signal, as well as limiting the excitotoxic potential of extracellular glutamate by rapid removal from the extracellular space. EAAT2 also supplies the astrocytes with glutamate that is converted into glutamine by the enzyme glutamine synthase. Glutamine is used by astrocytes for ammonia detoxification and is involved in signaling the metabolic needs of proximal neurons [340, 346]. EAAT1 and 2 are also expressed in the astrocytic processes that come into direct contact with brain capillaries, and as such play an important role in the transport of glutamate from the extracellular fluid into the blood [350]. EAAT3 and 4 are expressed in distinct neuronal subpopulations and have a more specialized function in limiting glutamate spillover to adjacent neurons [347]. These transporters are typically found in different population of cells, with EAAT3 enriched in the forebrain and EAAT4 in the cerebellum. Interestingly, EAATs that are expressed on inhibitory interneurons provide these cells with glutamate as an essential precursor to the synthesis of GABA, the main inhibitory neurotransmitter [351]. Given their critical role in the brain, it is not surprising that a variety of mechanisms exist for regulating EAAT expression and function. Short-term regulation of these transporters is achieved through posttranslational modifications, as well as through interaction with other membrane proteins, both of which can influence surface expression and intrinsic activity [352, 353]. Regulation on a longer time scale occurs through changes in gene expression, mediated through transcriptional regulation [354] as well as through alternative splicing [352]. The EAAT2 promoter has been shown to be regulated by physical astrocyte–neuron interactions, as well as through various growth factor signals.

Given the central role played by the EAATs, it is not surprising that abnormalities in their expression or function are associated with neuronal injury and death in a variety of acute and chronic neurodegenerative conditions [352, 353]. The mechanisms underlying EAAT dysfunction may vary from defects in intracellular trafficking, to altered mRNA splicing, to abnormal posttranslational modifications, but the result is typically reduced function. Several studies have been carried out in mice harboring a gene deletion of the murine homologs of EAAT1, EAAT2, or both. It was shown that mice lacking the GLAST gene (homolog of EAAT1), had an increased likelihood of cerebellar Purkinje cell loss following an ischemic event [355]. This is likely a reflection of a decreased capacity to remove the elevated extracellular glutamate and a corresponding increase in excitotoxic damage. Also, in mice lacking GLT1 (the mouse homolog of EAAT2), an increased propensity for seizures was observed, with animals not generally surviving past a few weeks of age [356]. These animal model studies emphasize the critical role of the EAATs in normal CNS function. Additional work has supported the role of these proteins in human disease. There are several examples of abnormalities in RNA splicing, which result in a truncated transporter protein of reduced function. This has been described in tissue from amyotrophic lateral sclerosis (ALS) patients [357, 358], as well as in tissue from patients with epilepsy and AD [359, 360]. However, the fact that these abnormally spliced variants are also found in tissue from control subjects indicates that while they may play a role in mediating disease pathology, they are clearly not the only factor involved. A role of EAAT2 in AD is further supported by decreased expression in AD brain [361], as well as a reduction in EAAT function in cultured astrocytes following treatment with the Aβ peptide [362].

Given the critical role of EAATs in regulating synaptic glutamate levels, increased expression or function of this protein may provide benefit in reducing excitotoxic damage. Interestingly, it was shown that EAAT1 and 2 are both upregulated in optic nerve from multiple sclerosis patients. In addition, glutamate uptake was shown to be increased in disease tissue, suggesting that glial cells upregulate expression of these transporters as a protective response to the excessive glutamate levels known to exist in this disease [363]. Other studies have shown that treatment of mice with the β-lactam antibiotic ceftriaxone leads to an increase in the expression of GLT1. Although the precise mechanism by which this increased expression occurs in unclear, it has been shown to increase glutamate uptake in functional assays, and to attenuate disease phenotypes in the R6/2 model of HD [364, 365]. Treatment with ceftriaxone also increases glutamate uptake and improves survival in a stroke model [366]. While the precise mechanism for these intriguing effects is unknown, it does suggest that upregulation of one or more of the EAATs may represent a promising therapeutic strategy for treating neurodegenerative diseases (Fig. 14).

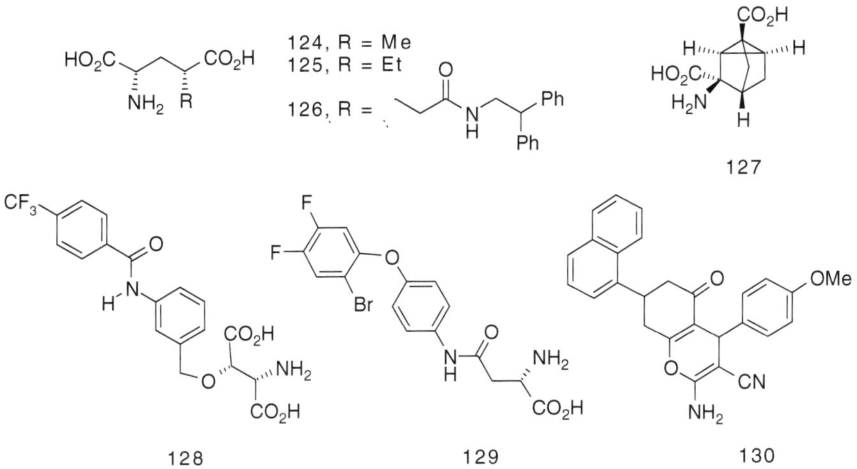

Fig. 14 EAAT inhibitors

5.2 *Glutamate Transporter Modulators*

Progress toward the discovery of potent and selective VGLUT and EAAT modulators has been slow. The majority of the reported VGLUT inhibitors are either conformationally restricted amino di-acid analogs or azo-dyes containing an amino di-acid motif [367]. The structures of known VGLUT inhibitors are not yet positioned to provide adequate brain exposure for in vivo evaluation (with respect to potency, PK, and brain penetration). Likewise, EAAT inhibitors have also been largely derived from conformationally restricted analogs of glutamic acid and aspartic acid. The SAR around EAATs was reviewed in 2003 [368, 369], and selected findings since then are reported here. Substitution of small alkyl groups at the 4-position of glutamic acid had effects on the EAAT1 pharmacology. 4-Methyl substitution (**124**) was found to be an EAAT1 substrate and an EAAT2/3 inhibitor (EAAT1 $K_m = 13$ μM, EAAT2 $K_i = 13$ μM, and EAAT3 $K_i = 6.6$ μM), whereas the 4-ethyl analog (**125**) was an inhibitor of EAAT1, 2, and 3 ($K_i = 23$, 14, and 39 μM, respectively) (Fig. 14) [370]. Further SAR development at the 4-position of glutamic acid led to **126**, a weak but selective inhibitor of EAAT2 (EAAT1–3, $IC_{50} > 1{,}000$, 89, and 1,000 μM, respectively) [371]. Efforts toward improving the potency and selectivity of EAAT inhibitors by confining the conformation of glutamic acid has led to a tricyclic analog, **127**, that was shown to be selective for EAAT2 ($IC_{50} = 2.2$ μM) compared to EAAT1 (50% at 100 μM) and EAAT3 (24.5 μM) [372]. Although not selective for any particular EAAT, benzyloxy analogs of aspartic acid such as **128** [373] showed improved potency for EAAT1–3 ($IC_{50} = 22$, 17, and 300 nM, respectively). Compound **128** dosed i.c.v. induced severe convulsive behaviors in mice suggesting an accumulation of glutamate in the brain. Selectivity for EAAT2 was subsequently obtained through the aryl ether aspartamide **129** (EAAT1–3 IC_{50} = 5,000, 85, and 3,800 nM, respectively) [374]. Finally, a novel, potent and selective inhibitor of EAAT1 was discovered through HTS screening, and subsequent SAR development led to compound **130** (EAAT1–3, IC_{50} = 0.66, >300, and >300 μM, respectively) [375]. Compound **130** is unique in that it is not related to glutamic acid or aspartic acid, and efforts to ascertain whether the binding is at the orthosteric site or an allosteric site are ongoing. More relevant to the treatment of neurodegenerative diseases would be positive modulators of EAATs, which theoretically could enhance the uptake of glutamate. Should compound **130** prove to bind at an allosteric site, this may provide a starting point for the discovery of positive allosteric modulators of the EAATs.

6 Conclusion

The central importance of glutamate signaling in both normal and pathological CNS function has prompted extensive research to better understand the biology and pharmacology of this neurotransmitter. In addition, the identification of drugs which modulate glutamate receptor function has been an area of focus for medicinal chemists working on diseases of the central nervous system. As a result of this work, significant progress has been made toward the identification of compounds that can serve as both research tools to better understand glutamate's role as a neurotransmitter, and potential therapeutic agents for the treatment of CNS diseases. Neurodegenerative diseases represent an especially challenging area for drug development. However, given the increasing prevalence of these diseases in an aging population, research to identify improved treatments for diseases such as Alzheimer's, Parkinson's, and Huntington's are likely to remain a priority for the pharmaceutical and biotech industries in the coming years. Identification of glutamatergic agents with improved potency, selectivity, and pharmaceutical properties will serve to advance our understanding of this complex area of CNS biology, and ultimately may open the door to safer and more efficacious therapies.

References

1. Liguz-Lecznar M, Skangiel-Kramska J (2007) Vesicular glutamate transporters (VGLUTs): the three musketeers of glutamatergic system. Acta Neurobiol Exp 67:207–218
2. Bunch L, Erichsen MN, Jensen AA (2009) Excitatory amino acid transporters as potential drug targets. Expert Opin Ther Targets 13:719–731
3. Bowie D (2008) Ionotropic glutamate receptors & CNS disorders. CNS Neurol Disord Drug Targets 7:129–143
4. Lodge D (2009) The history of the pharmacology and cloning of ionotropic glutamate receptors and the development of idiosyncratic nomenclature. Neuropharmacology 56:6–21

5. Niswender CM, Conn PJ (2010) Metabotropic glutamate receptors: physiology, pharmacology, and disease. Annu Rev Pharmacol Toxicol 50:295–322
6. Szydlowska K, Tymianski M (2010) Calcium, ischemia and excitotoxicity. Cell Calcium 47:122–129
7. Marek GJ, Behl B, Bespalov AY et al (2010) Glutamatergic (N-methyl-D-aspartate receptor) hypofrontality in schizophrenia: too little juice or a miswired brain? Mol Pharmacol 77:317–326
8. Cull-Candy S, Brickley S, Farrant M (2001) NMDA receptor subunits: diversity, development and disease. Curr Opin Neurobiol 11:327–335
9. Waxman EA, Lynch DR (2005) N-methyl-D-aspartate receptor subtypes: multiple roles in excitotoxicity and neurological disease. Neuroscientist 11:37–49
10. Burnashev N, Zhou Z, Neher E et al (1995) Fractional calcium currents through recombinant GluR channels of the NMDA, AMPA and kainate receptor subtypes. J Physiol 485 (Pt 2):403–418
11. Garaschuk O, Schneggenburger R, Schirra C et al (1996) Fractional Ca2+ currents through somatic and dendritic glutamate receptor channels of rat hippocampal CA1 pyramidal neurones. J Physiol 491(Pt 3):757–772
12. Schneggenburger R (1996) Simultaneous measurement of Ca2+ influx and reversal potentials in recombinant N-methyl-D-aspartate receptor channels. Biophys J 70:2165–2174
13. Dannhardt G, Kohl BK (1998) The glycine site on the NMDA receptor: structure-activity relationships and possible therapeutic applications. Curr Med Chem 5:253–263
14. Johnson JW, Ascher P (1987) Glycine potentiates the NMDA response in cultured mouse brain neurons. Nature 325:529–531
15. Wollmuth LP, Sobolevsky AI (2004) Structure and gating of the glutamate receptor ion channel. Trends Neurosci 27:321–328
16. Malenka RC, Nicoll RA (1999) Long-term potentiation–a decade of progress? Science 285:1870–1874
17. Ikonomidou C, Turski L (2002) Why did NMDA receptor antagonists fail clinical trials for stroke and traumatic brain injury? Lancet Neurol 1:383–386
18. Muir KW (2006) Glutamate-based therapeutic approaches: clinical trials with NMDA antagonists. Curr Opin Pharmacol 6:53–60
19. Hardingham GE, Bading H (2003) The Yin and Yang of NMDA receptor signalling. Trends Neurosci 26:81–89
20. Pohl D, Bittigau P, Ishimaru MJ et al (1999) N-Methyl-D-aspartate antagonists and apoptotic cell death triggered by head trauma in developing rat brain. Proc Natl Acad Sci USA 96:2508–2513
21. Lipton SA, Nakanishi N (1999) Shakespeare in love–with NMDA receptors? Nat Med 5:270–271
22. Collins MO, Husi H, Yu L et al (2006) Molecular characterization and comparison of the components and multiprotein complexes in the postsynaptic proteome. J Neurochem 97 (Suppl 1):16–23
23. Joyal JL, Burks DJ, Pons S et al (1997) Calmodulin activates phosphatidylinositol 3-kinase. J Biol Chem 272:28183–28186
24. Lafon-Cazal M, Perez V, Bockaert J et al (2002) Akt mediates the anti-apoptotic effect of NMDA but not that induced by potassium depolarization in cultured cerebellar granule cells. Eur J Neurosci 16:575–583
25. Wu GY, Deisseroth K, Tsien RW (2001) Activity-dependent CREB phosphorylation: convergence of a fast, sensitive calmodulin kinase pathway and a slow, less sensitive mitogen-activated protein kinase pathway. Proc Natl Acad Sci USA 98:2808–2813
26. Riccio A, Ahn S, Davenport CM et al (1999) Mediation by a CREB family transcription factor of NGF-dependent survival of sympathetic neurons. Science 286:2358–2361
27. Walton M, Woodgate AM, Muravlev A et al (1999) CREB phosphorylation promotes nerve cell survival. J Neurochem 73:1836–1842

28. Lonze BE, Riccio A, Cohen S et al (2002) Apoptosis, axonal growth defects, and degeneration of peripheral neurons in mice lacking CREB. Neuron 34:371–385
29. Larson J, Lynch G, Games D et al (1999) Alterations in synaptic transmission and long-term potentiation in hippocampal slices from young and aged PDAPP mice. Brain Res 840:23–35
30. Yoshiyama Y, Higuchi M, Zhang B et al (2007) Synapse loss and microglial activation precede tangles in a P301S tauopathy mouse model. Neuron 53:337–351
31. Harkany T, Abraham I, Timmerman W et al (2000) beta-amyloid neurotoxicity is mediated by a glutamate-triggered excitotoxic cascade in rat nucleus basalis. Eur J Neurosci 12:2735–2745
32. Miguel-Hidalgo JJ, Alvarez XA, Cacabelos R et al (2002) Neuroprotection by memantine against neurodegeneration induced by beta-amyloid(1-40). Brain Res 958:210–221
33. Lesne S, Ali C, Gabriel C et al (2005) NMDA receptor activation inhibits alpha-secretase and promotes neuronal amyloid-beta production. J Neurosci 25:9367–9377
34. Goto Y, Niidome T, Akaike A et al (2006) Amyloid beta-peptide preconditioning reduces glutamate-induced neurotoxicity by promoting endocytosis of NMDA receptor. Biochem Biophys Res Commun 351:259–265
35. Snyder EM, Nong Y, Almeida CG et al (2005) Regulation of NMDA receptor trafficking by amyloid-beta. Nat Neurosci 8:1051–1058
36. Tyszkiewicz JP, Yan Z (2005) beta-Amyloid peptides impair PKC-dependent functions of metabotropic glutamate receptors in prefrontal cortical neurons. J Neurophysiol 93: 3102–3111
37. Kim AH, Khursigara G, Sun X et al (2001) Akt phosphorylates and negatively regulates apoptosis signal-regulating kinase 1. Mol Cell Biol 21:893–901
38. Javitt DC (2006) Is the glycine site half saturated or half unsaturated? Effects of glutamatergic drugs in schizophrenia patients. Curr Opin Psychiatry 19:151–157
39. Shimazaki T, Kaku A, Chaki S (2010) D-Serine and a glycine transporter-1 inhibitor enhance social memory in rats. Psychopharmacology 209:263–270
40. Javitt DC (2009) Glycine transport inhibitors for the treatment of schizophrenia: symptom and disease modification. Curr Opin Drug Discov Dev 12:468–478
41. Johnson KA, Conn PJ, Niswender CM (2009) Glutamate receptors as therapeutic targets for Parkinson's disease. CNS Neurol Disord Drug Targets 8:475–491
42. Gotz T, Kraushaar U, Geiger J et al (1997) Functional properties of AMPA and NMDA receptors expressed in identified types of basal ganglia neurons. J Neurosci 17:204–215
43. Kaur S, Ozer H, Starr M (1997) MK 801 reverses haloperidol-induced catalepsy from both striatal and extrastriatal sites in the rat brain. Eur J Pharmacol 332:153–160
44. McAllister KH (1996) The competitive NMDA receptor antagonist SDZ 220-581 reverses haloperidol-induced catalepsy in rats. Eur J Pharmacol 314:307–311
45. Moore NA, Blackman A, Awere S et al (1993) NMDA receptor antagonists inhibit catalepsy induced by either dopamine D1 or D2 receptor antagonists. Eur J Pharmacol 237:1–7
46. Graham WC, Robertson RG, Sambrook MA et al (1990) Injection of excitatory amino acid antagonists into the medial pallidal segment of a 1-methyl-4-phenyl-1,2,3,6-tetrahydropyridine (MPTP) treated primate reverses motor symptoms of parkinsonism. Life Sci 47: PL91–PL97
47. Loschmann PA, De Groote C, Smith L et al (2004) Antiparkinsonian activity of Ro 25-6981, a NR2B subunit specific NMDA receptor antagonist, in animal models of Parkinson's disease. Exp Neurol 187:86–93
48. Steece-Collier K, Chambers LK, Jaw-Tsai SS et al (2000) Antiparkinsonian actions of CP-101, 606, an antagonist of NR2B subunit-containing N-methyl-d-aspartate receptors. Exp Neurol 163:239–243
49. Stauch Slusher B, Rissolo KC, Jackson PF et al (1994) Centrally-administered glycine antagonists increase locomotion in monoamine-depleted mice. J Neural Transm Gen Sect 97:175–185

50. Blanchet PJ, Konitsiotis S, Whittemore ER et al (1999) Differing effects of N-methyl-D-aspartate receptor subtype selective antagonists on dyskinesias in levodopa-treated 1-methyl-4-phenyl-tetrahydropyridine monkeys. J Pharmacol Exp Ther 290:1034–1040
51. Marin C, Papa S, Engber TM et al (1996) MK-801 prevents levodopa-induced motor response alterations in parkinsonian rats. Brain Res 736:202–205
52. Marti M, Paganini F, Stocchi S et al (2003) Plasticity of glutamatergic control of striatal acetylcholine release in experimental parkinsonism: opposite changes at group-II metabotropic and NMDA receptors. J Neurochem 84:792–802
53. Group THsDCR (1993) A novel gene containing a trinucleotide repeat that is expanded and unstable on Huntington's disease chromosomes. Cell 72:971–983
54. Gusella JF, MacDonald ME (1995) Huntington's disease. Semin Cell Biol 6:21–28
55. Vonsattel JP, Myers RH, Stevens TJ et al (1985) Neuropathological classification of Huntington's disease. J Neuropathol Exp Neurol 44:559–577
56. Cepeda C, Hurst RS, Calvert CR et al (2003) Transient and progressive electrophysiological alterations in the corticostriatal pathway in a mouse model of Huntington's disease. J Neurosci 23:961–969
57. Li L, Murphy TH, Hayden MR et al (2004) Enhanced striatal NR2B-containing N-methyl-D-aspartate receptor-mediated synaptic currents in a mouse model of Huntington disease. J Neurophysiol 92:2738–2746
58. Tang TS, Slow E, Lupu V et al (2005) Disturbed Ca2+ signaling and apoptosis of medium spiny neurons in Huntington's disease. Proc Natl Acad Sci USA 102:2602–2607
59. Heng MY, Detloff PJ, Wang PL et al (2009) In vivo evidence for NMDA receptor-mediated excitotoxicity in a murine genetic model of Huntington disease. J Neurosci 29:3200–3205
60. Zhang H, Li Q, Graham RK et al (2008) Full length mutant huntingtin is required for altered Ca2+ signaling and apoptosis of striatal neurons in the YAC mouse model of Huntington's disease. Neurobiol Dis 31:80–88
61. Fan J, Cowan CM, Zhang LY et al (2009) Interaction of postsynaptic density protein-95 with NMDA receptors influences excitotoxicity in the yeast artificial chromosome mouse model of Huntington's disease. J Neurosci 29:10928–10938
62. Sun Y, Savanenin A, Reddy PH et al (2001) Polyglutamine-expanded huntingtin promotes sensitization of N-methyl-D-aspartate receptors via post-synaptic density 95. J Biol Chem 276:24713–24718
63. Zeron MM, Chen N, Moshaver A et al (2001) Mutant huntingtin enhances excitotoxic cell death. Mol Cell Neurosci 17:41–53
64. Fan MM, Raymond LA (2007) N-methyl-D-aspartate (NMDA) receptor function and excitotoxicity in Huntington's disease. Prog Neurobiol 81:272–293
65. Milnerwood AJ, Gladding CM, Pouladi MA et al (2010) Early increase in extrasynaptic NMDA receptor signaling and expression contributes to phenotype onset in Huntington's disease mice. Neuron 65:178–190
66. Okamoto S, Pouladi MA, Talantova M et al (2009) Balance between synaptic versus extrasynaptic NMDA receptor act ivity influences inclusions and neurotoxicity of mutant huntingtin. Nat Med 15:1407–1413
67. Newcomer JW, Krystal JH (2001) NMDA receptor regulation of memory and behavior in humans. Hippocampus 11:529–542
68. Heresco-Levy U (2005) Glutamatergic neurotransmission modulators as emerging new drugs for schizophrenia. Expert Opin Emerg Drugs 10:827–844
69. Urwyler S, Floersheim P, Roy BL et al (2009) Drug design, in vitro pharmacology, and structure-activity relationships of 3-acylamino-2-aminopropionic acid derivatives, a novel class of partial agonists at the glycine site on the N-methyl-D-aspartate (NMDA) receptor complex. J Med Chem 52:5093–5107
70. Harsing LG Jr, Juranyi Z, Gacsalyi I et al (2006) Glycine transporter type-1 and its inhibitors. Curr Med Chem 13:1017–1044

71. Sur C, Kinney GG (2007) Glycine transporter 1 inhibitors and modulation of NMDA receptor-mediated excitatory neurotransmission. Curr Drug Targets 8:643–649
72. Duplantier AJ, Becker SL, Bohanon MJ et al (2009) Discovery, SAR, and pharmacokinetics of a novel 3-hydroxyquinolin-2(1H)-one series of potent D-amino acid oxidase (DAAO) inhibitors. J Med Chem 52:3576–3585
73. Williams M (2009) Commentary: genome-based CNS drug discovery: D-amino acid oxidase (DAAO) as a novel target for antipsychotic medications: progress and challenges. Biochem Pharmacol 78:1360–1365
74. Johnson JW, Kotermanski SE (2006) Mechanism of action of memantine. Curr Opin Pharmacol 6:61–67
75. Kotermanski SE, Johnson JW (2009) Mg2+ imparts NMDA receptor subtype selectivity to the Alzheimer's drug memantine. J Neurosci 29:2774–2779
76. Ferris SH (2003) Evaluation of memantine for the treatment of Alzheimer's disease. Expert Opin Pharmacother 4:2305–2313
77. Mobius HJ, Stoffler A, Graham SM (2004) Memantine hydrochloride: pharmacological and clinical profile. Drugs Today 40:685–695
78. Parsons CG, Danysz W, Quack G (1999) Memantine is a clinically well tolerated N-methyl-D-aspartate (NMDA) receptor antagonist–a review of preclinical data. Neuropharmacology 38:735–767
79. Cai SX (2006) Glycine/NMDA receptor antagonists as potential CNS therapeutic agents: ACEA-1021 and related compounds. Curr Top Med Chem 6:651–662
80. Catarzi D, Colotta V, Varano F (2006) Competitive Gly/NMDA receptor antagonists. Curr Top Med Chem 6:809–821
81. Nagata R, Katayama S, Ohtani K et al (2006) Tricyclic quinoxalinediones, aza-kynurenic acids, and indole-2-carboxylic acids as in vivo active NMDA-glycine antagonists. Curr Top Med Chem 6:733–745
82. Hargreaves RJ, Rigby M, Smith D et al (1993) Lack of effect of L-687, 414 ((+)-cis-4-methyl-HA-966), an NMDA receptor antagonist acting at the glycine site, on cerebral glucose metabolism and cortical neuronal morphology. Br J Pharmacol 110:36–42
83. Hawkinson JE, Huber KR, Sahota PS et al (1997) The N-methyl-D-aspartate (NMDA) receptor glycine site antagonist ACEA 1021 does not produce pathological changes in rat brain. Brain Res 744:227–234
84. Feng B, Morley RM, Jane DE et al (2005) The effect of competitive antagonist chain length on NMDA receptor subunit selectivity. Neuropharmacology 48:354–359
85. Morley RM, Tse HW, Feng B et al (2005) Synthesis and pharmacology of N1-substituted piperazine-2, 3-dicarboxylic acid derivatives acting as NMDA receptor antagonists. J Med Chem 48:2627–2637
86. Neyton J, Paoletti P (2006) Relating NMDA receptor function to receptor subunit composition: limitations of the pharmacological approach. J Neurosci 26:1331–1333
87. Borza I, Domany G (2006) NR2B selective NMDA antagonists: the evolution of the ifenprodil-type pharmacophore. Curr Top Med Chem 6:687–695
88. Mony L, Kew JN, Gunthorpe MJ et al (2009) Allosteric modulators of NR2B-containing NMDA receptors: molecular mechanisms and therapeutic potential. Br J Pharmacol 157: 1301–1317
89. Hadj Tahar A, Gregoire L, Darre A et al (2004) Effect of a selective glutamate antagonist on L-dopa-induced dyskinesias in drug-naive parkinsonian monkeys. Neurobiol Dis 15:171–176
90. Ouattara B, Belkhir S, Morissette M et al (2009) Implication of NMDA receptors in the antidyskinetic activity of cabergoline, CI-1041, and Ro 61-8048 in MPTP monkeys with levodopa-induced dyskinesias. J Mol Neurosci 38:128–142
91. Tahirovic YA, Geballe M, Gruszecka-Kowalik E et al (2008) Enantiomeric propanolamines as selective N-methyl-D-aspartate 2B receptor antagonists. J Med Chem 51:5506–5521

92. Marinelli L, Cosconati S, Steinbrecher T et al (2007) Homology modeling of NR2B modulatory domain of NMDA receptor and analysis of ifenprodil binding. ChemMedChem 2:1498–1510
93. Gitto R, De Luca L, Ferro S et al (2008) Computational studies to discover a new NR2B/NMDA receptor antagonist and evaluation of pharmacological profile. ChemMedChem 3:1539–1548
94. Wee XK, Ng KS, Leung HW et al (2010) Mapping the high-affinity binding domain of 5-substituted benzimidazoles to the proximal N-terminus of the GluN2B subunit of the NMDA receptor. Br J Pharmacol 159:449–461
95. Jiang SX, Zheng RY, Zeng JQ et al (2010) Reversible inhibition of intracellular calcium influx through NMDA receptors by imidazoline I(2) receptor antagonists. Eur J Pharmacol 629:12–19
96. Hollmann M, Heinemann S (1994) Cloned glutamate receptors. Annu Rev Neurosci 17: 31–108
97. Dingledine R, Borges K, Bowie D et al (1999) The glutamate receptor ion channels. Pharmacol Rev 51:7–61
98. Madden DR (2002) The structure and function of glutamate receptor ion channels. Nat Rev Neurosci 3:91–101
99. Rosenmund C, Stern-Bach Y, Stevens CF (1998) The tetrameric structure of a glutamate receptor channel. Science 280:1596–1599
100. Koike M, Tsukada S, Tsuzuki K et al (2000) Regulation of kinetic properties of GluR2 AMPA receptor channels by alternative splicing. J Neurosci 20:2166–2174
101. Bronson JR, Zhang Z, Vandenberghe W (1999) Ca(2+) permeation of AMPA receptors in cerebellar neurons expressing glu receptor 2. J Neurosci 19:9149–9159
102. Burnashev N, Monyer H, Seeburg PH et al (1992) Divalent ion permeability of AMPA receptor channels is dominated by the edited form of a single subunit. Neuron 8:189–198
103. Blackstone CD, Moss SJ, Martin LJ et al (1992) Biochemical characterization and localization of a non-N-methyl-D-aspartate glutamate receptor in rat brain. J Neurochem 58:1118–1126
104. Conti F, Weinberg RJ (1999) Shaping excitation at glutamatergic synapses. Trends Neurosci 22:451–458
105. Morari M, Sbrenna S, Marti M et al (1998) NMDA and non-NMDA ionotropic glutamate receptors modulate striatal acetylcholine release via pre- and postsynaptic mechanisms. J Neurochem 71:2006–2017
106. Patel DR, Croucher MJ (1997) Evidence for a role of presynaptic AMPA receptors in the control of neuronal glutamate release in the rat forebrain. Eur J Pharmacol 332:143–151
107. Schenk S, Matteoli M (2004) Presynaptic AMPA receptors: more than just ion channels? Biol Cell 96:257–260
108. Gill R, Lodge D (1997) Pharmacology of AMPA antagonists and their role in neuroprotection. Int Rev Neurobiol 40:197–232
109. Narayanan U, Chi OZ, Liu X et al (2000) Effect of AMPA on cerebral cortical oxygen balance of ischemic rat brain. Neurochem Res 25:405–411
110. Gill R (1994) The pharmacology of alpha-amino-3-hydroxy-5-methyl-4-isoxazole propionate (AMPA)/kainate antagonists and their role in cerebral ischaemia. Cerebrovasc Brain Metab Rev 6:225–256
111. Xue D, Huang ZG, Barnes K et al (1994) Delayed treatment with AMPA, but not NMDA, antagonists reduces neocortical infarction. J Cereb Blood Flow Metab 14:251–261
112. Rogawski MA, Donevan SD (1999) AMPA receptors in epilepsy and as targets for antiepileptic drugs. Adv Neurol 79:947–963
113. Tortorella A, Halonen T, Sahibzada N et al (1997) A crucial role of the alpha-amino-3-hydroxy-5-methylisoxazole-4-propionic acid subtype of glutamate receptors in piriform and perirhinal cortex for the initiation and propagation of limbic motor seizures. J Pharmacol Exp Ther 280:1401–1405

114. Kunig G, Niedermeyer B, Deckert J et al (1998) Inhibition of [3H]alpha-amino-3-hydroxy-5-methyl-4-isoxazole-propionic acid [AMPA] binding by the anticonvulsant valproate in clinically relevant concentrations: an autoradiographic investigation in human hippocampus. Epilepsy Res 31:153–157
115. Lees GJ, Leong W (1993) Differential effects of NBQX on the distal and local toxicity of glutamate agonists administered intra-hippocampally. Brain Res 628:1–7
116. Lees GJ, Leong W (1994) Synergy between diazepam and NBQX in preventing neuronal death caused by non-NMDA agonists. Neuroreport 5:2149–2152
117. Carroll RC, Lissin DV, von Zastrow M et al (1999) Rapid redistribution of glutamate receptors contributes to long-term depression in hippocampal cultures. Nat Neurosci 2: 454–460
118. Hayashi Y, Shi SH, Esteban JA et al (2000) Driving AMPA receptors into synapses by LTP and CaMKII: requirement for GluR1 and PDZ domain interaction. Science 287:2262–2267
119. Bredt DS, Nicoll RA (2003) AMPA receptor trafficking at excitatory synapses. Neuron 40: 361–379
120. Song I, Huganir RL (2002) Regulation of AMPA receptors during synaptic plasticity. Trends Neurosci 25:578–588
121. Kim JH, Anwyl R, Suh YH et al (2001) Use-dependent effects of amyloidogenic fragments of (beta)-amyloid precursor protein on synaptic plasticity in rat hippocampus in vivo. J Neurosci 21:1327–1333
122. Armstrong DM, Ikonomovic MD, Sheffield R et al (1994) AMPA-selective glutamate receptor subtype immunoreactivity in the entorhinal cortex of non-demented elderly and patients with Alzheimer's disease. Brain Res 639:207–216
123. Aronica E, Dickson DW, Kress Y et al (1998) Non-plaque dystrophic dendrites in Alzheimer hippocampus: a new pathological structure revealed by glutamate receptor immunocytochemistry. Neuroscience 82:979–991
124. Thorns V, Mallory M, Hansen L et al (1997) Alterations in glutamate receptor 2/3 subunits and amyloid precursor protein expression during the course of Alzheimer's disease and Lewy body variant. Acta Neuropathol 94:539–548
125. Chan SL, Griffin WS, Mattson MP (1999) Evidence for caspase-mediated cleavage of AMPA receptor subunits in neuronal apoptosis and Alzheimer's disease. J Neurosci Res 57:315–323
126. Szegedi V, Juhasz G, Budai D et al (2005) Divergent effects of Abeta1-42 on ionotropic glutamate receptor-mediated responses in CA1 neurons in vivo. Brain Res 1062:120–126
127. Tozaki H, Matsumoto A, Kanno T et al (2002) The inhibitory and facilitatory actions of amyloid-beta peptides on nicotinic ACh receptors and AMPA receptors. Biochem Biophys Res Commun 294:42–45
128. Zhao D, Watson JB, Xie CW (2004) Amyloid beta prevents activation of calcium/calmodulin-dependent protein kinase II and AMPA receptor phosphorylation during hippocampal long-term potentiation. J Neurophysiol 92:2853–2858
129. Almeida CG, Tampellini D, Takahashi RH et al (2005) Beta-amyloid accumulation in APP mutant neurons reduces PSD-95 and GluR1 in synapses. Neurobiol Dis 20:187–198
130. Louzada PR Jr, Paula Lima AC, de Mello FG et al (2001) Dual role of glutamatergic neurotransmission on amyloid beta(1-42) aggregation and neurotoxicity in embryonic avian retina. Neurosci Lett 301:59–63
131. Allen JW, Eldadah BA, Faden AI (1999) Beta-amyloid-induced apoptosis of cerebellar granule cells and cortical neurons: exacerbation by selective inhibition of group I metabotropic glutamate receptors. Neuropharmacology 38:1243–1252
132. Ting JT, Kelley BG, Lambert TJ et al (2007) Amyloid precursor protein overexpression depresses excitatory transmission through both presynaptic and postsynaptic mechanisms. Proc Natl Acad Sci USA 104:353–358
133. Ahmadian G, Ju W, Liu L et al (2004) Tyrosine phosphorylation of GluR2 is required for insulin-stimulated AMPA receptor endocytosis and LTD. EMBO J 23:1040–1050

134. Hayashi T, Huganir RL (2004) Tyrosine phosphorylation and regulation of the AMPA receptor by SRC family tyrosine kinases. J Neurosci 24:6152–6160
135. Papa SM, Engber TM, Boldry RC et al (1993) Opposite effects of NMDA and AMPA receptor blockade on catalepsy induced by dopamine receptor antagonists. Eur J Pharmacol 232:247–253
136. Zadow B, Schmidt WJ (1994) The AMPA antagonists NBQX and GYKI 52466 do not counteract neuroleptic-induced catalepsy. Naunyn Schmiedebergs Arch Pharmacol 349:61–65
137. Loschmann PA, Lange KW, Kunow M et al (1991) Synergism of the AMPA-antagonist NBQX and the NMDA-antagonist CPP with L-dopa in models of Parkinson's disease. J Neural Transm Park Dis Dement Sect 3:203–213
138. Loschmann PA, Kunow M, Wachtel H (1992) Synergism of NBQX with dopamine agonists in the 6-OHDA rat model of Parkinson's disease. J Neural Transm Suppl 38:55–64
139. Wachtel H, Kunow M, Loschmann PA (1992) NBQX (6-nitro-sulfamoyl-benzo-quinoxaline-dione) and CPP (3-carboxy-piperazin-propyl phosphonic acid) potentiate dopamine agonist induced rotations in substantia nigra lesioned rats. Neurosci Lett 142:179–182
140. Konitsiotis S, Blanchet PJ, Verhagen L et al (2000) AMPA receptor blockade improves levodopa-induced dyskinesia in MPTP monkeys. Neurology 54:1589–1595
141. Marin C, Jimenez A, Bonastre M et al (2001) LY293558, an AMPA glutamate receptor antagonist, prevents and reverses levodopa-induced motor alterations in Parkinsonian rats. Synapse 42:40–47
142. Silverdale MA, Nicholson SL, Crossman AR et al (2005) Topiramate reduces levodopa-induced dyskinesia in the MPTP-lesioned marmoset model of Parkinson's disease. Mov Disord 20:403–409
143. Black MD (2005) Therapeutic potential of positive AMPA modulators and their relationship to AMPA receptor subunits. A review of preclinical data. Psychopharmacology 179:154–163
144. Ward SE, Bax BD, Harries M (2010) Challenges for and current status of research into positive modulators of AMPA receptors. Br J Pharmacol 160:181–190
145. Marenco S, Weinberger DR (2006) Therapeutic potential of positive AMPA receptor modulators in the treatment of neuropsychiatric disorders. CNS Drugs 20:173–185
146. Fletcher EJ, Lodge D (1996) New developments in the molecular pharmacology of alpha-amino-3-hydroxy-5-methyl-4-isoxazole propionate and kainate receptors. Pharmacol Ther 70:65–89
147. Yamada KA (2000) Therapeutic potential of positive AMPA receptor modulators in the treatment of neurological disease. Expert Opin Invest Drugs 9:765–778
148. Mayer ML, Vyklicky L Jr (1989) Concanavalin A selectively reduces desensitization of mammalian neuronal quisqualate receptors. Proc Natl Acad Sci USA 86:1411–1415
149. Ito I, Tanabe S, Kohda A et al (1990) Allosteric potentiation of quisqualate receptors by a nootropic drug aniracetam. J Physiol 424:533–543
150. Ahmed AH, Oswald RE (2010) Piracetam defines a new binding site for allosteric modulators of alpha-amino-3-hydroxy-5-methyl-4-isoxazole-propionic acid (AMPA) receptors. J Med Chem 53:2197–2203
151. Yamada KA, Tang CM (1993) Benzothiadiazides inhibit rapid glutamate receptor desensitization and enhance glutamatergic synaptic currents. J Neurosci 13:3904–3915
152. Staubli U, Rogers G, Lynch G (1994) Facilitation of glutamate receptors enhances memory. Proc Natl Acad Sci USA 91:777–781
153. Arai A, Kessler M, Rogers G et al (1996) Effects of a memory-enhancing drug on DL-alpha-amino-3-hydroxy-5-methyl-4-isoxazolepropionic acid receptor currents and synaptic transmission in hippocampus. J Pharmacol Exp Ther 278:627–638
154. Goff DC, Lamberti JS, Leon AC et al (2008) A placebo-controlled add-on trial of the Ampakine, CX516, for cognitive deficits in schizophrenia. Neuropsychopharmacology 33: 465–472
155. Yamada KA, Rothman SM (1992) Diazoxide blocks glutamate desensitization and prolongs excitatory postsynaptic currents in rat hippocampal neurons. J Physiol 458:409–423

156. Francotte P, Tullio P, Goffin E et al (2007) Design, synthesis, and pharmacology of novel 7-substituted 3, 4-dihydro-2H–1, 2, 4-benzothiadiazine 1, 1-dioxides as positive allosteric modulators of AMPA receptors. J Med Chem 50:3153–3157
157. Phillips D, Sonnenberg J, Arai AC et al (2002) 5′-alkyl-benzothiadiazides: a new subgroup of AMPA receptor modulators with improved affinity. Bioorg Med Chem 10:1229–1248
158. Francotte P, Goffin E, Fraikin P et al (2010) New fluorinated 1, 2, 4-benzothiadiazine 1, 1-dioxides: discovery of an orally active cognitive enhancer acting through potentiation of the 2-amino-3-(3-hydroxy-5-methylisoxazol-4-yl)propionic acid receptors. J Med Chem 53:1700–1711
159. Francotte P, de Tullio P, Podona T et al (2008) Synthesis and pharmacological evaluation of a second generation of pyridothiadiazine 1, 1-dioxides acting as AMPA potentiators. Bioorg Med Chem 16:9948–9956
160. Marighetto A, Valerio S, Jaffard R et al (2008) The AMPA modulator S 18986 improves declarative and working memory performances in aged mice. Behav Pharmacol 19:235–244
161. Ptak CP, Ahmed AH, Oswald RE (2009) Probing the allosteric modulator binding site of GluR2 with thiazide derivatives. Biochemistry 48:8594–8602
162. Hald H, Ahring PK, Timmermann DB et al (2009) Distinct structural features of cyclothiazide are responsible for effects on peak current amplitude and desensitization kinetics at iGluR2. J Mol Biol 391:906–917
163. Sekiguchi M, Fleck MW, Mayer ML et al (1997) A novel allosteric potentiator of AMPA receptors: 4–2-(phenylsulfonylamino)ethylthio–2, 6-difluoro-phenoxyaceta mide. J Neurosci 17:5760–5771
164. Ornstein PL, Zimmerman DM, Arnold MB et al (2000) Biarylpropylsulfonamides as novel, potent potentiators of 2-amino-3- (5-methyl-3-hydroxyisoxazol-4-yl)- propanoic acid (AMPA) receptors. J Med Chem 43:4354–4358
165. Harpsoe K, Liljefors T, Balle T (2008) Prediction of the binding mode of biarylpropylsulfonamide allosteric AMPA receptor modulators based on docking, GRID molecular interaction fields and 3D-QSAR analysis. J Mol Graph Model 26:874–883
166. Chappell AS, Gonzales C, Williams J et al (2007) AMPA potentiator treatment of cognitive deficits in Alzheimer disease. Neurology 68:1008–1012
167. Fernandez MC, Castano A, Dominguez E et al (2006) A novel class of AMPA receptor allosteric modulators. Part 1: design, synthesis, and SAR of 3-aryl-4-cyano-5-substituted-heteroaryl-2-carboxylic acid derivatives. Bioorg Med Chem Lett 16:5057–5061
168. Zarrinmayeh H, Tromiczak E, Zimmerman DM et al (2006) A novel class of positive allosteric modulators of AMPA receptors: design, synthesis, and structure-activity relationships of 3-biphenyl-4-yl-4-cyano-5-ethyl-1-methyl-1H-pyrrole-2-carboxylic acid, LY2059346. Bioorg Med Chem Lett 16:5203–5206
169. Nikam SS, Kornberg BE (2001) AMPA receptor antagonists. Curr Med Chem 8:155–170
170. Pentikainen U, Settimo L, Johnson MS et al (2006) Subtype selectivity and flexibility of ionotropic glutamate receptors upon antagonist ligand binding. Org Biomol Chem 4:1058–1070
171. Varano F, Catarzi D, Colotta V et al (2008) Novel AMPA and kainate receptor antagonists containing the pyrazolo[1, 5-c]quinazoline ring system: Synthesis and structure-activity relationships. Bioorg Med Chem 16:2617–2626
172. Menniti FS, Chenard BL, Collins MB et al (2000) Characterization of the binding site for a novel class of noncompetitive alpha-amino-3-hydroxy-5-methyl-4-isoxazolepropionic acid receptor antagonists. Mol Pharmacol 58:1310–1317
173. Micale N, Colleoni S, Postorino G et al (2008) Structure-activity study of 2, 3-benzodiazepin-4-ones noncompetitive AMPAR antagonists: identification of the 1-(4-amino-3-methylphenyl)-3, 5-dihydro-7, 8-ethylenedioxy-4H–2, 3-benzodiaze pin-4-one as neuroprotective agent. Bioorg Med Chem 16:2200–2211

174. Ruel J, Guitton MJ, Puell JL (2002) Negative allosteric modulation of AMPA-preferring receptors by the selective isomer GYKI 53784 (LY303070), a specific non-competitive AMPA antagonist. CNS Drug Rev 8:235–254
175. Bialer M, Johannessen SI, Kupferberg HJ et al (2004) Progress report on new antiepileptic drugs: a summary of the Seventh Eilat Conference (EILAT VII). Epilepsy Res 61:1–48
176. Pelletier JC, Hesson DP, Jones KA et al (1996) Substituted 1, 2-dihydrophthalazines: potent, selective, and noncompetitive inhibitors of the AMPA receptor. J Med Chem 39:343–346
177. Pei XF, Sturgess MA, Valenzuela CF et al (1999) Allosteric modulators of the AMPA receptor: novel 6-substituted dihydrophthalazines. Bioorg Med Chem Lett 9:539–542
178. Gitto R, Caruso R, Pagano B et al (2006) Novel potent anticonvulsant agent containing a tetrahydroisoquinoline skeleton. J Med Chem 49:5618–5622
179. Gitto R, De Luca L, Pagano B et al (2008) Synthesis and anticonvulsant evaluation of N-substituted isoquinoline AMPA receptor antagonists. Bioorg Med Chem 16:2379–2384
180. Macchiarulo A, De Luca L, Costantino G et al (2004) QSAR study of anticonvulsant negative allosteric modulators of the AMPA receptor. J Med Chem 47:1860–1863
181. Ahmed AH, Wang Q, Sondermann H et al (2009) Structure of the S1S2 glutamate binding domain of GLuR3. Proteins 75:628–637
182. Karim F, Wang CC, RWt G (2001) Metabotropic glutamate receptor subtypes 1 and 5 are activators of extracellular signal-regulated kinase signaling required for inflammatory pain in mice. J Neurosci 21:3771–3779
183. Warwick HK, Nahorski SR, Challiss RA (2005) Group I metabotropic glutamate receptors, mGlu1a and mGlu5a, couple to cyclic AMP response element binding protein (CREB) through a common Ca2+ – and protein kinase C-dependent pathway. J Neurochem 93:232–245
184. Biber K, Laurie DJ, Berthele A et al (1999) Expression and signaling of group I metabotropic glutamate receptors in astrocytes and microglia. J Neurochem 72:1671–1680
185. Harrison PJ, Lyon L, Sartorius LJ et al (2008) The group II metabotropic glutamate receptor 3 (mGluR3, mGlu3, GRM3): expression, function and involvement in schizophrenia. J Psychopharmacol 22:308–322
186. Conn PJ, Pin JP (1997) Pharmacology and functions of metabotropic glutamate receptors. Annu Rev Pharmacol Toxicol 37:205–237
187. Pin JP, Duvoisin R (1995) The metabotropic glutamate receptors: structure and functions. Neuropharmacology 34:1–26
188. Miller EK, Cohen JD (2001) An integrative theory of prefrontal cortex function. Annu Rev Neurosci 24:167–202
189. Rao SG, Williams GV, Goldman-Rakic PS (2000) Destruction and creation of spatial tuning by disinhibition: GABA(A) blockade of prefrontal cortical neurons engaged by working memory. J Neurosci 20:485–494
190. Albasanz JL, Dalfo E, Ferrer I et al (2005) Impaired metabotropic glutamate receptor/phospholipase C signaling pathway in the cerebral cortex in Alzheimer's disease and dementia with Lewy bodies correlates with stage of Alzheimer's-disease-related changes. Neurobiol Dis 20:685–693
191. Bahr BA, Abai B, Gall CM et al (1994) Induction of beta-amyloid-containing polypeptides in hippocampus: evidence for a concomitant loss of synaptic proteins and interactions with an excitotoxin. Exp Neurol 129:81–94
192. Celsi F, Svedberg M, Unger C et al (2007) Beta-amyloid causes downregulation of calcineurin in neurons through induction of oxidative stress. Neurobiol Dis 26:342–352
193. Bahr BA, Hoffman KB, Yang AJ et al (1998) Amyloid beta protein is internalized selectively by hippocampal field CA1 and causes neurons to accumulate amyloidogenic carboxyterminal fragments of the amyloid precursor protein. J Comp Neurol 397:139–147
194. Bendiske J, Bahr BA (2003) Lysosomal activation is a compensatory response against protein accumulation and associated synaptopathogenesis–an approach for slowing Alzheimer disease? J Neuropathol Exp Neurol 62:451–463

195. Ure J, Baudry M, Perassolo M (2006) Metabotropic glutamate receptors and epilepsy. J Neurol Sci 247:1–9
196. Zhong J, Gerber G, Kojic L et al (2000) Dual modulation of excitatory synaptic transmission by agonists at group I metabotropic glutamate receptors in the rat spinal dorsal horn. Brain Res 887:359–377
197. Allen JW, Vicini S, Faden AI (2001) Exacerbation of neuronal cell death by activation of group I metabotropic glutamate receptors: role of NMDA receptors and arachidonic acid release. Exp Neurol 169:449–460
198. Bruno V, Copani A, Knopfel T et al (1995) Activation of metabotropic glutamate receptors coupled to inositol phospholipid hydrolysis amplifies NMDA-induced neuronal degeneration in cultured cortical cells. Neuropharmacology 34:1089–1098
199. Faden AI, O'Leary DM, Fan L et al (2001) Selective blockade of the mGluR1 receptor reduces traumatic neuronal injury in vitro and improvesoOutcome after brain trauma. Exp Neurol 167:435–444
200. Mukhin AG, Ivanova SA, Faden AI (1997) mGluR modulation of post-traumatic neuronal death: role of NMDA receptors. Neuroreport 8:2561–2566
201. Kohara A, Takahashi M, Yatsugi S et al (2008) Neuroprotective effects of the selective type 1 metabotropic glutamate receptor antagonist YM-202074 in rat stroke models. Brain Res 1191:168–179
202. Lyeth BG, Gong QZ, Shields S et al (2001) Group I metabotropic glutamate antagonist reduces acute neuronal degeneration and behavioral deficits after traumatic brain injury in rats. Exp Neurol 169:191–199
203. Mills CD, Johnson KM, Hulsebosch CE (2002) Group I metabotropic glutamate receptors in spinal cord injury: roles in neuroprotection and the development of chronic central pain. J Neurotrauma 19:23–42
204. Movsesyan VA, Stoica BA, Faden AI (2004) MGLuR5 activation reduces beta-amyloid-induced cell death in primary neuronal cultures and attenuates translocation of cytochrome c and apoptosis-inducing factor. J Neurochem 89:1528–1536
205. Vincent AM, TenBroeke M, Maiese K (1999) Metabotropic glutamate receptors prevent programmed cell death through the modulation of neuronal endonuclease activity and intracellular pH. Exp Neurol 155:79–94
206. Byrnes KR, Stoica B, Loane DJ et al (2009) Metabotropic glutamate receptor 5 activation inhibits microglial associated inflammation and neurotoxicity. Glia 57:550–560
207. Lea PM, Custer SJ, Vicini S et al (2002) Neuronal and glial mGluR5 modulation prevents stretch-induced enhancement of NMDA receptor current. Pharmacol Biochem Behav 73:287–298
208. Deng W, Wang H, Rosenberg PA et al (2004) Role of metabotropic glutamate receptors in oligodendrocyte excitotoxicity and oxidative stress. Proc Natl Acad Sci USA 101: 7751–7756
209. Awad H, Hubert GW, Smith Y et al (2000) Activation of metabotropic glutamate receptor 5 has direct excitatory effects and potentiates NMDA receptor currents in neurons of the subthalamic nucleus. J Neurosci 20:7871–7879
210. Conn PJ, Battaglia G, Marino MJ et al (2005) Metabotropic glutamate receptors in the basal ganglia motor circuit. Nat Rev Neurosci 6:787–798
211. Pisani A, Calabresi P, Centonze D et al (1997) Enhancement of NMDA responses by group I metabotropic glutamate receptor activation in striatal neurones. Br J Pharmacol 120: 1007–1014
212. Breysse N, Baunez C, Spooren W et al (2002) Chronic but not acute treatment with a metabotropic glutamate 5 receptor antagonist reverses the akinetic deficits in a rat model of parkinsonism. J Neurosci 22:5669–5678
213. Breysse N, Amalric M, Salin P (2003) Metabotropic glutamate 5 receptor blockade alleviates akinesia by normalizing activity of selective basal-ganglia structures in parkinsonian rats. J Neurosci 23:8302–8309

214. Ossowska K, Konieczny J, Wolfarth S et al (2001) Blockade of the metabotropic glutamate receptor subtype 5 (mGluR5) produces antiparkinsonian-like effects in rats. Neuropharmacology 41:413–420
215. Coccurello R, Breysse N, Amalric M (2004) Simultaneous blockade of adenosine A2A and metabotropic glutamate mGlu5 receptors increase their efficacy in reversing Parkinsonian deficits in rats. Neuropsychopharmacology 29:1451–1461
216. Kachroo A, Orlando LR, Grandy DK et al (2005) Interactions between metabotropic glutamate 5 and adenosine A2A receptors in normal and parkinsonian mice. J Neurosci 25:10414–10419
217. Diaz-Cabiale Z, Vivo M, Del Arco A et al (2002) Metabotropic glutamate mGlu5 receptor-mediated modulation of the ventral striopallidal GABA pathway in rats. Interactions with adenosine A(2A) and dopamine D(2) receptors. Neurosci Lett 324:154–158
218. Ferre S, Karcz-Kubicha M, Hope BT et al (2002) Synergistic interaction between adenosine A2A and glutamate mGlu5 receptors: implications for striatal neuronal function. Proc Natl Acad Sci USA 99:11940–11945
219. Fuxe K, Agnati LF, Jacobsen K et al (2003) Receptor heteromerization in adenosine A2A receptor signaling: relevance for striatal function and Parkinson's disease. Neurology 61: S19–S23
220. Nishi A, Liu F, Matsuyama S et al (2003) Metabotropic mGlu5 receptors regulate adenosine A2A receptor signaling. Proc Natl Acad Sci USA 100:1322–1327
221. Turle-Lorenzo N, Breysse N, Baunez C et al (2005) Functional interaction between mGlu 5 and NMDA receptors in a rat model of Parkinson's disease. Psychopharmacology 179: 117–127
222. Kearney JA, Frey KA, Albin RL (1997) Metabotropic glutamate agonist-induced rotation: a pharmacological, FOS immunohistochemical, and [14C]-2-deoxyglucose autoradiographic study. J Neurosci 17:4415–4425
223. Kearney JA, Becker JB, Frey KA et al (1998) The role of nigrostriatal dopamine in metabotropic glutamate agonist-induced rotation. Neuroscience 87:881–891
224. Wardas J, Pietraszek M, Wolfarth S et al (2003) The role of metabotropic glutamate receptors in regulation of striatal proenkephalin expression: implications for the therapy of Parkinson's disease. Neuroscience 122:747–756
225. Phillips JM, Lam HA, Ackerson LC et al (2006) Blockade of mGluR glutamate receptors in the subthalamic nucleus ameliorates motor asymmetry in an animal model of Parkinson's disease. Eur J Neurosci 23:151–160
226. De Leonibus E, Manago F, Giordani F et al (2009) Metabotropic glutamate receptors 5 blockade reverses spatial memory deficits in a mouse model of Parkinson's disease. Neuropsychopharmacology 34:729–738
227. Palucha A, Pilc A (2007) Metabotropic glutamate receptor ligands as possible anxiolytic and antidepressant drugs. Pharmacol Ther 115:116–147
228. Witkin JM, Marek GJ, Johnson BG et al (2007) Metabotropic glutamate receptors in the control of mood disorders. CNS Neurol Disord Drug Targets 6:87–100
229. Samadi P, Gregoire L, Morissette M et al (2008) mGluR5 metabotropic glutamate receptors and dyskinesias in MPTP monkeys. Neurobiol Aging 29:1040–1051
230. Levandis G, Bazzini E, Armentero MT et al (2008) Systemic administration of an mGluR5 antagonist, but not unilateral subthalamic lesion, counteracts l-DOPA-induced dyskinesias in a rodent model of Parkinson's disease. Neurobiol Dis 29:161–168
231. Mela F, Marti M, Dekundy A et al (2007) Antagonism of metabotropic glutamate receptor type 5 attenuates l-DOPA-induced dyskinesia and its molecular and neurochemical correlates in a rat model of Parkinson's disease. J Neurochem 101:483–497
232. Schiefer J, Sprunken A, Puls C et al (2004) The metabotropic glutamate receptor 5 antagonist MPEP and the mGluR2 agonist LY379268 modify disease progression in a transgenic mouse model of Huntington's disease. Brain Res 1019:246–254

233. Pinheiro PS, Mulle C (2008) Presynaptic glutamate receptors: physiological functions and mechanisms of action. Nat Rev Neurosci 9:423–436
234. Bradley SR, Marino MJ, Wittmann M et al (2000) Activation of group II metabotropic glutamate receptors inhibits synaptic excitation of the substantia Nigra pars reticulata. J Neurosci 20:3085–3094
235. Rouse ST, Marino MJ, Bradley SR et al (2000) Distribution and roles of metabotropic glutamate receptors in the basal ganglia motor circuit: implications for treatment of Parkinson's disease and related disorders. Pharmacol Ther 88:427–435
236. Dawson L, Chadha A, Megalou M et al (2000) The group II metabotropic glutamate receptor agonist, DCG-IV, alleviates akinesia following intranigral or intraventricular administration in the reserpine-treated rat. Br J Pharmacol 129:541–546
237. Murray TK, Messenger MJ, Ward MA et al (2002) Evaluation of the mGluR2/3 agonist LY379268 in rodent models of Parkinson's disease. Pharmacol Biochem Behav 73:455–466
238. Picconi B, Pisani A, Centonze D et al (2002) Striatal metabotropic glutamate receptor function following experimental parkinsonism and chronic levodopa treatment. Brain 125:2635–2645
239. Matsui T, Kita H (2003) Activation of group III metabotropic glutamate receptors presynaptically reduces both GABAergic and glutamatergic transmission in the rat globus pallidus. Neuroscience 122:727–737
240. Valenti O, Marino MJ, Wittmann M et al (2003) Group III metabotropic glutamate receptor-mediated modulation of the striatopallidal synapse. J Neurosci 23:7218–7226
241. Wittmann M, Marino MJ, Bradley SR et al (2001) Activation of group III mGluRs inhibits GABAergic and glutamatergic transmission in the substantia nigra pars reticulata. J Neurophysiol 85:1960–1968
242. MacInnes N, Messenger MJ, Duty S (2004) Activation of group III metabotropic glutamate receptors in selected regions of the basal ganglia alleviates akinesia in the reserpine-treated rat. Br J Pharmacol 141:15–22
243. Schoepp DD, Jane DE, Monn JA (1999) Pharmacological agents acting at subtypes of metabotropic glutamate receptors. Neuropharmacology 38:1431–1476
244. Pin JP, De Colle C, Bessis AS et al (1999) New perspectives for the development of selective metabotropic glutamate receptor ligands. Eur J Pharmacol 375:277–294
245. Layton ME (2005) Subtype-selective noncompetitive modulators of metabotropic glutamate receptor subtype 1 (mGluR1). Curr Top Med Chem 5:859–867
246. Litschig S, Gasparini F, Rueegg D et al (1999) CPCCOEt, a noncompetitive metabotropic glutamate receptor 1 antagonist, inhibits receptor signaling without affecting glutamate binding. Mol Pharmacol 55:453–461
247. Ott D, Floersheim P, Inderbitzin W et al (2000) Chiral resolution, pharmacological characterization, and receptor docking of the noncompetitive mGlu1 receptor antagonist (+/-)-2-hydroxyimino- 1a, 2-dihydro-1H-7-oxacyclopropa[b]naphthalene-7a-carboxylic acid ethyl ester. J Med Chem 43:4428–4436
248. Carroll FY, Stolle A, Beart PM et al (2001) BAY36-7620: a potent non-competitive mGlu1 receptor antagonist with inverse agonist activity. Mol Pharmacol 59:965–973
249. De Vry J, Horvath E, Schreiber R (2001) Neuroprotective and behavioral effects of the selective metabotropic glutamate mGlu(1) receptor antagonist BAY 36-7620. Eur J Pharmacol 428:203–214
250. Lavreysen H, Pereira SN, Leysen JE et al (2004) Metabotropic glutamate 1 receptor distribution and occupancy in the rat brain: a quantitative autoradiographic study using [3H]R214127. Neuropharmacology 46:609–619
251. Malherbe P, Kratochwil N, Knoflach F et al (2003) Mutational analysis and molecular modeling of the allosteric binding site of a novel, selective, noncompetitive antagonist of the metabotropic glutamate 1 receptor. J Biol Chem 278:8340–8347
252. Micheli F, Fabio RD, Cavanni P et al (2003) Synthesis and pharmacological characterisation of 2, 4-dicarboxy-pyrroles as selective non-competitive mGluR1 antagonists. Bioorg Med Chem 11:171–183

253. Mabire D, Coupa S, Adelinet C et al (2005) Synthesis, structure-activity relationship, and receptor pharmacology of a new series of quinoline derivatives acting as selective, noncompetitive mGlu1 antagonists. J Med Chem 48:2134–2153
254. Sekhar YN, Nayana MR, Ravikumar M et al (2007) Comparative molecular field analysis of quinoline derivatives as selective and noncompetitive mGluR1 antagonists. Chem Biol Drug Des 70:511–519
255. Noeske T, Jirgensons A, Starchenkovs I et al (2007) Virtual screening for selective allosteric mGluR1 antagonists and structure-activity relationship investigations for coumarine derivatives. ChemMedChem 2:1763–1773
256. Vanejevs M, Jatzke C, Renner S et al (2008) Positive and negative modulation of group I metabotropic glutamate receptors. J Med Chem 51:634–647
257. Wang X, Kolasa T, El Kouhen OF et al (2007) Rapid hit to lead evaluation of pyrazolo[3, 4-d]pyrimidin-4-one as selective and orally bioavailable mGluR1 antagonists. Bioorg Med Chem Lett 17:4303–4307
258. Wu WL, Burnett DA, Domalski M et al (2007) Discovery of orally efficacious tetracyclic metabotropic glutamate receptor 1 (mGluR1) antagonists for the treatment of chronic pain. J Med Chem 50:5550–5553
259. Sekhar YN, Nayana MR, Sivakumari N et al (2008) 3D-QSAR and molecular docking studies of 1, 3, 5-triazene-2, 4-diamine derivatives against r-RNA: novel bacterial translation inhibitors. J Mol Graph Model 26:1338–1352
260. Ito S, Satoh A, Nagatomi Y et al (2008) Discovery and biological profile of 4-(1-aryltriazol-4-yl)-tetrahydropyridines as an orally active new class of metabotropic glutamate receptor 1 antagonist. Bioorg Med Chem 16:9817–9829
261. Ito S, Hirata Y, Nagatomi Y et al (2009) Discovery and biological profile of isoindolinone derivatives as novel metabotropic glutamate receptor 1 antagonists: a potential treatment for psychotic disorders. Bioorg Med Chem Lett 19:5310–5313
262. Satoh A, Nagatomi Y, Hirata Y et al (2009) Discovery and in vitro and in vivo profiles of 4-fluoro-N-[4-[6-(isopropylamino)pyrimidin-4-yl]-1, 3-thiazol-2-yl]-N-methy lbenzamide as novel class of an orally active metabotropic glutamate receptor 1 (mGluR1) antagonist. Bioorg Med Chem Lett 19:5464–5468
263. Helton DR, Tizzano JP, Monn JA et al (1998) Anxiolytic and side-effect profile of LY354740: a potent, highly selective, orally active agonist for group II metabotropic glutamate receptors. J Pharmacol Exp Ther 284:651–660
264. Tizzano JP, Griffey KI, Schoepp DD (2002) The anxiolytic action of mGlu2/3 receptor agonist, LY354740, in the fear-potentiated startle model in rats is mechanistically distinct from diazepam. Pharmacol Biochem Behav 73:367–374
265. Monn JA, Valli MJ, Massey SM et al (1999) Synthesis, pharmacological characterization, and molecular modeling of heterobicyclic amino acids related to (+)-2-aminobicyclo[3.1.0] hexane-2, 6-dicarboxylic acid (LY354740): identification of two new potent, selective, and systemically active agonists for group II metabotropic glutamate receptors. J Med Chem 42:1027–1040
266. Jones CK, Eberle EL, Peters SC et al (2005) Analgesic effects of the selective group II (mGlu2/3) metabotropic glutamate receptor agonists LY379268 and LY389795 in persistent and inflammatory pain models after acute and repeated dosing. Neuropharmacology 49(Suppl 1):206–218
267. Monn JA, Massey SM, Valli MJ et al (2007) Synthesis and metabotropic glutamate receptor activity of S-oxidized variants of (−)-4-amino-2-thiabicyclo-[3.1.0]hexane-4, 6-dicarboxylate: identification of potent, selective, and orally bioavailable agonists for mGlu2/3 receptors. J Med Chem 50:233–240
268. Rorick-Kehn LM, Johnson BG, Burkey JL et al (2007) Pharmacological and pharmacokinetic properties of a structurally novel, potent, and selective metabotropic glutamate 2/3 receptor agonist: in vitro characterization of agonist (-)-(1R, 4S, 5S, 6S)-4-amino-2-sulfonylbicyclo [3.1.0]-hexane-4, 6-dicarboxylic acid (LY404039). J Pharmacol Exp Ther 321:308–317

269. Krystal JH, Abi-Saab W, Perry E et al (2005) Preliminary evidence of attenuation of the disruptive effects of the NMDA glutamate receptor antagonist, ketamine, on working memory by pretreatment with the group II metabotropic glutamate receptor agonist, LY354740, in healthy human subjects. Psychopharmacology 179:303–309
270. Perkins EJ, Abraham T (2007) Pharmacokinetics, metabolism, and excretion of the intestinal peptide transporter 1 (SLC15A1)-targeted prodrug (1S, 2S, 5R, 6S)-2-[(2′S)-(2-amino)propionyl]aminobicyclo[3.1.0.]hexen-2, 6-di carboxylic acid (LY544344) in rats and dogs: assessment of first-pass bioactivation and dose linearity. Drug Metab Dispos 35:1903–1909
271. Patil ST, Zhang L, Martenyi F et al (2007) Activation of mGlu2/3 receptors as a new approach to treat schizophrenia: a randomized Phase 2 clinical trial. Nat Med 13:1102–1107
272. Nakazato A, Kumagai T, Sakagami K et al (2000) Synthesis, SARs, and pharmacological characterization of 2-amino-3 or 6-fluorobicyclo[3.1.0]hexane-2, 6-dicarboxylic acid derivatives as potent, selective, and orally active group II metabotropic glutamate receptor agonists. J Med Chem 43:4893–4909
273. Dominguez C, Prieto L, Valli MJ et al (2005) Methyl substitution of 2-aminobicyclo[3.1.0] hexane 2, 6-dicarboxylate (LY354740) determines functional activity at metabotropic glutamate receptors: identification of a subtype selective mGlu2 receptor agonist. J Med Chem 48:3605–3612
274. Rudd MT, McCauley JA (2005) Positive allosteric modulators of the metabotropic glutamate receptor subtype 2 (mGluR2). Curr Top Med Chem 5:869–884
275. Fraley ME (2009) Positive allosteric modulators of the metabotropic glutamate receptor 2 for the treatment of schizophrenia. Expert Opin Ther Pat 19:1259–1275
276. Johnson MP, Baez M, Jagdmann GE Jr et al (2003) Discovery of allosteric potentiators for the metabotropic glutamate 2 receptor: synthesis and subtype selectivity of N-(4-(2-methoxyphenoxy)phenyl)-N-(2, 2, 2- trifluoroethylsulfonyl)pyrid-3-ylmethylamine. J Med Chem 46:3189–3192
277. Pinkerton AB, Cube RV, Hutchinson JH et al (2004) Allosteric potentiators of the metabotropic glutamate receptor 2 (mGlu2). Part 1: Identification and synthesis of phenyl-tetrazolyl acetophenones. Bioorg Med Chem Lett 14:5329–5332
278. Pinkerton AB, Vernier JM, Schaffhauser H et al (2004) Phenyl-tetrazolyl acetophenones: discovery of positive allosteric potentiatiors for the metabotropic glutamate 2 receptor. J Med Chem 47:4595–4599
279. Galici R, Jones CK, Hemstapat K et al (2006) Biphenyl-indanone A, a positive allosteric modulator of the metabotropic glutamate receptor subtype 2, has antipsychotic- and anxiolytic-like effects in mice. J Pharmacol Exp Ther 318:173–185
280. Pinkerton AB, Cube RV, Hutchinson JH et al (2004) Allosteric potentiators of the metabotropic glutamate receptor 2 (mGlu2). Part 2: 4-thiopyridyl acetophenones as non-tetrazole containing mGlu2 receptor potentiators. Bioorg Med Chem Lett 14:5867–5872
281. Bonnefous C, Vernier JM, Hutchinson JH et al (2005) Biphenyl-indanones: allosteric potentiators of the metabotropic glutamate subtype 2 receptor. Bioorg Med Chem Lett 15:4354–4358
282. Cube RV, Vernier JM, Hutchinson JH et al (2005) 3-(2-Ethoxy-4-{4-[3-hydroxy-2-methyl-4-(3-methylbutanoyl)phenoxy]butoxy}ph enyl)propanoic acid: a brain penetrant allosteric potentiator at the metabotropic glutamate receptor 2 (mGluR2). Bioorg Med Chem Lett 15:2389–2393
283. Govek SP, Bonnefous C, Hutchinson JH et al (2005) Benzazoles as allosteric potentiators of metabotropic glutamate receptor 2 (mGluR2): efficacy in an animal model for schizophrenia. Bioorg Med Chem Lett 15:4068–4072
284. Pinkerton AB, Cube RV, Hutchinson JH et al (2005) Allosteric potentiators of the metabotropic glutamate receptor 2 (mGlu2). Part 3: Identification and biological activity of indanone containing mGlu2 receptor potentiators. Bioorg Med Chem Lett 15:1565–1571

285. Zhang L, Rogers BN, Duplantier AJ et al (2008) 3-(Imidazolyl methyl)-3-aza-bicyclo[3.1.0] hexan-6-yl)methyl ethers: a novel series of mGluR2 positive allosteric modulators. Bioorg Med Chem Lett 18:5493–5496
286. D'Alessandro PL, Corti C, Roth A et al (2010) The identification of structurally novel, selective, orally bioavailable positive modulators of mGluR2. Bioorg Med Chem Lett 20: 759–762
287. Duplantier AJ, Efremov I, Candler J et al (2009) 3-Benzyl-1, 3-oxazolidin-2-ones as mGluR2 positive allosteric modulators: Hit-to lead and lead optimization. Bioorg Med Chem Lett 19:2524–2529
288. Lindsley CW, Niswender CM, Engers DW et al (2009) Recent progress in the development of mGluR4 positive allosteric modulators for the treatment of Parkinson's disease. Curr Top Med Chem 9:949–963
289. Marino MJ, Williams DL Jr, O'Brien JA et al (2003) Allosteric modulation of group III metabotropic glutamate receptor 4: a potential approach to Parkinson's disease treatment. Proc Natl Acad Sci USA 100:13668–13673
290. Maj M, Bruno V, Dragic Z et al (2003) (−)-PHCCC, a positive allosteric modulator of mGluR4: characterization, mechanism of action, and neuroprotection. Neuropharmacology 45:895–906
291. Williams R, Zhou Y, Niswender CM et al (2010) Re-exploration of the PHCCC scaffold: discovery of improved positive allosteric modulators of mGluR4. ACS Chem Neurosci 1:411–419
292. Niswender CM, Lebois EP, Luo Q et al (2008) Positive allosteric modulators of the metabotropic glutamate receptor subtype 4 (mGluR4): Part I. Discovery of pyrazolo[3, 4-d]pyrimidines as novel mGluR4 positive allosteric modulators. Bioorg Med Chem Lett 18:5626–5630
293. Niswender CM, Johnson KA, Weaver CD et al (2008) Discovery, characterization, and antiparkinsonian effect of novel positive allosteric modulators of metabotropic glutamate receptor 4. Mol Pharmacol 74:1345–1358
294. Williams R, Johnson KA, Gentry PR et al (2009) Synthesis and SAR of a novel positive allosteric modulator (PAM) of the metabotropic glutamate receptor 4 (mGluR4). Bioorg Med Chem Lett 19:4967–4970
295. Engers DW, Rodriguez AL, Williams R et al (2009) Synthesis, SAR and unanticipated pharmacological profiles of analogues of the mGluR5 ago-potentiator ADX-47273. ChemMedChem 4:505–511
296. Varney MA, Cosford ND, Jachec C et al (1999) SIB-1757 and SIB-1893: selective, noncompetitive antagonists of metabotropic glutamate receptor type 5. J Pharmacol Exp Ther 290:170–181
297. Gasparini F, Lingenhohl K, Stoehr N et al (1999) 2-Methyl-6-(phenylethynyl)-pyridine (MPEP), a potent, selective and systemically active mGlu5 receptor antagonist. Neuropharmacology 38:1493–1503
298. Maggos C (2010) http://www.addexpharma.com/press-releases/press-release-details/article/development-of-adx10059-ended-for-long-term-use
299. Porter RH, Jaeschke G, Spooren W et al (2005) Fenobam: a clinically validated nonbenzodiazepine anxiolytic is a potent, selective, and noncompetitive mGlu5 receptor antagonist with inverse agonist activity. J Pharmacol Exp Ther 315:711–721
300. Wallberg A, Nilsson K, Osterlund K et al (2006) Phenyl ureas of creatinine as mGluR5 antagonists. A structure-activity relationship study of fenobam analogues. Bioorg Med Chem Lett 16:1142–1145
301. Jaeschke G, Porter R, Buttelmann B et al (2007) Synthesis and biological evaluation of fenobam analogs as mGlu5 receptor antagonists. Bioorg Med Chem Lett 17:1307–1311
302. Hamill TG, Krause S, Ryan C et al (2005) Synthesis, characterization, and first successful monkey imaging studies of metabotropic glutamate receptor subtype 5 (mGluR5) PET radiotracers. Synapse 56:205–216

303. Ametamey SM, Treyer V, Streffer J et al (2007) Human PET studies of metabotropic glutamate receptor subtype 5 with 11C-ABP688. J Nucl Med 48:247–252
304. Hintermann S, Vranesic I, Allgeier H et al (2007) ABP688, a novel selective and high affinity ligand for the labeling of mGlu5 receptors: identification, in vitro pharmacology, pharmacokinetic and biodistribution studies. Bioorg Med Chem 15:903–914
305. Wyss MT, Ametamey SM, Treyer V et al (2007) Quantitative evaluation of 11C-ABP688 as PET ligand for the measurement of the metabotropic glutamate receptor subtype 5 using autoradiographic studies and a beta-scintillator. Neuroimage 35:1086–1092
306. Gasparini F, Bilbe G, Gomez-Mancilla B et al (2008) mGluR5 antagonists: discovery, characterization and drug development. Curr Opin Drug Discov Dev 11:655–665
307. Lindsley CW, Emmitte KA (2009) Recent progress in the discovery and development of negative allosteric modulators of mGluR5. Curr Opin Drug Discov Dev 12:446–457
308. Rodriguez AL, Williams R (2007) Recent progress in the development of allosteric modulators of mGluR5. Curr Opin Drug Discov Dev 10:715–722
309. Spooren W, Gasparini F (2004) mGlu5 receptor antagonists: a novel class of anxiolytics? Drug News Perspect 17:251–257
310. Rodriguez AL, Nong Y, Sekaran NK et al (2005) A close structural analog of 2-methyl-6-(phenylethynyl)-pyridine acts as a neutral allosteric site ligand on metabotropic glutamate receptor subtype 5 and blocks the effects of multiple allosteric modulators. Mol Pharmacol 68:1793–1802
311. Sharma S, Rodriguez AL, Conn PJ et al (2008) Synthesis and SAR of a mGluR5 allosteric partial antagonist lead: unexpected modulation of pharmacology with slight structural modifications to a 5-(phenylethynyl)pyrimidine scaffold. Bioorg Med Chem Lett 18:4098–4101
312. Sharma S, Kedrowski J, Rook JM et al (2009) Discovery of molecular switches that modulate modes of metabotropic glutamate receptor subtype 5 (mGlu5) pharmacology in vitro and in vivo within a series of functionalized, regioisomeric 2- and 5-(phenylethynyl)pyrimidines. J Med Chem 52:4103–4106
313. Iso Y, Grajkowska E, Wroblewski JT et al (2006) Synthesis and structure-activity relationships of 3-[(2-methyl-1, 3-thiazol-4-yl)ethynyl]pyridine analogues as potent, noncompetitive metabotropic glutamate receptor subtype 5 antagonists; search for cocaine medications. J Med Chem 49:1080–1100
314. Carroll FI, Kotturi SV, Navarro HA et al (2007) Synthesis and pharmacological evaluation of phenylethynyl[1, 2, 4]methyltriazines as analogues of 3-methyl-6-(phenylethynyl)pyridine. J Med Chem 50:3388–3391
315. Micheli F, Bertani B, Bozzoli A et al (2008) Phenylethynyl-pyrrolo[1, 2-a]pyrazine: a new potent and selective tool in the mGluR5 antagonists arena. Bioorg Med Chem Lett 18:1804–1809
316. Tehrani LR, Smith ND, Huang D et al (2005) 3-[Substituted]-5-(5-pyridin-2-yl-2H-tetrazol-2-yl)benzonitriles: identification of highly potent and selective metabotropic glutamate subtype 5 receptor antagonists. Bioorg Med Chem Lett 15:5061–5064
317. Bach P, Nilsson K, Wallberg A et al (2006) A new series of pyridinyl-alkynes as antagonists of the metabotropic glutamate receptor 5 (mGluR5). Bioorg Med Chem Lett 16:4792–4795
318. Ceccarelli SM, Jaeschke G, Buettelmann B et al (2007) Rational design, synthesis, and structure-activity relationship of benzoxazolones: new potent mglu5 receptor antagonists based on the fenobam structure. Bioorg Med Chem Lett 17:1302–1306
319. Kulkarni SS, Newman AH (2007) Design and synthesis of novel heterobiaryl amides as metabotropic glutamate receptor subtype 5 antagonists. Bioorg Med Chem Lett 17: 2074–2079
320. Ceccarelli SM, Schlotterbeck G, Boissin P et al (2008) Metabolite identification via LC-SPE-NMR-MS of the in vitro biooxidation products of a lead mGlu5 allosteric antagonist and impact on the improvement of metabolic stability in the series. ChemMedChem 3: 136–144

321. Galatsis P, Yamagata K, Wendt JA et al (2007) Synthesis and SAR comparison of regioisomeric aryl naphthyridines as potent mGlu5 receptor antagonists. Bioorg Med Chem Lett 17:6525–6528
322. Kulkarni SS, Newman AH (2007) Discovery of heterobicyclic templates for novel metabotropic glutamate receptor subtype 5 antagonists. Bioorg Med Chem Lett 17:2987–2991
323. Milbank JB, Knauer CS, Augelli-Szafran CE et al (2007) Rational design of 7-arylquinolines as non-competitive metabotropic glutamate receptor subtype 5 antagonists. Bioorg Med Chem Lett 17:4415–4418
324. Wendt JA, Deeter SD, Bove SE et al (2007) Synthesis and SAR of 2-aryl pyrido[2, 3-d] pyrimidines as potent mGlu5 receptor antagonists. Bioorg Med Chem Lett 17:5396–5399
325. Wang B, Vernier JM, Rao S et al (2004) Discovery of novel modulators of metabotropic glutamate receptor subtype-5. Bioorg Med Chem 12:17–21
326. Buttelmann B, Peters JU, Ceccarelli S et al (2006) Arylmethoxypyridines as novel, potent and orally active mGlu5 receptor antagonists. Bioorg Med Chem Lett 16:1892–1897
327. Hammerland LG, Johansson M, Malmstrom J et al (2006) Structure-activity relationship of thiopyrimidines as mGluR5 antagonists. Bioorg Med Chem Lett 16:2467–2469
328. Rodriguez AL, Williams R, Zhou Y et al (2009) Discovery and SAR of novel mGluR5 non-competitive antagonists not based on an MPEP chemotype. Bioorg Med Chem Lett 19:3209–3213
329. Spanka C, Glatthar R, Desrayaud S et al (2010) Piperidyl amides as novel, potent and orally active mGlu5 receptor antagonists with anxiolytic-like activity. Bioorg Med Chem Lett 20:184–188
330. Williams DL Jr, Lindsley CW (2005) Discovery of positive allosteric modulators of metabotropic glutamate receptor subtype 5 (mGluR5). Curr Top Med Chem 5:825–846
331. Lindsley CW, Wisnoski DD, Leister WH et al (2004) Discovery of positive allosteric modulators for the metabotropic glutamate receptor subtype 5 from a series of N-(1, 3-diphenyl-1H-pyrazol-5-yl)benzamides that potentiate receptor function in vivo. J Med Chem 47:5825–5828
332. Uslaner JM, Parmentier-Batteur S, Flick RB et al (2009) Dose-dependent effect of CDPPB, the mGluR5 positive allosteric modulator, on recognition memory is associated with GluR1 and CREB phosphorylation in the prefrontal cortex and hippocampus. Neuropharmacology 57:531–538
333. Zhao Z, Wisnoski DD, O'Brien JA et al (2007) Challenges in the development of mGluR5 positive allosteric modulators: the discovery of CPPHA. Bioorg Med Chem Lett 17:1386–1391
334. Ritzen A, Sindet R, Hentzer M et al (2009) Discovery of a potent and brain penetrant mGluR5 positive allosteric modulator. Bioorg Med Chem Lett 19:3275–3278
335. Yang ZQ (2005) Agonists and antagonists for group III metabotropic glutamate receptors 6, 7 and 8. Curr Top Med Chem 5:913–918
336. Mitsukawa K, Yamamoto R, Ofner S et al (2005) A selective metabotropic glutamate receptor 7 agonist: activation of receptor signaling via an allosteric site modulates stress parameters in vivo. Proc Natl Acad Sci USA 102:18712–18717
337. Greco B, Lopez S, van der Putten H et al (2010) Metabotropic glutamate 7 receptor subtype modulates motor symptoms in rodent models of Parkinson's disease. J Pharmacol Exp Ther 332:1064–1071
338. Nakamura M, Kurihara H, Suzuki G et al (2010) Isoxazolopyridone derivatives as allosteric metabotropic glutamate receptor 7 antagonists. Bioorg Med Chem Lett 20:726–729
339. Suzuki G, Tsukamoto N, Fushiki H et al (2007) In vitro pharmacological characterization of novel isoxazolopyridone derivatives as allosteric metabotropic glutamate receptor 7 antagonists. J Pharmacol Exp Ther 323:147–156
340. Beart PM, O'Shea RD (2007) Transporters for L-glutamate: an update on their molecular pharmacology and pathological involvement. Br J Pharmacol 150:5–17
341. Moriyama Y, Omote H (2008) Vesicular glutamate transporter acts as a metabolic regulator. Biol Pharm Bull 31:1844–1846
342. Takamori S (2006) VGLUTs: 'exciting' times for glutamatergic research? Neurosci Res 55:343–351

343. Kashani A, Lepicard E, Poirel O et al (2008) Loss of VGLUT1 and VGLUT2 in the prefrontal cortex is correlated with cognitive decline in Alzheimer disease. Neurobiol Aging 29:1619–1630
344. Kashani A, Betancur C, Giros B et al (2007) Altered expression of vesicular glutamate transporters VGLUT1 and VGLUT2 in Parkinson disease. Neurobiol Aging 28:568–578
345. Hinoi E, Takarada T, Tsuchihashi Y et al (2005) Glutamate transporters as drug targets. Curr Drug Targets CNS Neurol Disord 4:211–220
346. Shigeri Y, Seal RP, Shimamoto K (2004) Molecular pharmacology of glutamate transporters, EAATs and VGLUTs. Brain Res Brain Res Rev 45:250–265
347. Tzingounis AV, Wadiche JI (2007) Glutamate transporters: confining runaway excitation by shaping synaptic transmission. Nat Rev Neurosci 8:935–947
348. Rose EM, Koo JC, Antflick JE et al (2009) Glutamate transporter coupling to Na, K-ATPase. J Neurosci 29:8143–8155
349. Koch HP, Brown RL, Larsson HP (2007) The glutamate-activated anion conductance in excitatory amino acid transporters is gated independently by the individual subunits. J Neurosci 27:2943–2947
350. Teichberg VI, Cohen-Kashi-Malina K, Cooper I et al (2009) Homeostasis of glutamate in brain fluids: an accelerated brain-to-blood efflux of excess glutamate is produced by blood glutamate scavenging and offers protection from neuropathologies. Neuroscience 158:301–308
351. Stafford MM, Brown MN, Mishra P et al (2010) Glutamate spillover augments GABA synthesis and release from axodendritic synapses in rat hippocampus. Hippocampus 20:134–144
352. Maragakis NJ, Rothstein JD (2004) Glutamate transporters: animal models to neurologic disease. Neurobiol Dis 15:461–473
353. Sheldon AL, Robinson MB (2007) The role of glutamate transporters in neurodegenerative diseases and potential opportunities for intervention. Neurochem Int 51:333–355
354. Su ZZ, Leszczyniecka M, Kang DC et al (2003) Insights into glutamate transport regulation in human astrocytes: cloning of the promoter for excitatory amino acid transporter 2 (EAAT2). Proc Natl Acad Sci USA 100:1955–1960
355. Yamashita A, Makita K, Kuroiwa T et al (2006) Glutamate transporters GLAST and EAAT4 regulate postischemic Purkinje cell death: an in vivo study using a cardiac arrest model in mice lacking GLAST or EAAT4. Neurosci Res 55:264–270
356. Tanaka K, Watase K, Manabe T et al (1997) Epilepsy and exacerbation of brain injury in mice lacking the glutamate transporter GLT-1. Science 276:1699–1702
357. Lin CL, Bristol LA, Jin L et al (1998) Aberrant RNA processing in a neurodegenerative disease: the cause for absent EAAT2, a glutamate transporter, in amyotrophic lateral sclerosis. Neuron 20:589–602
358. Rothstein JD, Martin LJ, Kuncl RW (1992) Decreased glutamate transport by the brain and spinal cord in amyotrophic lateral sclerosis. N Engl J Med 326:1464–1468
359. Honig LS, Chambliss DD, Bigio EH et al (2000) Glutamate transporter EAAT2 splice variants occur not only in ALS, but also in AD and controls. Neurology 55:1082–1088
360. Hoogland G, van Oort RJ, Proper EA et al (2004) Alternative splicing of glutamate transporter EAAT2 RNA in neocortex and hippocampus of temporal lobe epilepsy patients. Epilepsy Res 59:75–82
361. Jacob CP, Koutsilieri E, Bartl J et al (2007) Alterations in expression of glutamatergic transporters and receptors in sporadic Alzheimer's disease. J Alzheimers Dis 11:97–116
362. Matos M, Augusto E, Oliveira CR et al (2008) Amyloid-beta peptide decreases glutamate uptake in cultured astrocytes: involvement of oxidative stress and mitogen-activated protein kinase cascades. Neuroscience 156:898–910
363. Vallejo-Illarramendi A, Domercq M, Perez-Cerda F et al (2006) Increased expression and function of glutamate transporters in multiple sclerosis. Neurobiol Dis 21:154–164
364. Miller BR, Dorner JL, Shou M et al (2008) Up-regulation of GLT1 expression increases glutamate uptake and attenuates the Huntington's disease phenotype in the R6/2 mouse. Neuroscience 153:329–337

365. Sari Y, Prieto AL, Barton SJ et al (2010) Ceftriaxone-induced up-regulation of cortical and striatal GLT1 in the R6/2 model of Huntington's disease. J Biomed Sci 17:62
366. Thone-Reineke C, Neumann C, Namsolleck P et al (2008) The beta-lactam antibiotic, ceftriaxone, dramatically improves survival, increases glutamate uptake and induces neurotrophins in stroke. J Hypertens 26:2426–2435
367. Thompson CM, Davis E, Carrigan CN et al (2005) Inhibitor of the glutamate vesicular transporter (VGLUT). Curr Med Chem 12:2041–2056
368. Campiani G, Fattorusso C, De Angelis M et al (2003) Neuronal high-affinity sodium-dependent glutamate transporters (EAATs): targets for the development of novel therapeutics against neurodegenerative diseases. Curr Pharm Des 9:599–625
369. Mennini T, Fumagalli E, Gobbi M et al (2003) Substrate inhibitors and blockers of excitatory amino acid transporters in the treatment of neurodegeneration: critical considerations. Eur J Pharmacol 479:291–296
370. Alaux S, Kusk M, Sagot E et al (2005) Chemoenzymatic synthesis of a series of 4-substituted glutamate analogues and pharmacological characterization at human glutamate transporters subtypes 1-3. J Med Chem 48:7980–7992
371. Sagot E, Jensen AA, Pickering DS et al (2008) Chemo-enzymatic synthesis of (2S, 4R)-2-amino-4-(3-(2, 2-diphenylethylamino)-3-oxopropyl)pentanedioic acid: a novel selective inhibitor of human excitatory amino acid transporter subtype 2. J Med Chem 51:4085–4092
372. Dunlop J, Eliasof S, Stack G et al (2003) WAY-855 (3-amino-tricyclo[2.2.1.02.6]heptane-1, 3-dicarboxylic acid): a novel, EAAT2-preferring, nonsubstrate inhibitor of high-affinity glutamate uptake. Br J Pharmacol 140:839–846
373. Shimamoto K, Sakai R, Takaoka K et al (2004) Characterization of novel L-threo-beta-benzyloxyaspartate derivatives, potent blockers of the glutamate transporters. Mol Pharmacol 65:1008–1015
374. Greenfield A, Grosanu C, Dunlop J et al (2005) Synthesis and biological activities of aryl-ether-, biaryl-, and fluorene-aspartic acid and diaminopropionic acid analogs as potent inhibitors of the high-affinity glutamate transporter EAAT-2. Bioorg Med Chem Lett 15: 4985–4988
375. Jensen AA, Erichsen MN, Nielsen CW et al (2009) Discovery of the first selective inhibitor of excitatory amino acid transporter subtype 1. J Med Chem 52:912–915

Top Med Chem 6: 149–176
DOI: 10.1007/7355_2010_9

Published online: 10 September 2010

Modulation of the Kynurenine Pathway for the Potential Treatment of Neurodegenerative Diseases

Stephen Courtney and Andreas Scheel

Abstract Modulation of tryptophan metabolism and in particular the kynurenine pathway is of considerable interest in the discovery of potential new treatments for neurodegenerative diseases. A number of small molecule inhibitors of the kynurenine metabolic pathway enzymes have been identified over recent years; a summary of these and their utility has been reviewed in this chapter. In particular, inhibitors of kynurenine monooxygenase represent an opportunity to develop a therapy for Huntington's disease; progress in the optimization of small molecule inhibitors of this enzyme is also described.

Keywords Enzyme inhibitors, Huntington's disease, Kynurenine monooxygenase (KMO), Kynuremine pathway, Neurodegenerative disease, Tryptophan Metabolism

Contents

S. Courtney (✉)
Evotec (UK) Ltd, 114 Milton Park, Abingdon, Oxfordshire OX14 4SA, UK

A. Scheel
Evotec AG, Schnackenburgallee 114, 22525, Hamburg Germany
Prosidion Ltd, Windrush Court, Watlington Road, Oxford OX4 6LT, UK

1 Introduction

The role of tryptophan metabolism in human biology has been studied for many years; more recently, the significance of the kynurenine pathway (KP), the major breakdown pathway of tryptophan, has been widely examined [1–5]. These studies specifically relate to the role of several kynurenine catabolic products in immunomodulation and CNS function. It is believed that modulation of the levels of the key metabolites of tryptophan catabolism represents a potential new approach to developing a treatment for neurodegenerative and inflammatory diseases [2]. A number of small molecule enzyme inhibitors have become available, which modulate different stages of the pathway and can be used to further study its role in disease. These compounds will initially serve as tools in the pursuit of developing a clearer understanding of the underlying mechanisms of a variety of diseases. In this review, we have focused on the KP and in particular on kynurenine monooxygenase (KMO) due to its potential as a new target for the treatment of Huntington's disease (HD).

2 Tryptophan Metabolism and the Kynurenine Pathway

L-tryptophan is an essential amino acid. In addition to being used for protein synthesis, it serves as a precursor for several biologically active substances. Non-proteinogenic tryptophan is used to produce bioactive substances such as serotonin, the hormone melatonin, and tryptamine. The majority of tryptophan, however, is catabolized through the so-called kynurenine pathway (KP), a cascade of enzymatic reactions that yields important cofactors such as NAD^+ and $NADP^+$ and a number of important kynurenine metabolites on the way [6]. Although the KP has been known for more than five decades, primarily from its function in peripheral tissues [7], it has attracted a particular interest in the last 15 years from a drug discovery perspective as a number of the KP metabolites have immunomodulatory and neuroactive properties and may thus be involved in normal brain function and might contribute to human disease. The link between KP metabolism, the immune system, and CNS diseases is increasingly appreciated, and many reports in the literature have recently focused on its role in brain physiology and pathology [1, 3].

3 Introduction of KP Enzymes and Key Metabolites

The first step of tryptophan catabolism is the oxidative cleavage of the indole ring of L-tryptophan, which is catalyzed by members of the family of pyrrole dioxygenases. A key member of this family, indoleamine-2,3-dioxygenase (IDO, see Fig. 1, EC 1.13.11.17), is expressed in all tissues except in the liver and produces the central metabolite kynurenine (KYN). Two different and competing branches of the pathway then further metabolize KYN: the first pathway includes a family of enzymes called kynurenine aminotransferases (KATs), which produce kynurenic acid (KYNA) in a terminal branch. In a second arm, KMO (or kynurenine

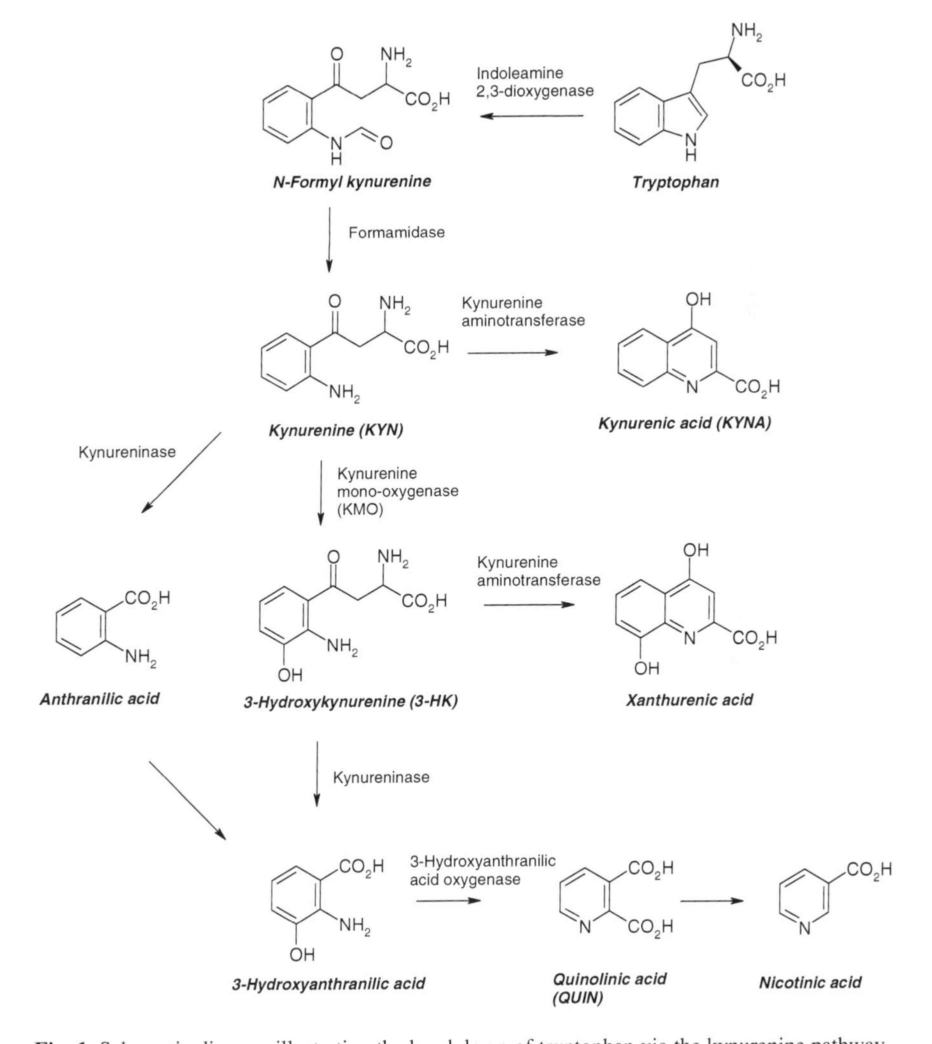

Fig. 1 Schematic diagram illustrating the breakdown of tryptophan via the kynurenine pathway

hydroxylase, EC 1.14.13.9) metabolizes KYN into 3-hydroxykynurenine (3-HK). In the third branch of the pathway, kynureninase transforms KYN to give anthranilic acid, which subsequently provides a further route to generate 3-HK. 3-HK, in turn, is a substrate for kynureninase, which produces 3-hydroxyanthranilic acid (3HANA). 3HANA is the substrate of 3-hydroxyanthanilic acid oxygenase that produces quinolinic acid (QUIN). After further enzymatic steps, the final product, NAD^+, is formed (Fig. 1).

Recent pharmacological interest in KYN metabolites with respect to CNS diseases has mainly focused on three brain-active molecules: 3-HK and QUIN, two molecules with neurotoxic properties, and KYNA, a presumed neuroprotective metabolite. These will be discussed in more detail in the following sections.

All of the enzymes of the KP can be found in the periphery, especially in the liver and in immune system cells of the monocyte, macrophages lineage [8], where increased KP activity is believed to play an important role in immunomodulatory activities [9] (see Sect. 4). In addition, components of the KP have been shown to be expressed in endothelial cells and pericytes of the blood–brain barrier (BBB) [10]. In the central nervous system, all components of the KP pathway are also expressed although at much lower levels than in the periphery. Although KYN can be produced in the brain, it seems that the cerebral KP levels can be also driven by kynurenine uptake from the blood through the BBB [11]. Further processing of KYN in the brain is primarily carried out after uptake of KYN by glia cells which express the relevant downstream KP enzymes, whereas no or little expression has been found in neurons. Astrocytes are described to contain no or little KMO but express KAT and therefore favor the KYNA arm of the pathway [12]. In contrast, resident and reactive microglia (i.e., in situations of injury or inflammation) and also infiltrating macrophages harbor very little KAT activity but express KMO and are thus responsible for the QUIN-producing branch of the KP pathway [13].

Interestingly, no clear and specific physiological function of the KP metabolites has emerged in the brain. Several reports demonstrate the ability of KYNA and QA to modulate glutamate and acetylcholine receptor functions [1, 3], although the endogenous contributions to neurotransmission remain poorly understood in normal CNS physiology. In fact, the activity of kynurenine catabolizing enzymes may be harmful in the CNS since an increase of the KP pathway leads to the accumulation of potentially neurotoxic KP metabolites, which might contribute to neurodegeneration in a variety of chronic disorders.

4 Physiological and Pathological Features of KP Pathway

4.1 Physiological Role

The KP pathway has a number of different and important physiological roles, especially in immunomodulation in peripheral tissues. Metabolizing Trp via the

KP pathway is believed to reduce the growth of intracellular pathogens, to play a role in tumor escape mechanisms (immune surveillance) and to preserve immune tolerance at the fetal–maternal interface, by preventing rejection of the embryo [9]. The key enzyme in this process is IDO, whose expression is induced by IFN-γ and other immune-active molecules [14]. The immunosuppressive effects of IDO-mediated tryptophan metabolism in this context were initially believed to be solely due to Trp depletion, which suppresses T cell proliferation. However, there is increasing evidence that various KP metabolites directly modulate T cell biology, by inhibiting proliferation of T cells undergoing activation and inducing T cell apoptosis [15]. Thus, the KP and its metabolites are believed to play a key role in tolerizing T cells, a process that is highly relevant for the maintenance of immune tolerance.

4.2 Kynurenines and CNS Dysfunction

Being at the center of a complex network linking immune response and inflammatory reactions, it is not surprising that imbalances in KP metabolites lead to pathophysiological consequences. In Drosophila (*Drosophila melanogaster*), genetic deletions of individual KP enzymes resulted in neuronal abnormalities [16]. Already two decades ago, it was found that QUIN promotes excitotoxicity *in vivo*, and it was later suggested that these excitotoxic effects are mediated by QUIN's weak agonistic activity on the NMDA receptor, with an IC_{50} of 30–100 μM [1]. The observed lesions, as mediated by intracerebral injection of QUIN, can be prevented through the application of NMDA receptor antagonists, so it is indeed likely that the observed effects are mediated through NMDA receptor activation [17]. Additionally, QUIN has also been shown to stimulate lipid peroxidation and to produce radical oxygen species (ROS) [18].

QUIN is present in the brain at basal levels that usually are in the range of 0.01 μM, rarely exceeding 1 μM [4, 19]. In the diseased brain, the QUIN concentration may increase considerably. It has been demonstrated that the cerebral KP seems to be stimulated in response to local injury and/or inflammation, by activating the KP pathway in microglial cells [14, 20]. Also, infiltrating macrophages following central inflammatory reactions are believed to produce 20- to 30-fold more QUIN than brain glial cells. In cell culture, reports have demonstrated that micromolar concentrations of QUIN are toxic after a few hours, and chronic exposure of as little as 100 nM concentrations of QUIN to organotypic cortico-striatal cultures produces damage after several weeks [21]. It is thus conceivable that QUIN levels after pathological activation of the KP may indeed reach local concentrations that are sufficient to substantially activate NMDA receptors, especially under chronic conditions, and thus cause neurotoxicity although this has not been adequately addressed *in vivo*. Interestingly, QUIN-induced lesions are significantly attenuated by reagents that scavenge reactive oxygen species [22, 23]. Thus, the remarkable toxicity of QUIN may be explained by its ability to activate NMDA receptors and cause oxidative damage at the same time.

In addition, QUIN is likely not to act on its own as a neurotoxic agent. 3-HK is known to produce reactive oxygen radicals which can cause oxidative cell damage and ultimately lead to apoptosis *in vitro* [24]. Its cytotoxic properties seem to be due to auto-oxidation which subsequently leads to the production of hydrogen peroxide and other reactive products. It has also been shown that 3-HK potentiates QUIN-mediated cell damage: co-injection of both substances in low doses into the rat brain, which alone would not cause any damage, results in substantial neuronal loss [25].

Despite the evidence mentioned above that QUIN may indeed exert its main neurotoxic properties *in vivo* through the NMDA receptor, and despite synergistic properties with 3-HK, there is still a controversial discussion within the field if the concentration of these metabolites reached *in vivo* is indeed sufficient to substantially stimulate the NMDA receptor and thus lead to a toxic calcium overload, or whether alternative pathophysiological mechanisms are at work [26].

In contrast, KYNA, which is formed from kynurenine by KAT, has shown to be a neuroprotective agent. It is a competitive blocker of the glycine site of the NMDA receptor, albeit with a low potency of ~7 μM [27]. However, the affinity of KYNA for this site is weaker than that of glycine; so it is unclear whether under normal physiological conditions the glycine site is occupied at all by KYNA. KYNA is also a blocker of the α,7-nicotinic acetylcholine receptor with a similar potency [28] and has been shown to modulate dopamine and glutamate release presynaptically *in vivo* [29]. More recently, KYNA has also been shown to be an agonist of an orphan GPCR with unknown function, GPR35 [30], albeit at low potency (30–100 μM). Accordingly, high concentrations of KYNA are anticonvulsant and provide protection against excitotoxic lesions caused by QUIN and protect against various conditions such as ischemia and traumatic brain injury [2]. Perhaps not surprisingly, it has been shown that KYNA levels are altered in various CNS disorders. However, local concentrations of KYNA in the brain have consistently been found to be in the low nanomolar range (10–150 nM) [31], and it is questionable if the concentrations necessary to modulate NMDA and nicotinic receptors can actually be reached *in vivo,* even under pathological conditions. Some data imply inhibitory actions of KYNA on presynaptic glutamate release as a mechanism for its anti-excitotoxic activity, even at nanomolar concentrations [32]; however, these data need to be substantiated further. Application of KYNA itself has been considered for therapeutic intervention, but the poor ability to cross the BBB has limited its applicability. An additional finding relates to the ability of KYNA at lower (nM) concentrations to modulate synaptic glutamate release when infused directly into the brain through a dialysis probe [31, 33]. Therefore, it is likely that KYNA plays a synaptic role through unknown mechanisms under physiological conditions. This is an area that requires further investigation.

4.3 Kynurenine Pathway and Disease

Not surprisingly, disturbance in the KP has been implicated in a number of diseases, and pharmacological intervention has thus potential for treatment. The immunosuppressive effects of the KP metabolites in the periphery are used in the treatment of multiple sclerosis (MS), where synthetic kynurenines are undergoing clinical development. Furthermore, IDO inhibitors are in preclinical development to treat cancers, for example, ovarian and colorectal [34]. This has been reviewed extensively elsewhere [3, 35], and the focus of the discussion in this review will be on the role of KP metabolites in brain pathophysiology.

As illustrated above, changes in the KP have been implicated in a large number of diseases including neurodegenerative diseases (Huntington's, Parkinson's, Alzheimer's disease), psychiatric diseases (schizophrenia, depression), and AIDS dementia [2, 4].

It is unclear at this point if increased levels of neurotoxic metabolites (QUIN and 3-HK) and a reduction in KYNA, the neuroprotective metabolite, may be a direct cause for some of these diseases. Alternatively, kynurenines may simply play a secondary role in disease progression: after focal physical injury and in a state of neuroinflammation, the cerebral KP is stimulated, through activation of brain microglial cells or infiltration of macrophages. Under those conditions, KP metabolites can be increased substantially over a longer period and may thus prolong and exacerbate disease symptoms.

However, many conclusions are purely based on the imbalance of the KP metabolites as determined in animal models or patients [36]. Although this does point toward a link between the KP and nervous system diseases, it is difficult to distinguish disease-causing effects from secondary effects purely happening because of chronic inflammatory processes accompanying the primary disease progress. Indeed, clear-cut target validation studies using genetic animal models or small molecule enzyme modulators in relevant *in vivo* disease models, to validate enzymes within the KP pathway as suitable targets for therapeutic interventions, are scarce. It is thus not clear at this point which of these diseases may benefit from blocking the KP pathway in an *in vivo* situation.

5 Enzymes in the KP as Drug Targets and Their Inhibitors

As discussed above, inhibition of key enzymes of the KP (Fig. 1) may represent a viable opportunity to develop therapeutic agents for the treatment of a number of inflammatory, neurodegenerative, and psychiatric disorders. In this section, we will review the available chemical inhibitors of the different enzymes in the KP, with a broader focus on KMO; however, we will also briefly review the current status of inhibitors of IDO, kynurenine aminotransferase II (KAT II), kynureninase, and 3-hydroxyanthranilic acid oxygenase (HAO).

5.1 Indoleamine 2,3-Dioxygenase

IDO is a key enzyme in the degradation of tryptophan in extra-hepatic tissues [37], through the generation of *N*-formyl kynurenine which is further degraded to kynurenine (L-KYN) by formamidase. In addition to its potential role in neurodegeneration, inhibition of IDO has been implicated as an important new therapeutic target for the treatment of cancer through tumor immunosuppression [3, 38].

One of the earliest inhibitors of IDO reported in the literature is the carboline (**1**) with a K_i of 120 μM [39]; several other carboline-based inhibitors were subsequently reported with improved activity, for example, the 3-butyl derivative (**2**, $K_i = 3.3$ μM) [40]. In fact, until recently the most commonly available competitive inhibitors were tryptophan-based analogs, for example, *N*-methyl tryptophan (**3**, $K_i = 34$ μM) [41, 42]. In 2006, the natural product Brassinin (**4**) was identified as a weakly active inhibitor of IDO ($K_i = 97.7$ μM). Investigation of structure–activity relationships (SAR) of this series identified further analogs with improved potency (**5**, $K_i = 11.5$ μM), and examples that more importantly did not require the core indole ring (**6**, $K_i = 42$ μM), however, do retain the dithiocarbamate functionality [43].

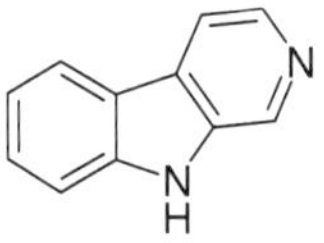

(**1**) 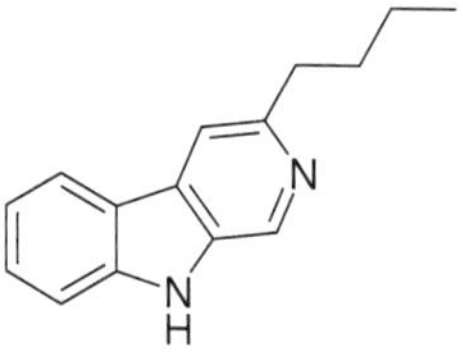 (**2**) 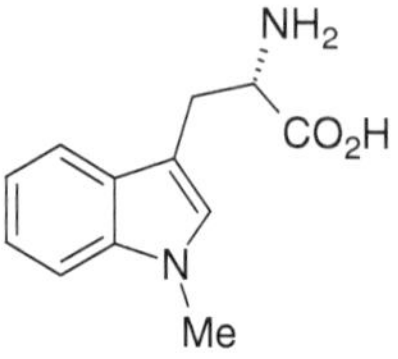 (**3**)

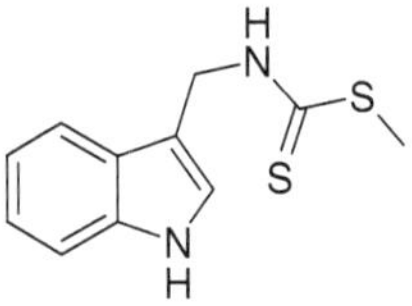

(**4**) 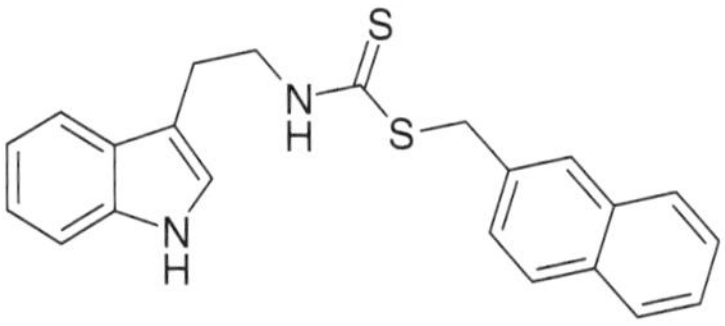 (**5**) 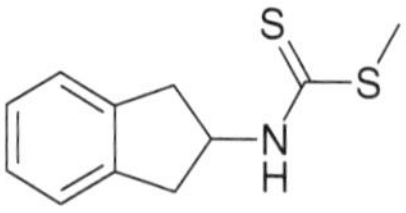 (**6**)

The naphthoquinone natural product Annulin B (**7**), isolated from marine hydroid, has also been reported to have relatively potent IDO inhibitory activity ($K_i = 0.12$ μM); screening other natural product analogs and commercial compounds revealed Menadione as a potent inhibitor of IDO with *in vivo* efficacy in mouse tumor models. Further optimization was carried out with the quinone scaffold to reveal compounds with improved potency (**8**, $K_i = 0.055$ μM) [34]. Related quinone marine natural products have also been identified as potent IDO inhibitors (Exiguamine A; **9**, $K_i = 41$ nM); simplification of this natural product has resulted in the analog (**10**, $K_i = 200$ nM) as a potential tool compound for further investigations [44]. The role of the quinone redox system and kinetic analysis for these classes of compounds remains to be explored. However, it is likely that the activity of these compounds involves the iron bound to the heme of IDO.

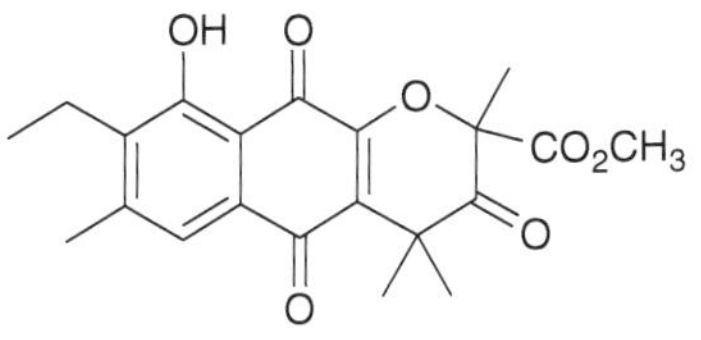

(**7**) 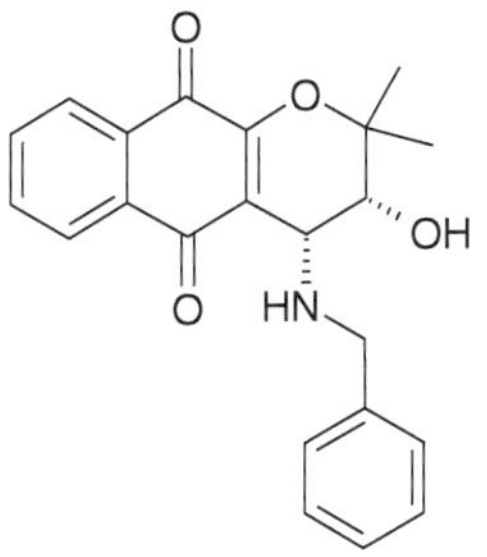 (**8**)

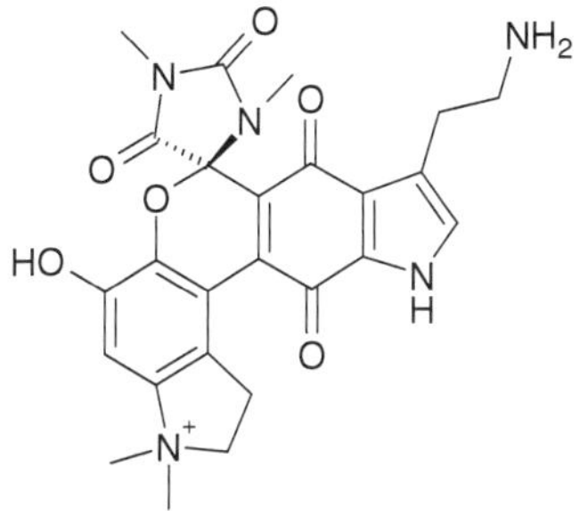

(**9**) 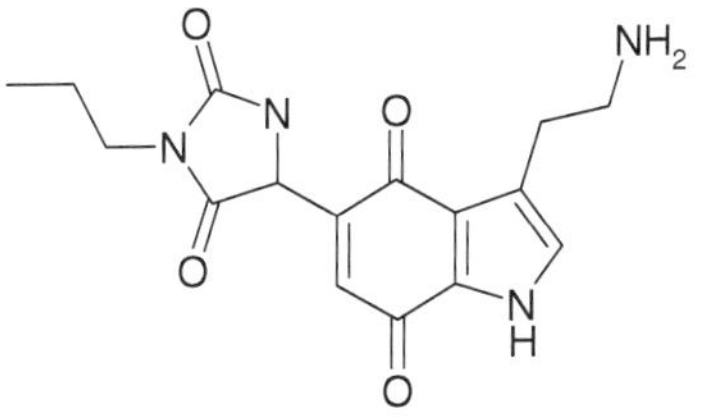 (**10**)

In 1989, 4-phenylimidazole (**11**) was reported as a weak inhibitor with an IC_{50} of 48 μM against IDO [45]. Recently, the reported crystal structure of IDO was used to design and optimize this structure to generate a new series of inhibitors [46], the most potent compound reported being (**12**, $IC_{50} = 7.6$ μM). Incyte corporation [47] have also identified through high-throughput screening a new, non-indole, non-quinone redox system with good competitive IDO inhibition (**13**, $K_i = 1.5$ μM) and tenfold selectivity over tryptophan 2,3-dioxygenase (TDO), a related enzyme predominantly present in the liver, responsible for maintaining the balance of dietary tryptophan. Simple substitution of the phenyl ring resulted in improvement in activity both in a biochemical IDO assay and in a HeLa cellular assay as exemplified with compound **14**, IC_{50} (biochemical) = 67 nM, IC_{50} (cellular) = 19 nM. Further *in vivo* studies were used to illustrate the effectiveness of this compound at decreasing kynurenine levels in plasma as well as inhibiting tumor growth.

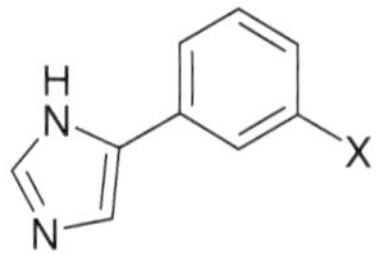

(**11**; X = H, **12**; X = SH) 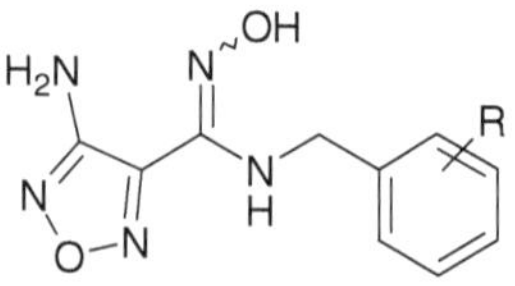 (**13**; R = H, **14**; R = 3-Cl, 4-F)

The tryptophan and quinone classes of IDO inhibitors have been limited by their potency and poor physical properties and to date have not progressed into preclinical setting. However, the oxadiazoles identified by Incyte (above, **13–14**) represent a new class of competitive IDO inhibitor, which have progressed further toward preclinical development for the treatment of a variety of cancers.

5.2 KAT II Inhibitors

To date, four KAT isoenzymes (KAT I, II, III, and IV) have been identified as present in the mammalian brain [48, 49]; however, only KAT I and KAT II have been widely associated with the transamidation of KYN into KYNA. Indeed KAT II [a pyridoxal-5′-phosphate (PLP)-dependant enzyme] accounts for the majority of KYNA formation in the rat and human brain and as such represents a key transformation in the KP [50, 51].

Inhibition of KAT II would result in a decrease in the synthesis of KYNA and consequently enhanced NMDA receptor activity and glutamate release [52]. Therefore, KAT II blockade is thought to be useful in the treatment of disorders with implicated glutamatergic and cholinergic hypofunction (such as Alzheimer's disease and schizophrenia).

Reported inhibitors of KAT enzymes in the literature are very limited and until relatively recently centered around chlorinated substrates [53]. (S)-4-Ethylsulfonylbenzoylalanine (S-ESBA, **15**) was the first reported synthetic and selective inhibitor of KAT II with an IC_{50} of 6.1 μM and no inhibition of KAT I. Effects of (**15**) on the reduction of extracellular KYNA concentrations in the rat hippocampus *in vivo* using microdialysis were investigated [51]. In this study, the levels of KYNA were successfully decreased by approximately 30% from basal levels. Interestingly (**15**) does not inhibit the human enzyme as effectively. In a follow-up publication, Pellicciari illustrated a 10- to 20-fold reduction in potency against human KAT II; this was speculated to result from sequence variants in the enzyme catalytic sites [54]. S-ESBA continues to be a valuable tool compound in the *in vivo* investigations of the KP to investigate the significance of KYNA levels in neurodegenerative disease.

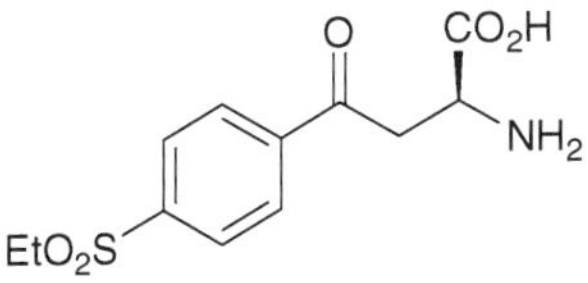

(15)

5.3 Kynureninase Inhibitors

Kynureninase catalyzes the hydrolytic cleavage of both kynurenine and 3-hydroxykynurenine to generate anthranilic acid and 3-hydroxyanthranilic acid, respectively [55]. The majority of inhibitors of kynureninase are substrate based and are designed based on the postulated transition state intermediate where water attacks the benzoyl group carbonyl through a PLP-dependant mechanism. The hydroxy (**16**; $K_i = 0.3\ \mu M$) and sulfone (**17**; $IC_{50} = 11\ \mu M$) derivatives have been reported as inhibitors of human kynureninase with moderate to good potency [56, 57]. The methoxy derivative (**18**; $IC_{50} = 3\ \mu M$) was identified as a moderately potent and selective inhibitor of kynureninase [58]. *In vivo* studies were also carried out using this inhibitor to demonstrate that 3-hydroxylation is the preferred route of KYN metabolism in the brain. As expected, this study demonstrated that inhibition of kynureninase resulted in an increase in the levels of the neurotoxic 3-HK.

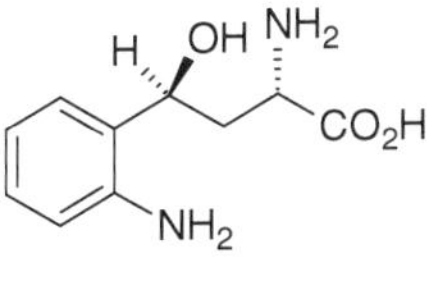

(16) 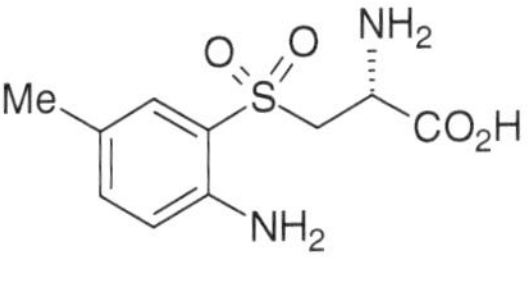 (17) 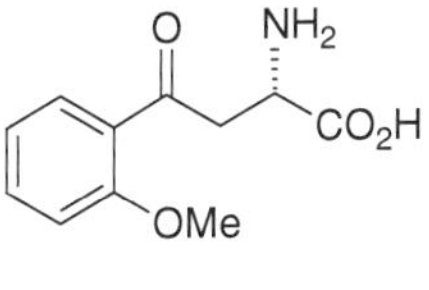 (18)

A series of novel bicyclic analogs of kynurenine were subsequently reported exhibiting moderate potency against human kynureninase; for example, the napthyl derivative (**19**) exhibits a K_i of 22 μM [59]. Further enhancements in the potency against human kynureninase were achieved through the synthesis of the di-hydroxy compound (**20**; $K_i = 100$ nM) which demonstrated improved selectivity (1,000-fold) over the bacterial enzyme [60]. In 2009, the co-crystal structure of human kynureninase with 3-hydroxyhippurate (**21**; $K_i = 60\ \mu M$) was solved [61]. Subsequently, a series of mutants were designed to establish the preliminary binding residues contributing to substrate specificity. Needless to say, selectivity against kynureninase is an essential prerequisite in the progression of a neurodegenerative disease treatment.

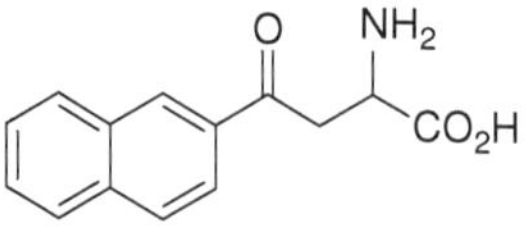

(19) 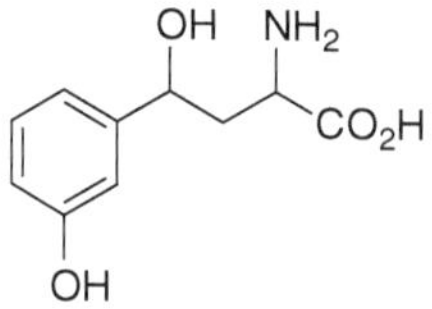 (20) 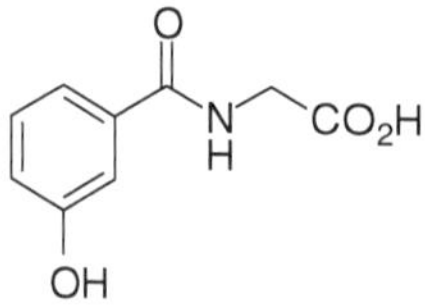 (21)

5.4 3-Hydroxyanthranilic Acid Oxygenase Inhibitors

3-Hydroxyanthranilic acid oxygenase (HAO) catalyzes the final transformation of the pathway's conversion of tryptophan into QUIN [62]. To date, the only inhibitors of HAO reported are halogenated substrates, and these are represented by the compounds **22**, **23**, and **24**, below. These compounds have been shown to be highly potent (IC_{50}s of **22**, **23**, and **24** are 6, 0.3, and 5.8 nM, respectively), and their utility as potential therapeutic agents has been limited, possibly due to the oxidative instability of these compounds [63]. Compound **22** has, however, been shown to attenuate QUIN accumulation (following immune activation) in brain and blood following systemic administration to mice [64], thus illustrating that targeting HAO may have the ability to potentially provide a neuroprotective agent. Compound **24** have also shown the ability to inhibit cerebral HAO *in vivo* following intracerebroventricular administration to rats resulting in reduction of QUIN production, as measured by GC/MS [65].

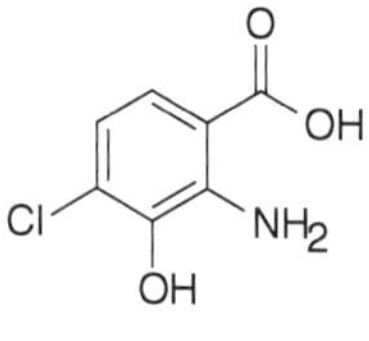

(22) 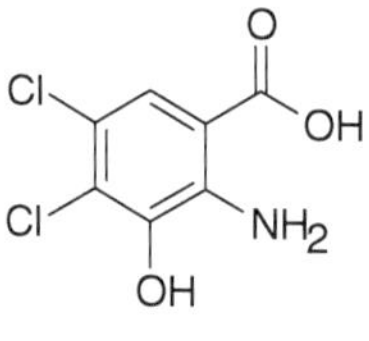 (23) 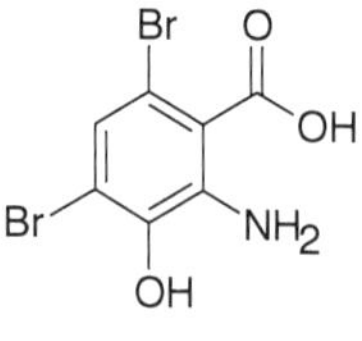 (24)

5.5 KMO Inhibitors

KMO is a NADPH-dependant flavin monooxygenase which is localized to the outer membrane of mitochondria [66]. KMO is expressed at high levels in the liver, endothelial cells, and monocytic cells and to a lower extent in the brain. Here, its expression is mainly found in cells of glial nature, specifically in microglial cells and in infiltrating macrophages, whereas little or no expression has been found in astrocytes or in neurons [12, 67]. However, the lack of good antibodies to detect endogeneous KMO has limited the verification of this expression pattern in the brain.

As described above, KMO catalyzes the hydroxylation at the third position of kynurenine. The KMO enzyme is thus at a key position of the pathway as its activity determines the level of flux through the two arms of the pathway. KMO inhibition is expected to be beneficial in neurodegenerative disease as this would increase the availability of KYN to KATII, and thus achieves a shift away from QUIN and 3-HK production to an increase in KYNA production. Thus, KMO has been considered the most relevant target for therapeutic intervention in the KP for CNS disease [1].

The first reported competitive inhibitor of KMO was nicotinoylalanine (NAL, **25**). NAL is a simple analog of the natural substrate (KYN) where the amino ring substituent has been removed and a pyridyl ring nitrogen has been placed to block the hydroxylation position [68]. The potency of this analog is low (IC_{50} = 900 μM) and is equipotent against kynureninase; however, it has been shown to increase the concentration of KYNA in brain tissue and has anticonvulsant activity at high doses. Further optimization of this template was carried out by and resulted in the identification of (*m*-nitrobenzoyl)alanine (*m*-NBA, **26**). NBA has a 100-fold selectivity over kynureninase with an IC_{50} of 0.9 μM for KMO. *In vivo* NBA was shown to increase the concentration of KYN and KYNA in brain, blood, and liver of rats [69, 70]. Molecular modeling studies were also carried out [71] to rationalize the potency and selectivity obtained for NBA. A protein structure of KMO is currently not available; however, investigation of the mechanism of action of KMO and generation of a "pseudo active site" model was carried out. These studies are the first reported investigations of quantitative structure–activity relationships (QSAR) for KMO inhibitors.

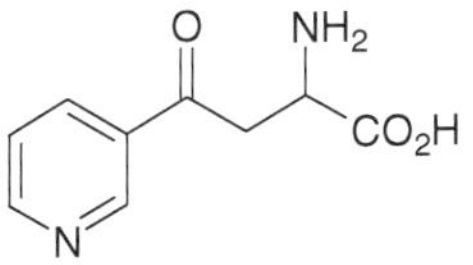

(25) 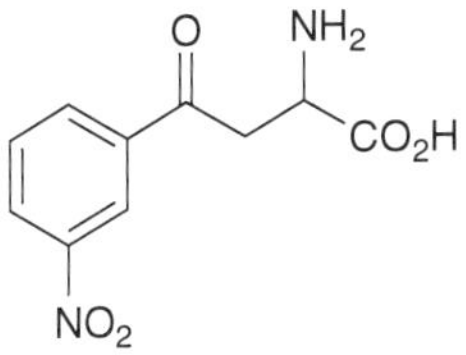 **(26)**

Pharmacia and Upjohn subsequently carried out an SAR study of the phenyl ring of **26** and highlighted that 3,4-dichloro substitution was preferred [72] giving rise to PNU-156561 (**27**, IC_{50} = 0.2 μM). The same group [73] then explored the SAR around the benzoylalanine side chain of this series and highlighted that the carboxylic acid was essential for activity, whereas the amino group may be removed with limited effect on the activity (**28**, R = H; IC_{50} = 0.9 μM). This was an important discovery as this moved the inhibitor series away from the natural substrate and opened the door for exploration of side chain substitutions. Modification of the second position (i.e., replacement for the amino group) highlighted the hydroxy and benzyl analogs (**28**, R = OH, CH_2Ph, respectively) as potent enzyme inhibitors, IC_{50} = 0.3 and 0.18 μM, respectively. Substitution with a methyl group gave slightly lower activity

(**28**, R = CH_3; IC_{50} = 2.2 μM). The stereochemistry of these inhibitors was also investigated through a stereoselective synthesis, as perhaps expected it was found that the S-(−) isomer was favored in all cases. In the same study, the 4-oxo-butenoic acid analogs were also reported (**29**, R = H, OH, CH_3) with similar activity against KMO (IC_{50} = 0.9, 0.95, 2.2 μM, respectively) [73].

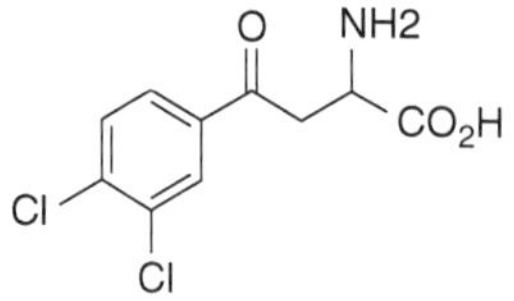

(**27**) 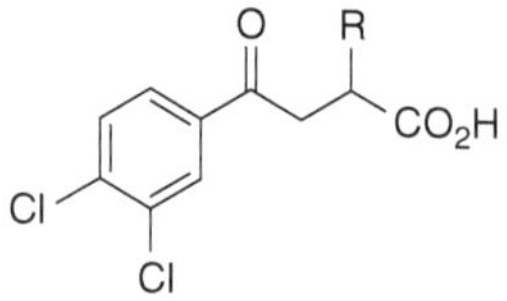 (**28**, R = H, CH_3, OH, CH_2Ph) 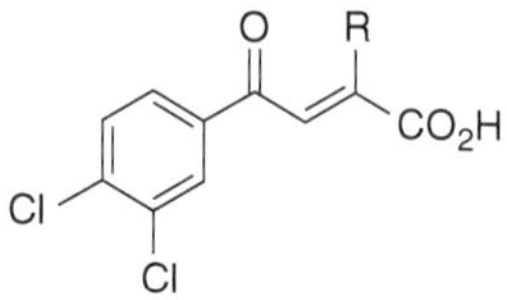 (**29**, R = H, OH, CH_3)

An identical template was also explored by Glaxo Wellcome [74] 2 years later, again showing a preference for the 3,4-dichloro substituents on the phenyl ring. The ability of these compounds to inhibit the production of QUIN was performed using macrophage cultures stimulated with interferon-γ as a model for QUIN formation in inflammatory disease. Pharmacia and Upjohn continued to develop the benzoyl carboxylic acid series of inhibitors, subsequently identifying that conformational restriction of the side chain through incorporation of a cyclopropyl ring (**30**, UPF-648) gave an improvement in potency to the nanomolar level [5, 75]. UPF-648 has become a valuable tool for academics and industry and is widely used as a benchmark compound for studies investigating the utility of KMO inhibitors. The strategy of rigidifying the side chain was further explored by the same industrial group [76]; through molecular overlays, they designed the quinoline-based inhibitor (**31**) which had moderate activity against KMO and good selectivity against kynureninase and KAT. Although the best reported IC_{50} was only 24 μM for (**31**), this illustrates the possibility of modification of the central core to generate new inhibitor series with the goal to improve the activity profile and CNS penetration.

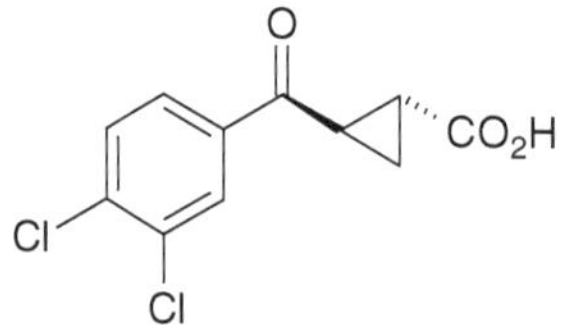

(**30**) 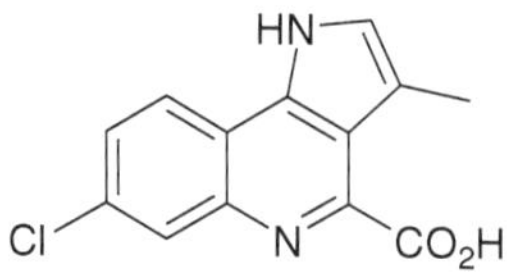 (**31**)

In addition to the "ketoacids" described above, an interesting sulfonamide screening hit (**32**) was described by Roche [77]. The SAR around this compound was explored resulting in the identification of many compounds with improved potency against KMO; these are illustrated by compounds (**33**, Ro61-8048) and (**34**), with IC_{50}s of 37 and 39 nM, respectively. As well as being highly potent inhibitors of KMO *in vitro*, these compounds were shown to inhibit KMO following oral dosing to gerbils. Compound **33** (100 μmol/kg, p.o.) also increased KYNA concentrations (7.5-fold) in the extracellular hippocampal fluid of rats. Many studies have been carried out with these compounds; however, there remains a question mark over the brain penetrability of this series.

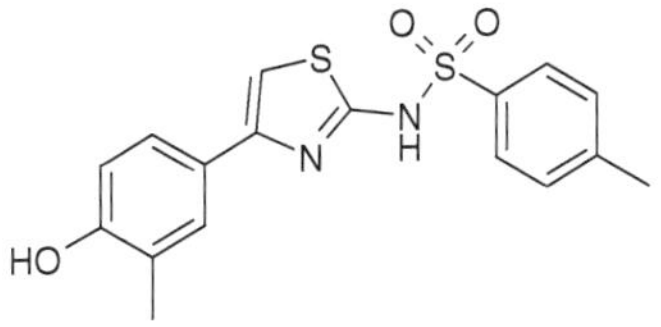

(**32**) 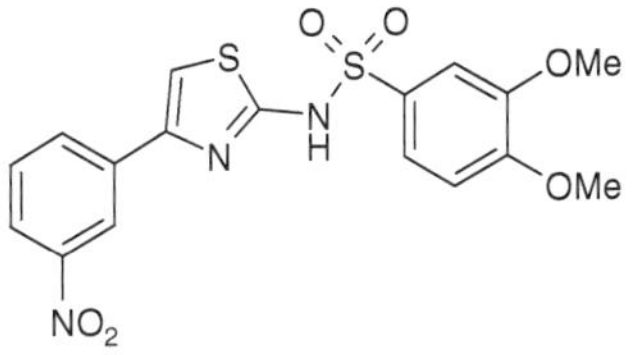 (**33**)

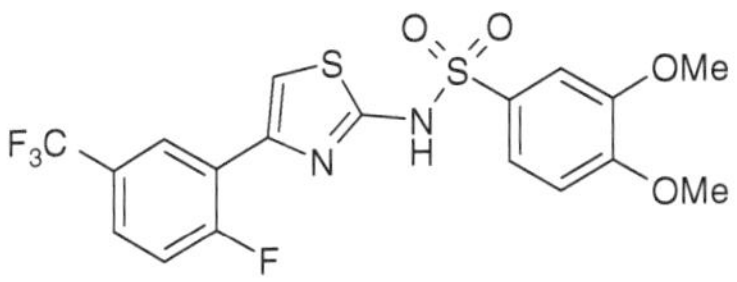

(**34**)

Although there have been a number of potent and selective inhibitors of KMO reported in the literature, to date there have been no clinical trials initiated. The most likely explanation for this lack of progress is due to the fact that reported chemotypes have suffered from the same problem of limited penetration into the CNS.

5.6 KMO Inhibition and Stroke/Ischemia

Sustained upregulation of the KP enzyme system is observed after brain injury. After a cerebral insult, glia cells become activated and secrete cytokines and kynurenines [20]. It has been shown that in gerbils, after induced transient ischemia in the brain, kynurenine metabolite levels, especially QUIN, increase dramatically in the ischemic region of the brain for several days [20]. It is assumed this is a secondary, inflammatory response to brain damage due to induction of microglia cells and infiltrating macrophages, and may further prolong or enhance the damage. Application of a KMO inhibitor has been pursued as a strategy to reduce potentially neurotoxic metabolites such as 3-HK and QUIN, and to increase neuroprotective KYNA levels through an increase in kynurenine levels. Ro61-8048 (**33**, see above) was applied to hippocampal slice cultures exposed to oxygen and glucose deprivation, and to an *in vivo* model of stroke in gerbils [77, 78]. In these experiments, KYNA levels were shown to increase up to 7.5-fold after application of the KMO inhibitor, accompanied by a substantial reduction of cell death. In a model of rat cerebral ischemia, both Ro61-8048 and another KMO inhibitor, mNBA (**26**), substantially reduced the level of hippocampal cell death [79]. The administration of 3-HK or QUIN prevents the neuroprotective effects of these inhibitors, suggesting that the neuroprotective mechanism may occur via 3-HK/QUIN [78]. Surprisingly, Ro61-8048 is claimed to be brain impermeable [1], which has been confirmed by our own studies (data not shown). The central effects achieved *in vivo* after oral administration can thus only be attributed to increases of kynurenine levels in the periphery and subsequent changes of KP metabolites in the brain. Taken together, these data suggest that modulating the KP pathway through KMO inhibition may help to modulate the outcome of neuronal cell damage after ischemic conditions; however, the mechanism underlying these pharmacological effects needs to be elucidated.

5.7 KMO Inhibition and Huntington's Disease

HD is a fatal progressive neurodegenerative disorder that is characterized by a triad of motor, cognitive, and psychiatric dysfunctions, typically starting in midlife and progressing relentlessly to death. It is caused by a mutation in the Huntingtin gene, where a stretch of normally up to 35 CAG repeats are elongated from 35 to >100 CAG. The corresponding Huntingtin protein (*Htt*) is expressed ubiquitously in the body, but the elongated poly-glutamine stretch leads mainly to neuronal cell loss and brain dysfunction. Medium spiny neurons in the striatum most severely suffer from degeneration in HD; however, other brain regions are also affected, leading to brain dysfunction and eventually to death [80, 81].

In terms of the tryptophan pathway and HD, it is known that patients with HD display an activated immune system and, in particular, decreased Trp levels [82]. Furthermore, the KP has been shown to be activated in HD animal models and human patients, thus pointing at a potential link between dysregulated KP and disease [83]. One KP metabolite, in particular, QUIN, has been linked to HD. Intra-striatal injection of QUIN in rodents has been shown to cause lesions in the striatum reminiscent of HD striatum [84]. This system hence became an animal model to study HD in animals, before genetic models became available. In a genetic mouse model of HD, R6/2, 3-HK levels are elevated in certain brain regions such as the cortex and striatum at 1–4 months of age, and KMO activity is increased [85]. In other animal models that show a more modest phenotype such as YAC128 or knock-in mice, QUIN and 3-HK were also elevated in striatum and cortex, although at a later stage [86]. It has also been found that both 3-HK and QUIN levels are significantly increased – by three- to fourfold – in low-grade human HD brain, but remain unchanged in higher grade cases [87], leading to the hypothesis that these metabolites may participate in the early phases of neurodegeneration.

Additional evidence directly pointing at KMO as a potential target for the treatment of HD comes from functional genomics studies in yeast (*Saccharomyces cerevisiae*), where a large-scale genetic screen was designed to identify gene deletions that suppress the toxicity of mutant Huntingtin (mHtt). The most efficient rescue was achieved by deletion of the yeast homologue of the human KMO protein. It was subsequently shown that further suppressor genes identified do not encode for KP enzymes but have indirect effects on KP metabolite levels [88].

Although final genetic validation of KMO as a target for HD is still outstanding, the evidence summarized above indicates KMO as a viable target. Inhibition of KMO activity is predicted not only to attenuate the flux through the QUIN branch of the pathway but also to shunt the KP metabolism toward enhanced KYNA levels and thus enhance neuroprotection.

5.7.1 Optimization of KMO Inhibitors for the Potential Treatment of HD

To identify KMO inhibitors with the potential to treat Huntington's disease, CHDI, a not-for-profit research organisation, has embarked on a discovery program with drug discovery company Evotec. The initial goal of the program is to identify potent, selective BBB-permeable KMO inhibitors for proof of concept experiments in HD *in vivo* studies; however, the ultimate goal of the collaboration is to progress a clinical candidate for the treatment of HD. As a first step in this process, a full evaluation of compounds (**30**) and (**34**) was undertaken to establish a benchmark for both these series of inhibitors. This included an evaluation of the potency in both biochemical and cellular assays as well as profiling the adsorption, distribution, metabolism, excretion, and toxicity (ADMET) properties.

To evaluate the inhibitors, Evotec has developed a robust *in vitro* KMO inhibition assay based on monitoring the enzyme reaction by means of LC/MS/MS (Fig. 2), which enables a direct quantification of the substrate and product and has been successfully applied to both biochemical (mouse, rat, and human KMO enzymes) and cellular (using both a stable CHO cell line over-expressing human, mouse, or rat KMO enzymes, and a rat microglia cellular system) assays.

The profiles of compounds (**30**) and (**34**) are shown in Table 1. As can be seen in the above table, the potency (biochemical and cellular) of keto acid **30** is superior to the sulfonamide **34**. However, due to difficulties with solubility and detection of the compound in LC/MS, the full ADMET profile could not be gathered for compound **30**. Compound **34** although more readily profiled in the ADMET assays was shown to have several liabilities, including microsome instability, inhibition of

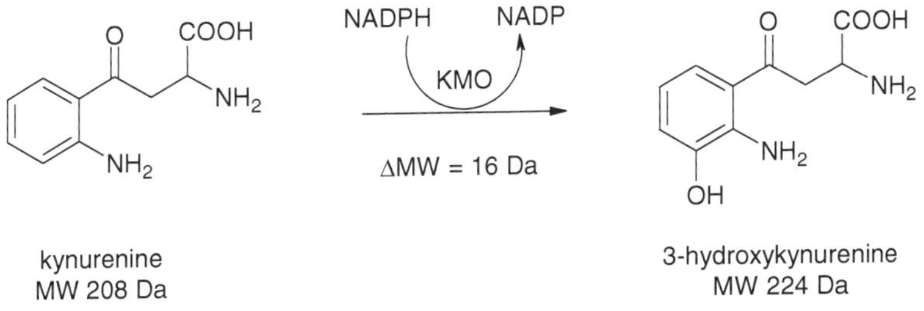

kynurenine
MW 208 Da

NADPH NADP
KMO
ΔMW = 16 Da

3-hydroxykynurenine
MW 224 Da

Fig. 2 KMO reaction monitored by LC/MS/MS

Table 1 Profiling data for compounds **30** and **34**

Assay	Compound **34**	Compound **30**
IC_{50} mouse enzyme (μM)	0.205	0.001
IC_{50} rat enzyme (μM)	ND	0.0022
IC_{50} human enzyme (μM)	0.1705	0.08
IC_{50} cellular CHO cells (human) (μM)	1.128	0.30
IC_{50} cellular CHO cells (rat) (μM)	ND	0.035
IC_{50} cellular microglia (rat) (μM)	0.349	0.22
Aqueous solubility (mg/ml)	0.1	<0.01
Half-life human liver microsomes (min)	10.7	NS
Half-life mouse liver microsomes (min)	8.8	NS
Plasma protein binding (% unbound)	1%	NS
Caco-2 (nM/s) [A–B]/[B–A]	421/164	NS
CYP450 enzyme inhibition (μM) 1A2/2C9/2C19/2D6/3A4	23/>50/1.8/>50/>50	>50/>50/>50/>50/>50
LogP	3.4 (pH 2)	2.62
Cytotoxicity (μM)	>50	>50
PK comment	No brain penetration	No brain penetration

NS no signal; ND not determined

cytochrome P450s, and high plasma protein binding. The *in vivo* pharmacokinetics of both the above compounds was evaluated in mice (intravenous and oral dosing); unfortunately, neither compound showed any appreciable BBB penetration. It is clear that to fully evaluate KMO as a potential target for diseases such as HD, there must be an appreciable level of penetration of the inhibitor through the BBB into the CNS.

Further evaluation of compound **30** was also carried out in an *in vivo* microdialysis study of kynurenine metabolites (study carried out by Brains OnLine[1]). The extracellular levels of anthranilic acid, KYNA, and 3-hydroxykynurenine were determined through a microdialysis probe inserted into the medial prefrontal cortex (mPFC) of mouse brains after oral dosing (10 mg/kg) of compound **30**.

As can be seen in Fig. 3, the changes in metabolite level are consistent with inhibition of KMO *in vivo*, even with the caveat that compound **30** does not penetrate the brain to an appreciable level.

To overcome the liabilities outlined in Table 1 and to establish a new series of potent inhibitors of KMO with BBB penetration and selectivity against kynureninase and KAT II, optimization of the keto acid inhibitors (e.g., **30**) was undertaken.

Initial analogs of **30** focused on replacement of the carbonyl (keto) group, one such analog was the oxime ether **35**. Unfortunately, this analog was less potent than the parent keto derivative **30** (Table 2); however, the compound did exhibit improved solubility and microsome stability. Interestingly the "parent" oxime analog where the cyclopropyl group has not been introduced (**36**) has a much improved potency while retaining the improved ADMET profile of **35** (Table 2). Encouraged by these results, we prepared a series of amide derivatives of **36** to remove the carboxylic acid functionality, one such example was compound **37**.

[1] http://www.brainsonline.org

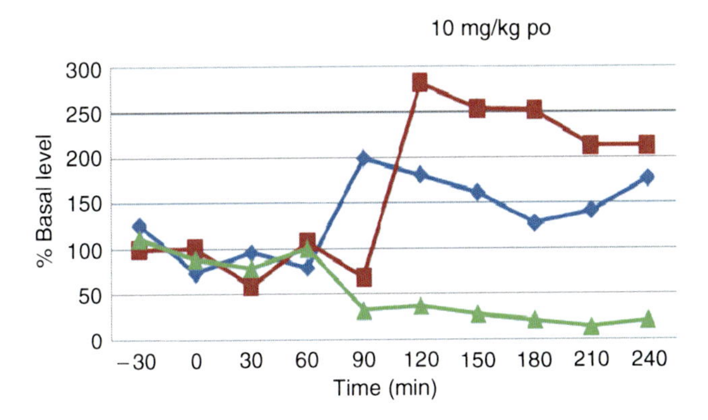

Fig. 3 Effects on extracellular levels of Anthranilic acid (*blue squares*), KYNA (*red squares*), and 3-HK (*green triangles*) in rat prefrontal cortex (PFC) ($n = 1$) after oral dosing (10 mg/kg) of compound **30**

Table 2 Profile of selected oxime ethers

Assay	Compound			
	30	**35**	**36**	**37**
IC_{50} mouse enzyme (μM)	0.001	4.5	0.014	0.3
IC_{50} cellular CHO cells (human) (μM)	0.3	5.1	0.45	2.0
Aqueous solubility (mg/ml)	<0.01	0.27	0.33	0.05
Half-life human liver microsomes (min)	NS	>60	>60	12
Half-life mouse liver microsomes (min)	NS	>60	>60	5
Cytotoxicity (μM)	>50	30	>50	>50

(35) **(36)** **(37)**

Although this particular amide was less potent in the KMO inhibition assays and a degree of microsomal instability was reintroduced, both **36** and **37** were submitted for pharmacokinetic evaluation in mice to investigate the potential improvement from removal of the carboxylic acid. It was found that both **36** and **37** have a good level of oral bioavailability (77 and 128%, respectively), and as expected **37** has a high clearance rate (59 l/h/kg) compared to **36** (0.3 l/h/kg). However, the major difference between these two compounds comes from the brain penetration. It was found that compound **36** had only a 5% penetration across the BBB, whereas **37** gave 49% brain penetration. To date, these KMO inhibitors have been shown to be selective for KMO over kynureninase and KAT II. Further optimization of this oxime scaffold to identify KMO inhibitors that meet the program goals is underway at CHDI and Evotec and will be reported in the near future.

To assist in the design of new inhibitors, CHDI and Evotec have also used a variety of both structure- and ligand-based computer-aided drug design (CADD) tools. For example, in terms of structure-based design, a number of homology models have been developed to investigate the possible binding modes of the inhibitors, to enable clear evaluation of SAR and to assist in the design of new inhibitor structures (through docking/scoring procedures). One such model is illustrated in Fig. 4. Here, the postulated key interactions of the keto acid template can be seen alongside the accommodation of the methyloxime derivative.

The above homology model (Fig. 4) has also been used to carry out a virtual screen of commercially available compounds. From catalog suppliers, 330,000 virtual compounds were selected using CNS drug-like property values [89] subsequently docked into the homology model and scored using a variety of scoring functions [90]. From these studies, a selection of 1,000 compounds was purchased

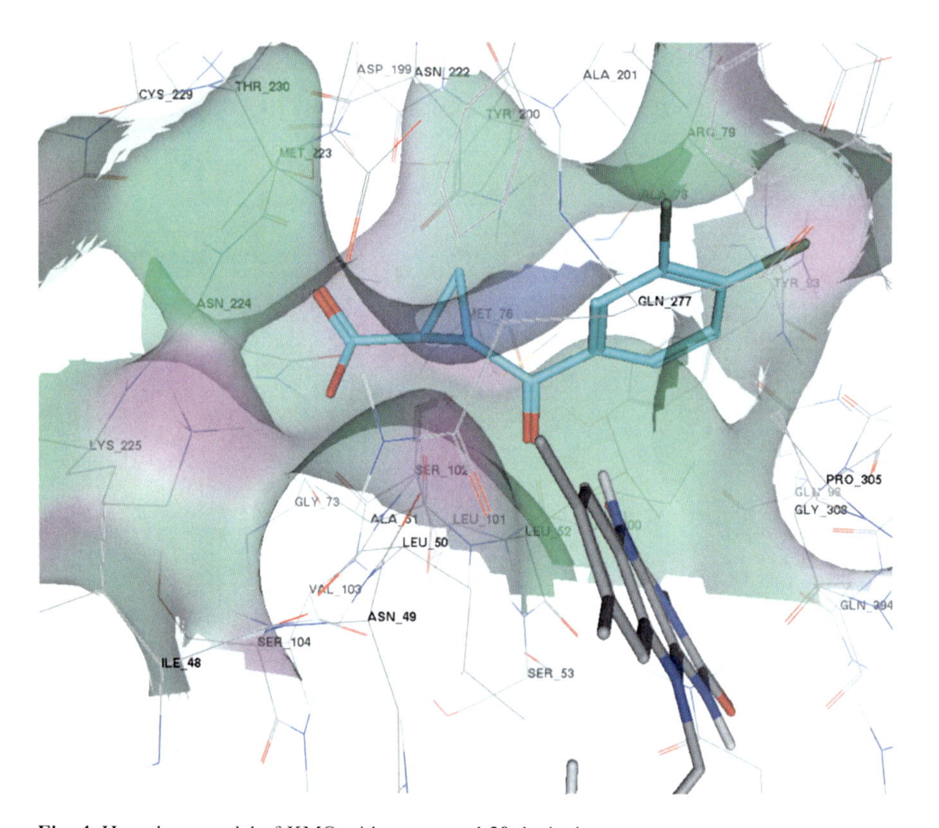

Fig. 4 Homology model of KMO with compound **30** docked

and screened for KMO inhibition. This successfully identified 27 hits with 17 having an $IC_{50} < 10\ \mu M$. These compounds are currently under optimization as part of our ongoing KMO program.

A variety of QSAR models have also been investigated. Comparative molecular field analysis (CoMFA) and comparative molecular similarity indices analysis (CoMSIA) [91, 92] models have been used for the substrate-based and keto acid inhibitors (represented by compounds **27** and **30**). Molecular overlays were initially developed for each class, and by comparison of activity data with these overlays, a model can be developed to take into account both steric and electrostatic fields around the inhibitors; this is illustrated in Fig. 5 [91].

Using a set of 50 training molecules, the model was tested with 21 test compounds. In this example, the model correctly predicted 83% of compounds with $IC_{50} < 1\ \mu M$, 81% of those molecules between 1 and 10 μM, and 75% of those with $IC_{50} > 10\ \mu M$. Figure 6 illustrates the correlation of this model (CoMFA correlation coefficient = 0.965, CoMSIA correlation coefficient = 0.704).

QSAR tools such as these are currently being used to rationalize SAR and to predict the potency of newly designed analogs currently undergoing evaluation before initiation of synthesis.

6 Perspective and Future Challenges

Manipulation of the tryptophan pathway continues to provide a range of potential therapeutic targets for disease intervention. Inhibition of the KP and the control of excitotoxic and neuroprotective metabolites are of particular interest in the treatment of HD. Although the discovery of a potent and selective inhibitor of the KP capable of efficacy *in vivo* in animals is challenging, the true hurdle for researchers is the translation of these lead compounds into viable therapies for the treatment and control of HD.

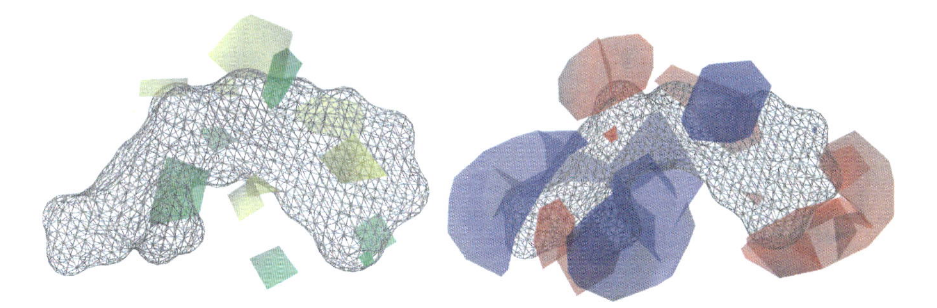

Fig. 5 CoMFA overlays for KMO inhibitors, *Green* and *yellow* – Steric fields. *Green areas* show where bulk is favorable; *yellow areas* indicate (*green*) where bulk is unfavorable. *Blue* and *red* – electrostatic fields. *Red areas* show are favorable for negative charge, i.e., oxygens, whereas *blue* are areas unfavorable for negative charge

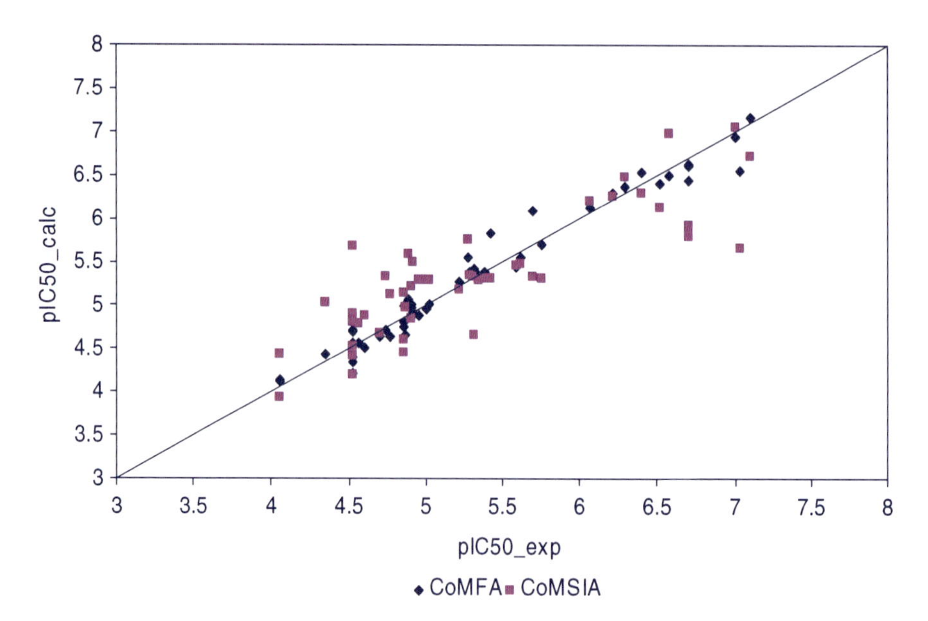

Fig. 6 Correlation plot for CoMFA and CoMSIA models with experimental enzyme inhibition data on human KMO enzyme

The need to discover an effective treatment for HD is paramount. Currently, the available therapies are limited to treatment of the most widely recognized symptoms of HD such as chorea, depression, irritability/aggression, sleep disturbance, and psychosis. Only one drug (tetrabenazine, a dopamine transport inhibitor) is currently available that has been licensed specifically for HD; this drug that has been licensed in Europe for the treatment of chorea for a number of years, however, was only recently approved in the USA due to its poor side effect profile. Through the targeted interception of key pharmacological mechanisms implicated in HD, the intention is that the onset and progression of HD may be delayed. To this end, the development of transgenic mouse models mimicking the disease has been extremely important milestones for HD studies. These models allow the role of new targets such as KMO to be evaluated in a relevant disease model to fully explore the significance of inhibition on the neurodegenerative effects of mHtt. Models such as the widely used R6/2 mice do, however, have some limitations in that they are expensive, time consuming, and of course have yet to be proven in terms of the predictability for progressive human HD.

One of the major challenges of inhibition of the KP is the presumed interplay between peripheral and central effects of KP metabolites. TRP, KYN, and 3-HK readily penetrate the BBB, whereas the acidic metabolites QUIN and KYNA cannot. Indeed, peripheral administration of TRP increases brain levels of KP metabolites. Considering the large capacity of the KP in the periphery, it is important but very challenging to determine by what extent changes in the levels of peripheral KP metabolites affect brain function and also dysfunction.

Inhibition of KMO specifically has challenges. For example, the understanding of the pharmacological mechanism of KMO inhibitors (QUIN-mediated excitotoxicy, kynurenic acid-mediated neuroprotection versus anti-inflammatory effects) although not critical for the optimization of a lead series requires further clarification before progression of compounds to clinical development. In addition, the differences in glia cell biology between rodents and humans and the potential cognitive side effects due to increasing KYNA levels during chronic treatment are key challenges for any ongoing research program.

Progression of a KMO inhibitor into an HD clinical trial will require evidence that the inhibition of the enzyme *in vitro* safely translates to both the anticipated modulation of tryptophan metabolites *in vivo* in the brain as well as positive phenotypical responses within relevant transgenic mouse models.

HD clinical trials present many challenges by themselves, and to date these trials have been varied in their design and quality of data produced. One of the most apparent challenges is the assessment of disease progression. Currently, the most widely recognized system is the Unified Huntington's Disease Rating Scale (UHDRS). This scale encompasses six subscales, namely motor, cognitive, behavioral, total functional capacity, total functional assessment, and independence score, and was developed by the Huntington's Study Group (HSG). Although beneficial in terms of unification of a single scale, the UHDRS must still suffer from the problem of interpretation of a disease that is complex both in terms of onset point and the vast array of symptoms that may be presented. Thus, the identification of validated and measurable biomarkers and clinical endpoints will be vital for such costly trials. Further work is needed in this area to ensure that future clinical trials are as effective as possible. One key factor for future success is the effective recruitment of patients into clinical trials. CHDI are coordinating a new initiative that will increase awareness and capture valuable information relating to disease onset and symptomatic presentation, including early-stage cognitive impairment. This information will be extremely valuable for future clinical trial design and management and ultimately in the search for an effective treatment for HD.

Acknowledgment The authors thank Andreas Ebneth, Celia Dominguez, and Ignacio Munoz-Sanjuan for their help.

References

1. Schwarcz R, Guidetti P, Sathyasaikumar KV, Muchowski PJ (2009) Of mice, rats and men: revisiting the quinolinic acid hypotheses of Huntington's disease. Prog Neurobiol. doi:10.1016/j.pneurobio.2009.04.005
2. Vamos E, Pardutz A, Klivenyi P, Toldi J, Vescei L (2009) The role of kynurenines in disorders of the central nervous system: possibilites for neuroprotection. J Neurol Sci 283:21–27
3. Costantino G (2009) New promises for manipulation of kynurenine pathway in cancer and neurological diseases. Expert Opin Ther Targets 13:247–258

4. Stone TW, Darlington LG (2002) Endogenous Kynurenines as targets for drug discovery and develop-ment. Nat Rev Drug Disc 1:609–620
5. Schwarcz R, Pellicciari R (2002) Manipulation of brain kynurenines: glial targets, neuronal effects, and clinical opportunities. J Pharm Exp Ther 303:1–10
6. Ruddick JP, Evans AK, Nutt DJ, Lightman SL, Rook GA, Lowry CA (2006) Tryptophan metabolism in the central nervous system: medical implications. Expert Rev Mol Med 8:1–27
7. Schlossberger HG, Kochen W, Linzen B, Steinhart H (1984) Progress in tryptophan and serotonin research. Walter de Gruyter, Berlin
8. Heyes MP, Saito K, Markey SP (1992) Human macrophages convert L-tryptophan into the neurotoxin quinolinic acid. Biochem J 283:633–635
9. Grohmann U, Fallarino F, Puccetti P (2003) Tolerance, DCs and tryoptophan: much ado about IDO. Trends Immunol 24:242–248
10. Owe-Young R, Webster NL, Mukhtar M, Pomerantz RJ, Smythe G, Walker D, Armati PJ, Crowe SM, Brew BJ (2008) Kynurenine pathway metabolism in human blood-brain-barrier cells: implications for immune tolerance & neurotoxicity. J Neurochem 105:1346–1357
11. Fukui S, Schwarcz R, Rapoport SI, Takada Y, Smith QR (1991) Blood-brain barrier transport of kynurenines: implications for brain synthesis and metabolism. J Neurochem 56:2007–2017
12. Guillemin GJ, Kerr SJ, Smythe GA, Smith DG, Kapoor V, Armati PJ, Croitoru J, Brew BJ (2001) Kynurenine pathway metabolism in human astrocytes: a paradox for neuronal protection. J Neurochem 78:1–13
13. Guillemin GJ, Smith DG, Smythe GA, Armati PJ, Brew BJ (2003) Expression of the kynurenine path-way enzymes in human microglia and macrophages. Adv Exp Med Biol 527:105–112
14. Alberati-Giani D, Ricciardi-Castagnoli P, Kohler C, Cesura AM (1996) Regulation of the kynurenine metabolic pathway by interferon-gamma in murine cloned macrophages and microglial cells. J Neurochem 66:996–1004
15. Frumento G, Rotondo R, Tonetti M, Damonte G, Benatti U, Ferrara GB (2002) Tryptophan-derived catabolites are responsible for inhibition of T and natural killer cell proliferation induced by indoleamine 2, 3-dioxygenase. J Exp Med 196:459–468
16. Savvateeva E, Popov A, Kamyshev N, Bragina J, Heisenberg M, Senitz D, Kornhuber J, Riederer P (2000) Age-dependent memory loss, synaptic pathology and altered brain plasticity in the Drosophila mutant cardinal accumulating 3-hydroxykynurenine. J Neural Transm 107:581–601
17. Foster AC, Collins JF, Schwarcz R (1983) On the excitotoxic properties of quinolinic acid, 2, 3-piperidine dicarboxylic acids and structurally related compounds. Neuropharmacology 22:1331–1342
18. Rios C, Santamaria A (1991) Quinolinic acid is a potent lipid peroxidant in rat brain homogenates. Neurochem Res 16:1139–1143
19. During MJ, Heyes MP, Freese A, Markey SP, Martin JB, Roth RH (1989) Quinolinic acid concentrations in strial extracellular fluid reach potentially neurotoxic levels following systemic L-tryptophan loading. Brain Res 476:384–387
20. Saito K, Nowak TS Jr, Markey SP, Heyes MP (1993) Mechanism of delayed increases in kynurenine pathway metabolism in damaged brain regions following transient cerebral ischemia. J Neurochem 60:180–192
21. Whetsell WO Jr, Schwarcz R (1989) Prolonged exposure to submicromolar concentrations of quinolinic acid causes excitotxic damage in organotypic cultures of rat corticostriatal system. Neurosci Lett 97:271–275
22. Nakai M, Qin ZH, Wang Y, Chase TN (1999) Free radical scavenger OPC-14117 attenuates quinolinic acid-induced NF-kappaB activation and apoptosis in rat striatum. Brain Res Mol Brain Res 22:59–68
23. Santamaría A, Salvatierra-Sánchez R, Vázquez-Román B, Santiago-López D, Villeda-Hernández J, Galván-Arzate S, Jiménez-Capdeville ME, Ali SF (2003) Protective effects of

the antioxidant selenium on quinolinic acid-induced neurotoxicity in rats: *in vitro* and *in vivo* studies. J Neurochem 86:479–488
24. Okuda S, Nishiyama N, Saito H, Katsuki K (1998) 3-Hydroxykynurenine, an endogenous oxidative stress generator, causes neuronal death with apoptotic features and region selectivity. J Neurosci 21:7463–7473
25. Guidetti P, Schwarcz R (1999) 3-Hydroxykynurenine potentiates quinolinate but not NMDA toxicity in the rat striatum. Eur J Neurosc 11:3857–3863
26. Obrenovitch TP, Urenjak J (2003) Accumulation of quinolinic acid with neuroinflammation: does it mean excitotoxicity? Exp Med Biol 527:147–154
27. Kessler M, Terramani T, Lynch G, Baudry M (1989) A glycine site associated with N-methyl-D-aspartic acid receptors: characterization and identification of a new class of antagonists. J Neurochem 52:1319–1328
28. Hilmas C, Pereira EF, Alkondon M, Rassoulpor A, Schwarcz R, Albuquerque EX (2001) The brain me-tabolite kynurenic acid inhibits alpha7 nicotinic receptor activity and increases non-alpha7 nicotinic re-ceptor expression: pathophysiological implications. J Neurosci 21:7463–7473
29. Rassoulpour A, Wu H-Q, Ferre S, Schwarcz R (2005) Nanomolar concentrations of kynurenic acid re-duce extracellular dopamine levels in the striatum. J Neurochem 93:762–765
30. Wang J, Simonavicius N, Wu X, Swaminath G, Reagan J, Tian H, Ling L (2006) Kynurenic acid as a ligand for orphan G protein-coupled receptor GPR35. J Biol Chem 281:22021–22028
31. Moroni F, Russi P, Lombardi G, Beni M, Carla V (1988) Presence of kynurenic acid in the mammalian brain. J Neurochem 51:177–180
32. Carpenedo R, Pittaluga A, Cozzi A, Attucci S, Galli A, Raiteri M, Moroni F (2001) Presynaptic kynurenate-sensitive receptors inhibit glutamate release. Eur J Neurosci 13: 2141–2147
33. Wu H-Q, Pereira EFR, Bruno JP, Pellicciari R, Albuquerque EX, Schwarcz R (2010) The astrocyte-derived α7 nicotinic receptor antagonist kynurenic acid controls extracellular glutamate levels in the prefrontal cortex. J Mol Neurosci 40:204–210
34. Kumar S, Malachowski WP, DuHadaway JB, LaLonde JM, Carroll PJ, Jaller D, Metz R, Prendergast GC, Muller AJ (2008) Indoleamine 2, 3-dioxygenase is the anticancer target for a series of potent naph-thoquinone-based inhibitors. J Med Chem 51:1706–1718
35. Platten M, Ho PG, Sreinmann L (2009) Anti-inflammatory strategies for the treatment of multiple sclerosis – tryptophan catabolites may hold the key. Drug Disc Today 3:401–408
36. Schwarcz R (2004) The kynurenine pathway of tryptophan degredation as a drug target. Curr Opin Pharmacol 4:12–17
37. Shimizu T, Nomiyama S, Hirata F, Hayaishi O (1978) Indoleamine 2, 3-dioxygenase: purification and some properties. J Biol Chem 253:4700–4706
38. Muller AJ, Scherle PA (2006) Targeting the mechanisms of tumoral immune tolerance with small-molecule inhibitors. Nat Rev Cancer 6:613–625
39. Eguchi N, Watanabe Y, Kawanishi K, Hashimoto Y, Hayaishi O (1984) Inhibition of indoleamine 2, 3-dioxygenase and tryptophan 2, 3-dioxygenase by beta-carbolines and indole derivatives. Arch Biochem Biophys 232:602–609
40. Peterson AC, La Loggia AJ, Hamaker LK, Arend RA, Fisette PL, Ozaki Y, Will JA, Brown RR, Cook JM (1993) Evaluation of substituted b-carbolines as noncompetitive indoleamine 2, 3-dioxygenase in-hibitors. Med Chem Res 3:473–482
41. Cady SG, Sono M (1991) 1-Methyl-DL-tryptophan, beta-(3-benzofuranyl)-DL-alanine (the oxygen ana-log of tryptophan), and beta-[3-benzo. (b)thienyl]-DL-alanine (the sulfur analog of tryptophan) are com-petitive inhibitors for indoleamine 2, 3-dioxygenase. Arch Biochem Biophys 291:326–333
42. Peterson AC, Migawa MT, Martin MJ, Hamaker LK, Czerwinski KM, Zhang W, Arend RA, Fisette PL, Ozaki Y, Will JA, Brown RR, Cook JM (1994) Evaluation of functionalized tryptophan derivatives and related compounds as competitive inhibitors of indoleamine 2, 3-dioxygenase. Med Chem Res 3:531–544

43. Gaspari P, Banerjee T, Malachowski WP, Muller AJ, Prendergast GC, DuHadaway J, Bennett S, Dono-van AM (2006) Structure-activity study of Brassin derivatives as indoleamine 2, 3-dioxygenase inhibi-tors. J Med Chem 49:684–692
44. Carr G, Chung MKW, Mauk G, Anderson RJ (2008) Synthesis of indoleamine 2, 3-dioxgenase inhibitory analogues of the sponge alkaloid Exiguamine A. J Med Chem 51: 2634–2637
45. Sono M, Cady SG (1989) Enzyme kinetic and spectroscopic studies of inhibitor and effector interactions with indoleamine 2, 3-dioxygenase. 1. Norharman and 4-phenylimidazole binding to the enzyme as inhibitors and heme ligands. Biochemistry 28:5392–5399
46. Kumar S, Jaller D, Patel B, LaLonde JM, DuHadaway JB, Malachowski WP, Prendergast GC, Muller AJ (2008) Structure based development of phenylimidazole-derived inhibitors of indoleamine 2, 3-dioxygenase. J Med Chem 51:4968–4977
47. Yue EW, Douty B, Wayland B, Bower M, Liu X, Leffet L, Wang Q, Bowman KJ, Hansbury MJ, Liu C, Wei M, Li Y, Wynn R, Burn TC, Koblish HK, Fridman JS, Metcalf B, Scherle PA, Combs AP (2009) Discovery of potent competitive inhibitors of indoleamine 2,3-dioxygenase with *in vivo* pharmacody-namic activity and efficacy in a mouse melanoma model. J Med Chem 52:7364–7367. doi:10.1021/jm900518f
48. Guidetti P, Amori L, Sapko MT, Okuno E, Schwarcz R (2007) Mitochondrial aspartate aminotrans-ferase: a third kynurenate-producing enzyme in the mammalian brain. J Neurochem 102:103–111
49. Yu P, Li Z, Zhang L, Tagle DA, Cai T (2006) Characterization of kynurenine aminotransferase III, a novel member of a phylogenetically conserved KAT family. Gene 365:111–118
50. Yu P, DiProspero NA, Sapko MT, Cai T, Chen A, Melendez-Ferro M, Du F, Whetsell WO, Guidetti P, Schwarcz R, Tagle DA (2004) Biochemical and phenotypic abnormalities in kynurenine aminotrans-ferase II-deficient mice. Mol Cell Biol 24:6919–6930
51. Pellicciari R, Rizzo RC, Costantino G, Marinozzi M, Amori L, Guidetti P, Wu HQ, Schwarcz R (2006) Modulators of the kynurenine pathway of tryptophan metabolism: synthesis and preliminary biological evaluation of (S)-4-(ethylsulfonyl)benzoylalanine, a potent and selective kynurenine aminotransferase II (KAT II) inhibitor. Chem Med Chem 1:528–531
52. Alkondon M, Pereira EFR, Yu P, Arruda EZ, Almeida LEF, Guidetti P, Fawcett WP, Sapko MT, Randall WR, Schwarcz R, Tagle DA, Albuquerque EX (2004) Targeted deletion of the kynurenine aminotrans-ferase II gene reveals a critical role of endogenous kynurenic acid in the regulation of synaptic transmis-sion via a7 nicotinic receptors in the hippocampus. J Neurosci 24:4635–4648
53. Varasi M, Della Torre A, Heidempergher F, Pevarello P, Speciale C, Guidetti P, Wells DR, Schwarcz R (1996) Derivatives of kynurenine as inhibitors of rat brain kynurenine aminotransferase. Eur J Med Chem 31:11–21
54. Pellicciari R, Venturoni F, Bellocchi D, Carotti A, Marinozzi M, Macchiarulo A, Amori L, Schwarcz R (2008) Sequence variants in kynurenine aminotransferase II (KAT II) orthologs determine different po-tencies of the inhibitor S-ESBA. Chem Med Chem 3:1199–1202
55. Soda K, Tanizawa K (1979) The mechanism of kynurenine hydrolysis catalyzed by kynureninase. Biochem J 86:1199–1209
56. Phillips RS, Dua RK (1991) Stereochemistry and mechanism of Aldol reactions catalyzed by kynureni-nase. J Am Chem Soc 113:7385–7388
57. Drysdale MJ, Reinhard JF (1998) S-aryl cysteine S, S-dioxides as inhibitors of mammalian kynureninase. Bioorg Med Chem Lett 8:133–138
58. Chiarugi A, Carpenedo R, Molina MT, Mattoli L, Pellicciari R, Moroni F (1995) Comparison of the neurochemical and behavioural effects resulting from the inhibition of kynurenine hydroxylase and/or kyureninase. J Neurochem 65:1176–1183
59. Fitzgerald DH, Muirhead KM, Botting NP (2001) A comparative study on the inhibition of human and bacterial kynureninase by novel bicyclic kynurenine analogues. Bioorg Med Chem 9:983–989

60. Walsh HA, Leslie PL, O'Shea KC, Botting NP (2002) 2-Amino-4[3'-ydroxyphenyl]-4-hydroxybutanoic acid; a potent inhibitor of rat and recombinant human kynureninase. Bioorg Med Chem Lett 12:361–363
61. Lima S, Kumar S, Gawandi V, Momany C, Phillips RS (2009) Crystal structure of the Homo sapiens kynureninase-3-hydroxhippuric acid inhibitor complex: insights into the molecular basis of kynureninase substrate specificity. J Med Chem 52:389–396
62. Bokman AH, Schweigert BS (1951) 3-Hydroxyanthranilic acid metabolism. IV. Spectrophotometric evi-dence for the formation of an intermediate. Arch Biochem Biophys 33:270–276
63. Manthey MK, Pyne SG, Truscott RJW (1988) The autoxidation of 3-hydroxyanthranilic acid. J Org Chem 53:1486–1488
64. Saito K, Markey SP, Heyes MP (1994) 6-Chloro-D, L-tryptophan, 4-chloro-3-hydroxyyanthranilate and dexamethasone attenuate quinolinic acid accumulation in brain and bllod following systemic immune activation. Neurosci Lett 178:211–215
65. Linderberg M, Hellberg S, Björk S, Gotthammer B, Högberg T, Persson K, Schwarcz R, Luthman J, Jo-hansson R (1999) Synthesis and QSAR of substituted 3-hydroxyanthranilic acid derivatives as inhibitors of 3-hydroxyanthranilic acid dioxygenase (3-HAO). Eur J Med Chem 34:729–744
66. Okamoto H, Yamamoto S, Nozaki M, Hayashi O (1967) On the submitochondrial localization of L-Kynurenine-3-hydroxylase. Biochem Biophys Res Commun 26:309–314
67. Amori L, Guidetti P, Pelliciari R, Kajii Y, Schwarcz R (2009) On the relationship between the two branches of the kynurenine pathway in the rat brain *in vivo*. J Neurochem 109:316–325
68. Moroni F, Russi P, Gallo-Mezo MA, Moneti G, Pellicciari R (1991) Modulation of quinolinic acid and kynurenic acid content in the rat brain: effects of endotoxins and nicotinylalanine. J Neurochem 57:1630–1635
69. Pellicciari R, Natalini B, Costantino G, Mahmoud MR, Mattoli L, Sadeghpour BM (1994) Modulation of the kynurenine pathway in search for new neuroprotective agents: synthesis and preliminary evaluation of (m-nitrobenzoyl)alanine, a potent inhibitor of kynurenine-3-hydroxylase. J Med Chem 37:647–655
70. Moroni F, Carpenedo R, Chiarugi A (1996) Kynurenine hydroxylase and kynureninase inhibitors as tools to study the role of kynurenine metabolites in the central nervous system. Adv Exp Med Biol 398:203–210
71. Costantino G, Mattoli L, Moroni F, Natalini B, Pellicciari R (1996) Kynurenine-3-hydroxylase asnd its selective inhibitors: molecular modelling studies. Adv Exp Med Biol 398: 493–497
72. Giordani A, Corti L, Cini M, Marconi M, Pillan A, Ferrario R, Schwarcz R, Guidetti P, Speciale C, Varasi M (1996) Benzoylalanine analogues as inhibitors of rat brain kynureninase and kynurenine 3-hydroxylase. Adv Exp Med Biol 398:499–505
73. Giordani A, Pevarello P, Cini M, Bormetti R, Greco F, Toma S, Speciale C, Varasi M (1998) 4-Phenyl-4-oxo-butanoic acid derviatives inhibitors of kynurenine 3-hydroxylase. Bioorg Med Chem Lett 8:2907–2912
74. Drysdale M, Hind SL, Jansen M, Renhard JF (2000) Synthesis and SAR of 4-aryl-2-hydroxy-4-oxobut-2enoic acids and esters and 2-amino-4-aryl-4-oxobut-2-enoic acids and esters: potent inhibitors of kynurenine-3-hydroxylase as potential neuroprotective agents. J Med Chem 43:123–127
75. Pellicciari R, Amori L, Costantino G, Giordani A, Macchiarulo A, Mattoli L, Pevarello P, Speciale C (2003) Modulation of the kynurine pathway of tryptophan metabolism in search for neuroprotective agents. Focus on kynurenine-3-hydroxylase. Adv Exp Med Biol 527:621–628
76. Heidempergher F, Pevarello P, Pillan A, Pinciroli V, Della Torre A, Speciale C, Marconi M, Cini M, Toma S, Greco F, Varasi M (1999) Pyrrolo[3, 2-c]quinoline derivatives: a new class of kynurenine-3-hydroxylase inhibitors. Il Farmaco 54:152–160
77. Rover S, Cesura AM, Huguenin P, Kettler R, Szente A (1997) Synthesis and biological evaluation of N-(4-phenyl-2-yl)benzenesulfonamides as high-affinity inhibitors of kynurenine 3-hydroxylase. J Med Chem 40:4378–4385

78. Carpenedo R, Meli E, Peruginelli F, Pellegrini-Giampietro DE, Moroni F (2002) Kynurenine 3-mono-oxygenase inhibitors attenuate post-ischemic neuronal death in organotypic hippocampal slice cultures. J Neurochem 82:1465–1471
79. Cozzi A, Carpenedo R, Moroni F (1999) Kynurenine hydroxylase inhibitors reduce ischemic brain damage: studies with (m-nitrobenzoyl)-alanine (mNBA) and 3, 4-dimethoxy-[-N-4-(nitrophenyl)thiazol-2yl]-benzenesulfonamide (Ro 61–8048) in models of focal or global brain ischemia. J Cereb Blood Flow Metab 19:771–777
80. Roze E, Saudou F, Caboche J (2008) Pathophysiology of Huntington's disease: from huntingtin functions to potential treatments. Curr Opin Neurol 21:497–503
81. Cowan CM, Raymond LA (2006) Selective neuronal degeneration in Huntington's disease. Curr Top Dev Biol 75:25–71
82. Leblhuber F, Walli J, Jellinger K, Tilz GP, Widner B, Laccone F, Fuchs D (1998) Activated immune system in patients with Huntington's disease. Clin Chem Lab Med 36:747–750
83. Stoy N, Mackay GM, Forrest CM, Stone CCJ, TW DLG (2005) Tryptophan metabolism and oxidative stress in patients with Huntington's disease. J Neurochem 93:611–623
84. Beal MF, Ferrante RJ, Swartz KJ, Kowall NW (1990) Chronic quinolinic acid lesions in rates closely resemble Huntingdon's disease. J Neurosci 11:1649–1659
85. Guidetti P, Bates GP, Graham RK, Hayden MR, Leavitt BR, MacDonald ME, Slow EJ, Wheeler VC, Woodman B, Schwarcz R (2006) Neurobiol Dis 23:190–197
86. Slow EJ, van Raamsdonk J, Rogers D, Coleman SH, Graham RK, Deng Y, Oh R, Bissada N, Hossain SM, Yang YZ, Li XJ, Simpson EM, Gutekunst CA, Leavitt BR, Hayden MR (2003) Selective striatal neuronal loss in a YAC128 mouse model of Huntrington disease. Hum Mol Get 12:1555–1567
87. Guidetti P, Luthi-Carter RE, Augood SJ, Schwarcz R (2004) Neostriatal and cortical quinolinate levels are increased in early grade Huntington's disease. Neurobiol Dis 17:455–461
88. Giorgini F, Guidetti P, Nguyen O, Bennet SC, Muchowski PJ (2005) A genomic screen in yeast implicates kynurenine 3-monooxygenase as a therapeutic target for Huntington disease. Nat Gen 37:526–531
89. Norinder U, Haeberlein M (2002) Computational approaches to the prediction of the blood-brain distribution. Adv Drug Del Rev 54:291–313
90. Cheng T, Li X, Li Y, Liu Z, Wang RJ (2009) Comparative assessment of scoring functions on a diverse test set. Chem Inf Model 49:1079–1093
91. Cramer RD III, Patterson DE, Bunce JD (1988) Comparative molecular field analysis (CoMFA) 1. Effect of shape on binding of steroids to carrier proteins. J Am Chem Soc 110:595959–595967
92. Klebe G (1998) Comparative molecular similarity indices, CoMSIA. In: Kubinyi H, Folkers G, Martin YC (eds) 3D QSAR in drug design. Kluwer Academic Publishers, Great Britain

Top Med Chem 6: 177–192
DOI: 10.1007/7355_2010_7

Published online: 22 September 2010

Spinal Muscular Atrophy: Current Therapeutic Strategies

Alex S. Kiselyov and Mark E. Gurney

Abstract Proximal spinal muscular atrophy (SMA) is an autosomal recessive disorder characterized by death of motor neurons in the spinal cord. SMA is caused by deletion and/or mutation of the survival motor neuron gene (*SMN1*) on chromosome *5q13*. There are variable numbers of copies of a second, related gene named *SMN2* located in the proximity to *SMN1*. Both genes encode the same protein (Smn). Loss of *SMN1* and incorrect splicing of *SMN2* affect cellular levels of Smn triggering death of motor neurons. The severity of SMA is directly related to the normal number of copies of *SMN2* carried by the patient. A considerable effort has been dedicated to identifying modalities including both biological and small molecule agents that increase SMN2 promoter activity to upregulate gene transcription and produce increased quantities of full-length Smn protein. This review summarizes recent progress in the area and suggests potential target product profile for an SMA therapeutic.

Keywords Spinal muscular atrophy, SMN2, mRNA stability, DcpS inhibitor, Structure-based design

Contents

A.S. Kiselyov (✉) and M.E. Gurney
deCODE Chemistry & Biostructures, 2501 Davey Road, Woodridge, IL 60517, USA
e-mail: akiselyov@decode.com

1 Introduction

Spinal muscular atrophy (SMA) is a group of juvenile autosomal recessive disorders. The general feature shared by all forms of SMA is progressive muscle weakness. It results from degeneration and eventual loss of the anterior horn cells in the spinal cord and the brain stem nuclei. Low muscle mass, inadequate weight gain, respiratory infections including pneumonia, scoliosis, and joint contractures are common complications. SMA is the second most common inherited disease in humans (after cystic fibrosis) affecting approximately 1 in every 6,000 newborns [1]. The carrier frequency for SMA is about 1 in 40 individuals. Internationally, the incidence of SMA is 7.8–10 cases per 100,000 live births [2].

2 SMA Genetics

The genetic cause underlying SMA is mutation of the survival motor neuron 1 (*SMN1*) gene, encoding the protein survival motor neuron (Smn). SMA is caused by mutation or homozygous deletion of the telomeric copy of the *SMN1* gene on chromosome 5q13 [3]. The majority (ca. 95%) of SMA patients carry homozygous deletions [4–9]. The region on chromosome 5 surrounding *SMN1* has an inverted duplication that includes variable numbers of copies of a second, related gene named *SMN2*. The clinical course of SMA varies from mild to severe depending upon the number of copies of *SMN2* carried by the patient [3]. The most common *SMN1* mutation is deletion, but other mutations, including gene conversion of *SMN1* to *SMN2*, may occur. Smn functions in the synthesis and trafficking of small nuclear ribonucleoproteins or snRNPs required for exon splicing to create mRNA. The protein heterodimerizes with multiple cellular targets including SIP1, GEMIN4, and others involved in the production of snRNPs, as exemplified by hnRNP U and the small nucleolar RNA-binding proteins.

SMN2 differs from *SMN1* by eight nucleotides, one of which results in skipping of exon 7 in *SMN2* mRNA processing. The *SMN2* gene is transcribed, but a mutation in a splice enhancer causes missplicing such that most transcripts lack exon 7 and produce a truncated protein product that is rapidly destroyed [10]. Because the *SMN2* gene does produce a small percentage of correctly spliced transcripts, a small amount of Smn protein is produced. Copy number variation in *SMN2* modifies disease severity [11]. A majority of babies with b SMA are severely affected with survival only 1–2 years after birth. Such patients typically have one to two copies of the *SMN2* gene as some patients may have chromosome 5 deletions covering both *SMN1* and *SMN2*. SMA patients with milder disease may carry three to six copies of the *SMN2* gene [12].

3 SMA Diagnosis and Categorization

Diagnosis of SMA includes a blood test which looks for the presence or absence of the *SMN1* gene. The DNA diagnostic test is combined with both physical examination and assessment of family history [13]. A carrier is identified from the number of exon 7-containing *SMN1* gene copies present. Due to variability within this region of human chromosome 5q13, some carriers may appear normal by DNA test as they may have two copies of *SMN1* on one copy of chromosome 5 and deletion of *SMN1* on the other [14]. Although the number of *SMN2* copies is related to the severity of the disease, it does not reliably predict the outcome. Therefore, in addition to the genetic testing, clinical evaluation includes assessing extent of weakness and motor abilities. These include electromyography (EMG) and nerve conduction velocity (NCV).

Although genetic evidence suggests that mutation of *SMN1* combined with variable *SMN2* copy number places patients on a continuum, pediatricians and neurologists historically categorized the disease into types based on clinical evidence. This classification is currently in use for proper treatment regimen and prognosis. There are four different types of SMAs (Type I–IV) recognized by the medical community. An additional Type 0 disease has been proposed to categorize SMA diagnosed within 30–36 weeks of gestation. Newborns with Type I SMA (SMA1, acute infantile SMA, Werdnig–Hoffman disease) feature facial weakness, low muscle tone (flaccidity), have serious issues with breathing, rolling over, and sitting without support, and eventually succumb to the disease within the first 2 years [15]. Type II SMA (SMA2, chronic infantile SMA, Dubowitz disease) is normally diagnosed at 6–12 months. Although the affected children do not require breathing aid and can sit, they fail to stand or walk unaided. They do survive into adulthood, but with significant motor disability and are vulnerable to respiratory infections [16]. Patients affected by Type III SMA (SMA3, Kugelberg–Welander disease, chronic juvenile SMA) are diagnosed in childhood (>age 1) or adolescence but remain mobile into adulthood. For these patients, it is difficult to rise from a sitting position. Usual complaints include trembling fingers and weakness of proximal limbs. SMA2 and SMA3 are detected in 1 of 24,000 births [16]. Type IV SMA usually occurs after age 30–35. This is the mildest form of the disease manifesting itself in moderate muscle weakness, tremors, and twitching [17].

4 Current Treatments

4.1 *Preclinical Evaluation*

4.1.1 *In Vitro* and *Ex Vivo* Assays

In this chapter, we will summarize several ex vivo and *in vivo* assay systems used to prioritize small molecules for preclinical development. Much work has focused on identifying compounds that increase *SMN2* promoter activity. Given that *SMN2* gene copy number modifies disease severity, a small molecule activator could mimic an increase in copy number. Histone deactylase inhibitors, for example, which elevate gene expression by modifying chromatin structure, have been shown to increase *SMN2* gene expression in cellular assays. Moreover, multiple groups have used cellular systems to screen for additional compounds with potential therapeutic activity. For example, Jarecki et al. constructed a cell-based *SMN2* gene reporter assay by transforming the NSC34 mouse neuroblastoma × motor neuron hybrid cell line with a fragment of the human *SMN2* gene promoter that is functionally linked to a bacterial β-lactamase gene. This assay is both robust and amenable to high-throughput screening (HTS) [18]. In a subsequent test, primary hits are confirmed by their ability to increase *SMN* mRNA, as measured by real-time PCR, in fibroblasts collected from SMA patients. A 1.3- to 2-fold increase in *SMN* mRNA at sub-micromolar to low micromolar concentration is generally observed for active compounds regardless of their molecular mechanism of action. A patient-derived skin cell assay is further used to assess the effect of an agent on Smn protein level and the number of nuclear organelles termed Gems or Cajal bodies (intranuclear concentrations of Smn found in most cells). The latter is a functional read-out of Smn protein levels [19].

4.1.2 *In Vivo* Models

Efforts at generating mouse models of SMA have been encouraging but not entirely successful. The mouse *SMN* gene is present as a single copy and does not undergo alternative splicing. Therefore, there is no counterpart in mice to the *SMN2* gene present in humans. $SMN^{-/-}$ mice are pre-implantation lethal and underscore the importance of the Smn protein for cellular and organismal survival. Since a milder phenotype allowing survival past birth is a desirable feature of an SMA model, several groups have introduced human *SMN2* BAC transgenes or mutant forms of *SMN* into mice. The combination of an *SMN2* BAC transgene on the mouse $SMN^{-/-}$ background results in neonatal lethality at low *SMN2* copy number or complete rescue at high *SMN2* copy number. Burghes and coworkers additionally introduced a transgene producing a truncated *SMN1* transgene lacking exon 7 (often referred to as the SMNΔ7 model) [20]. Lifespan of the SMNΔ7 mice is generally about 15 days. The role of the truncated Δ7 Smn protein in extending lifespan at least until 15 days is not well understood. More recently, DiDonato et al. has described a model in which the single nucleotide polymorphism causing missplicing of *SMN2* mRNA lacking exon 7 was knocked into the mouse *SMN* locus (referred to as the 2B/- model) [21]. These mice live until about 25 days of age. To further our understanding of *SMN* in terms of structure/function relationships and disease pathogenesis, it would be ideal if a panel of animals with intermediate/mild phenotypes of SMA existed.

The Jackson Laboratory received support from the SMA Foundation to make available the first group of mouse models for SMA. Each of the models includes a targeted mutation of the endogenous mouse *SMN* gene, combined with transgenes involving various forms of human *SMN2* and/or *SMN1*. Mice that are both homozygous for the targeted mutant Smn1 and carry the *SMN2* transgene exhibit symptoms and neuropathology similar to humans with type I proximal SMA. These mutants are either stillborn or survive for only 4–6 days. Homozygotes bearing the Smn-targeted mutation without a copy of the *SMN2* transgene die embryonically [22]. Mice homozygous for the Smn-targeted mutation and hemizygous for the *SMN2* transgene are viable, fertile, and have short thickened tails. There is a strong correlation between estimated copy number of the transgene and severity of the phenotype. These mice exhibit a molecular and progressive neurodegenerative phenotype similar to Type III SMA [23].

4.2 Supportive Treatments

SMA patients are regularly assessed for nutritional state, respiratory function, and orthopedic status. Currently available treatment is aimed at improvement of the patients' quality of life and addressing disability(ies). Examples of possible treatments, depending on type and severity of condition, include dietary assessment (i.e., recommendations for dealing with swallowing issues) and/or nutritional support via tube feeding; physical therapy to improve or maintain mobility and flexibility; wheelchair assistance for independent mobility; orthoses to prevent/minimize spinal curvature and/or to support walking; spinal fusion surgery; and respiratory therapies (e.g., supplementary oxygen, mechanical ventilation, and chest physiotherapy) [24].

4.3 Investigational Therapies

SMA is considered one of the better validated diseases for therapeutic intervention for a number of reasons. While SMA is manifested in a broad clinical spectrum, a single gene is responsible for all clinical forms of the disease (e.g., severe, intermediate, and mild). Loss of *SMN1* and *SMN2* is lethal, therefore essentially all SMA patients typically retain one or more copies of *SMN2*. *SMN2* encodes a fully functional Smn protein. Molecules that stimulate full-length Smn protein expression from the *SMN2* gene are of interest to a broad range of SMA patient populations. It has been suggested that SMA may be the first inherited disorder in which the activation/splicing correction of a copy of the gene may cure or ameliorate the disease. The level of Smn protein could be increased by either (1) enhancing *SMN2* gene transcription or (2) suppressing defective splicing of the *SMN2* mRNA, thereby increasing the number of full-length transcripts.

4.3.1 Biological Strategies for the Treatment of SMA

Antisense nucleotides hybridizing to an exon 7 reportedly promoted its inclusion and increased full-length Smn protein levels. Notably, the treatment did not interfere with either mRNA export or translation [25]. A development of bifunctional RNAs that modulate *SMN2* pre-mRNA splicing and could be delivered via a gene therapy vector has been reported [26]. These agents feature two distinct domains: *SMN*-complimentary RNA sequence and RNA segment recognized by various cellular splicing factors, such as SR and SR-like proteins. Further developments of this approach have been published recently [27].

A trans-splicing strategy is based on combining mutant and therapeutic RNAs to restore parent RNA sequence. A recent report described a specific system that reduced the competition between the splice sites and hence enhanced the efficiency of the process. Trans-splicing RNAs were shown to redirect splicing from the *SMN2* mini-gene as well as from endogenous transcripts. In the next step, trans-splicing RNAs were successfully delivered to SMA patient fibroblasts via recombinant adeno-associated viral vectors to yield increased levels of full-length SMN mRNA and total Smn protein levels. Notably, this treatment also restored snRNP assembly, a critical function of Smn. Authors concluded that the alternatively spliced *SMN2* exon 7 was a viable target for replacement by trans-splicing [28].

A series of vectors have been designed that express modified U7 snRNAs containing antisense sequences complementary to the 3V splice site of *SMN*. Over 20 anti-SMN U7 snRNAs were tested for their ability to promote inclusion of exon 7 in the *SMN2*. Transient expression of anti-SMN U7 snRNAs in HeLa cells enhanced *SMN2* splicing by ca. 70% yielding anticipated exon 7 inclusion in a sequence-specific and dose-dependent manner. The administration of anti-SMN U7 snRNPs also resulted in the increased concentrations of Smn protein [29]. Two novel recombinant splicing factors, namely hnRNP-G and its paralogue RBM, promoted inclusion of *SMN2* exon 7 via the specific protein–protein interaction involving hnRNP-G/RBM and Htra2-B1 [30].

Several symptomatic therapies aimed at muscle mass maintenance have been tested. For example, administration of follistatin to SMA mouse models resulted in increased muscle mass, gross motor function improvement, and a 30% increase in average lifespan. It has been suggested that follistatin targets the pathways that affect muscle maintenance and growth, specifically myostatin, a protein that limits muscle tissue growth. Skeletal muscle is a viable therapeutic target that may reduce the severity of some SMA symptoms. It is conceivable that the most effective treatment would combine strategies that directly address the genetic defect in SMA and SMN-independent strategies that enhance skeletal muscles. Similarly, thyrotropin-releasing hormone (TRH) has been recently shown to enhance peronal nerve conductancy in Type II–III SMA patients. The agent was delivered via percutaneous intravenous catheters at a dose of 0.1 mg/kg (in 50 ml of normal saline) for a total of 29 days. Improvements lasted 6–12 months [31].

4.3.2 Small Molecule Agents

Small molecule activators of the *SMN2* gene promoter, which enhance Smn expression, represent a promising strategy for the treatment of SMA. A number of mechanistically and chemically distinct agents were found to enhance the transcription, amend *SMN2* splicing, and stabilize or increase levels of Smn protein in patients. However, the unambiguous clinical proof-of-concept studies of these agents are still absent. The following are some agents under advanced investigation [32–34].

1. *Histone deacetylase* (HDAC) inhibitors can increase the level of fl-SMN [35]. Although *valproic acid* (VPA) has been discovered to affect a multitude of pathways, its activity in the SMA models was associated with HDAC inhibition. This molecule was shown to increase SMN protein in skin fibroblasts [36, 37]. A similar mechanistic hypothesis has been suggested for both *phenylbutyrate* and *hydroxyurea*. Phenylbutyrate is an approved agent for the treatment of urea acid cycle disorders. It was found to increase full-length SMN2 transcripts in skin fibroblasts [38]. Oral administration of phenylbutyrate increased SMN expression in white blood cells [39]. In a pilot trial, phenylbutyrate featured a short-term function improvement in ten SMA patients [40]. Hydroxyurea is an agent that enhances the expression of human fetal hemoglobin. Similar to VPA and phenylbutyrate, it increased SMN levels in skin fibroblasts from individuals with SMA [41, 42].
2. *Indoprofen*, a nonsteroidal anti-inflammatory drug (NSAID), increased SMN2 levels in fibroblasts of SMA patients [43].
3. A Phase I trial of *Rilutek* (*Riluzole*) in infants with SMA showed the molecule to be safe but not sufficiently beneficial for the treatment [44].
4. Studies of *gabapentin* in individuals with SMA II and III showed improvement in muscle strength but not in motor or respiratory function [45, 46].

A detailed discussion of these agents is summarized below.

HDAC Inhibitors

HDAC inhibitors were found to increase both the expression of *SMN2* and Smn protein in various cell types. The initial evidence regarding their potential therapeutic utility in SMA came from the studies of a weak, non-specific HDAC inhibitor phenylbutyrate [47]. Both this agent along and sodium butyrate showed promise in a mouse model and in an open-label pilot study, but was not effective in a human Phase II clinical trial [40, 48, 49].

Although an increase in the levels of full-length *SMN2* mRNA/protein production is generally regarded as beneficial for the treatment of SMA, there is no consensus regarding its desired magnitude. For example, butyrates and their derivatives featured improved survival of a mouse model of SMA but did not increase SMN levels in the spinal cord of these mice. On the other hand, fibroblast cultures derived from SMA patients were treated with therapeutic doses (0.5–500 μM) of VPA to result in a 2- to 4-fold increase in the levels of full-length *SMN2* mRNA/protein. VPA was discovered to elevate Smn protein levels through transcription activation in organotypic hippocampal brain slices from rats. VPA-treated animals featured both higher body weights and significant increase in lifespan. They were able to reorient more quickly and showed improved limb strength and motor function [50]. Sodium valproate has induced motor function improvement in patients presumably via enhancing transcription and reversing *SMN2* splicing pattern [51].

Hydroxyurea was described to increase the number of gems encapsulating functional Smn protein in cells from Type I to Type IV SMA patients. This agent was proposed to act via the inclusion of the missing exon from already existing RNA or/and increase in the expression of other transcription factors. Authors further speculated that these events may allow cell to bypass the *SMN2* mutation, making it work as a surrogate *SMN1* [41, 42].

Suberoylanilide hydroxamic acid (SAHA) was introduced as both potent and safe molecule for the treatment of SMA. SAHA increased Smn protein levels at low micromolar concentrations in rat hippocampal brain slices, motoneurone-rich cell fractions, and in a human brain slice culture assay. This agent was more efficient than VPA in both activating *SMN2* and inhibiting HDACs. SAHA also features good oral bioavailability. It was well tolerated and reported to cross the blood–brain barrier [52]. A close analog of SAHA, M344 increased full-length *SMN2* mRNA, Smn protein level, and number of gems in fibroblast cultures derived from SMA patients. The molecule was noted to be cytotoxic at high concentration featuring therapeutic index of ca. 2 [53].

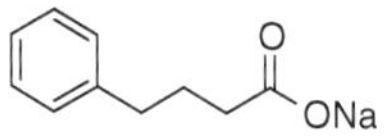

Phenyl butyrate

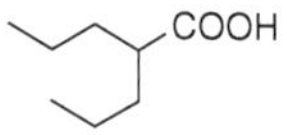

Valproic acid

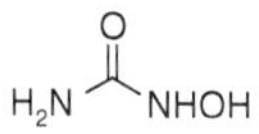

Hydroxyurea

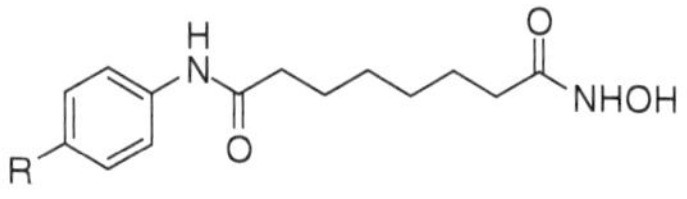

SAHA, R = H
M344, R = NMe_2

Trichostatin A, a more potent HDAC inhibitor compared to SAHA, increased *SMN2* expression in cultured cells and *in vivo*. It ameliorated neuromuscular abnormalities and improved the clinical phenotype of an SMA mouse model [54]. Daily treatment of the 5-day-old SMA mice with trichostatin A enhanced the level of *SMN2* expression and production of *SMN2* mRNA. In addition, it facilitated the assembly of *SMN*/RNA complexes. Treatment with trichostatin A restored normal size to anterior horn cells and increased both total muscle area and myofiber diameter. Treated mice lived longer and had better motor function. The median increase in survival was 3 days (20%). In discussing the mode of action of trichostatin A, authors suggested that the inhibition of diverse HDACs makes *SMN2* (and, perhaps other genes) more accessible to a frequent transcription. However, the main issue associated with HDAC inhibitors is their effect on off-target genes, especially in the chronic setting. A potential solution may be identification of the key HDAC enzymes involved in the *SMN2* expression and their inhibition with highly specific small molecules.

Agent TRO19622 featuring steroid template is under development by Trophos. It has successfully completed phase I studies in healthy volunteers and phase Ib studies in SMA patients. The compound was well tolerated, featured good safety profile and PK suitable for once-daily oral dosing based on preclinical models. Unfortunately, very limited biochemical data are available on this molecule in the literature [55].

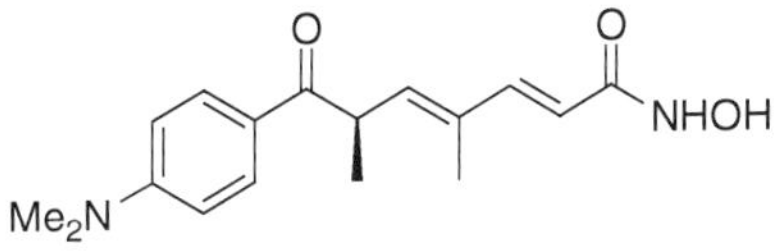

Trichostatin A

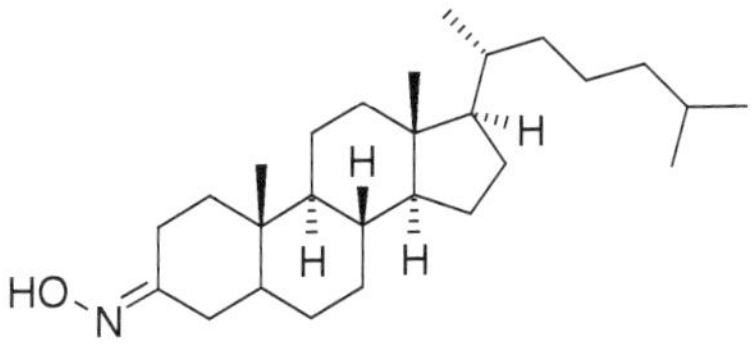

TRO19622

Antibiotics and Their Derivatives

Antibiotic *aclarubicin* was reported to facilitate the retention of exon 7 into the *SMN2* transcript. It promoted incorporation of exon 7 into the *SMN2* transcripts in Type I SMA fibroblasts, bringing the number of *SMN* gems to normal levels [56].

A synthetic derivative of antibiotic *PTK-SMA-01* has been developed by Paratek. Similar to aclarubicin, the compound increased exon 7 inclusion by greater than sixfold above background (19.2% vs. 3.1%) at 10 μM concentration. Cell-based assays using patient fibroblasts revealed that the agent increased the expression of both Smn protein (40% increase at 10 μM) and gems (2.8-fold increase at 2.5 μM). Adult *SMN2* transgenic mice were treated with PTK-SMA-01 to determine whether the compound increases full-length *SMN2* mRNA levels in non-CNS tissues *in vivo*. Four daily administrations (i.v. or p.o.) of the compound at 25 and 50 mg/kg furnished increased expression of *SMN2* mRNA by 23% in the kidney and by 74% in the liver compared to mice dosed with vehicle. Unfortunately, the molecule displayed very limited blood–brain barrier permeability [57].

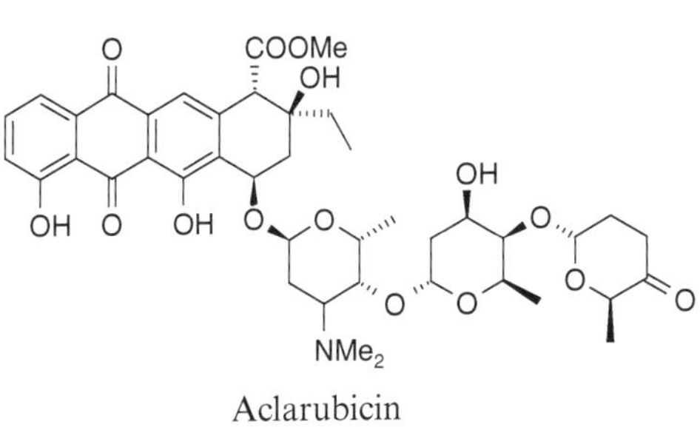

Aclarubicin

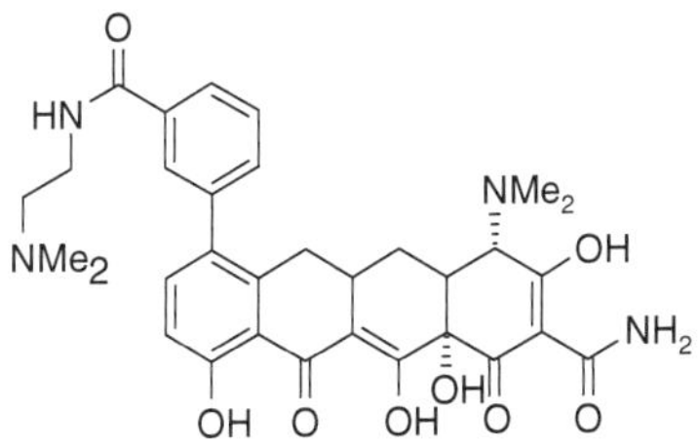

PTK-SMA-01

Agents that Inhibit the Decapping Scavenger Enzyme

A cell-based assay for *SMN2* promoter activation and the details of an uHTS screening campaign of 558,000 compounds was described by Jarecki et al. [18]. This effort led to the identification of several small molecule hits representing nine scaffolds. Derivatives of *2,4-diaminoquinazoline* were prioritized as potent activators of the *SMN2* promoter. Structure–activity relationship studies culminated in a lead compound X featuring high potency (EC_{50} = 4 nM) and 2.3-fold induction of the *SMN2* gene [58].

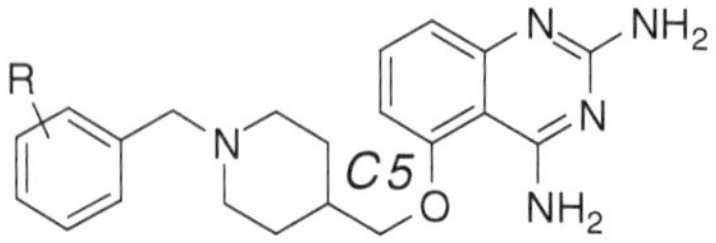

2,4-Diaminoquinazolines
R = 2-F (D156844), 2-Cl, 3-Cl

The optimized molecules featuring a 5-C substituent also upregulated expression of the mouse *SMN* gene in a mouse motor neuron hybrid cell line NSC-34. In Type I SMA patient fibroblasts, these compounds induced *SMN* in a dose-dependent manner and restored the number of intranuclear gems to levels corresponding to unaffected genetic carriers of SMA. Smn protein concentration has also been increased throughout the cell. In addition to favorable *ex vivo* functional activity, selected 2,4-diaminoquinazolines afforded high brain exposure levels and long brain half-life following oral dosing in mice. The decapping scavenger enzyme (DcpS) was recently reported as a potential molecular target of these compounds [59]. Specifically, screening of ~5,000 human proteins arrayed on a glass slide with an I-125 labeled C5-substituted quinazoline ligand identified DcpS as the sole interactor. DcpS binds and hydrolyzes the ^{7}methyl guanine cap structure of mRNA (^{7}mGpppG) in a two-step reaction. There was a tight correlation between DcpS inhibition and *SMN2* promoter induction [59]. D156844, for example, inhibited DcpS with an IC_{50} = 8 nM and activated the SMN2 promoter at an EC_{50} = 4 nM. Co-crystallization of D156844 and other C5-quinazolines with DcpS revealed that the compounds trapped the enzyme in a non-productive, open conformation. Structural data further suggested that the 2,4-diaminoquinazoline motif occupies the ^{7}mG-binding pocket of DcpS with the SAR around the quinazoline core [58].

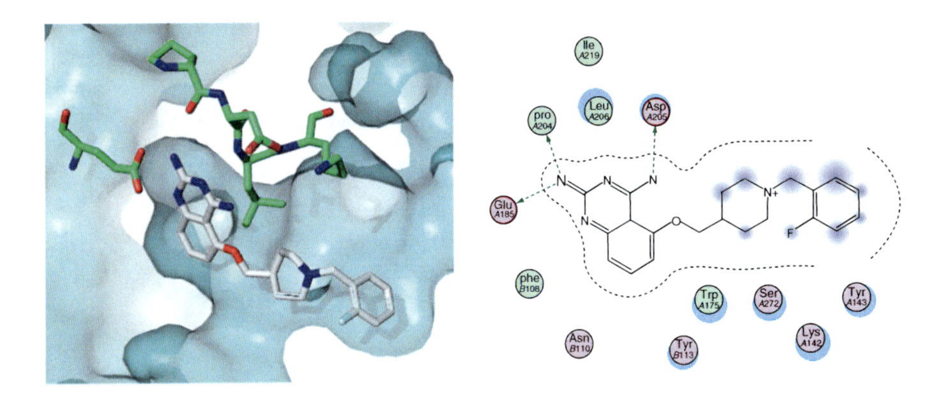

Natural Polyphenols

A red wine component, *(E)-resveratrol*, was evaluated in SMA Type-1 fibroblasts. Treatment with the molecule (100 μM) resulted in a 1.2- to 1.3-fold increase in the levels of full-length *SMN2* mRNA and Smn protein. However, variable results were obtained from other cell lines [60]. Similarly, other natural polyphenols including *curcumin* and *epigallocatechin galate* (EGCG) moderately increased exon 7 inclusion of *SMN2* transcripts, stimulated the production of full-length *SMN2* mRNA, Smn protein, and enhanced the formation of Smn-containing nuclear Gems [61].

(*E*)-Resveratrol Curcumin EGCG

Other Chemical Agents

Indoprofen: In a high-throughput screen of ca. 47,000 compound library, authors [43] have converged on indoprofen as a molecule that enhanced production of an *SMN2*- vs. *SMN1*-luciferase reporter protein. Indoprofen, a NSAID and cyclooxygenase (COX) inhibitor, afforded a 13% enhancement of Smn protein and a fivefold increase in the number of nuclear gems in fibroblasts from SMA patients. Notably, other tested NSAIDs or COX inhibitors were inactive in the assay.

Salbutamol, a β2-adrenoceptor agonist, was shown to promote both expeditious and significant increase in *SMN2* full-length mRNA and Smn protein in SMA fibroblasts [62].

Amidine [5-(*N*-ethyl-*N*-isopropyl)-amiloride, EIPA] has been introduced a potent and efficaceous Na^+/H^+ exchanger inhibitor. This agent afforded comparable enhancement of *SMN2* exon 7 inclusion, Smn protein production, and a number of nuclear gems across six lymphoid cell lines derived from Type I–III SMA patients. Proposed mode of action involves EIPA-induced upregulation of the splicing factor SRp20 in the nucleus [63].

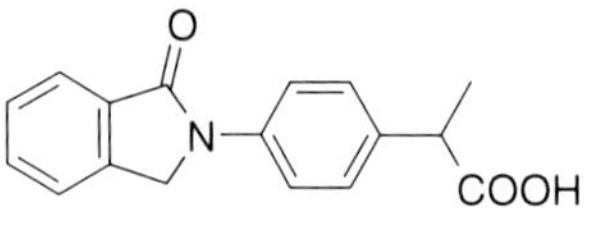

Indoprofen 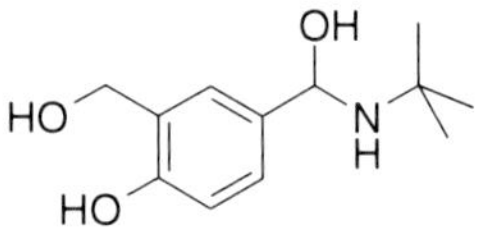 Salbutamol 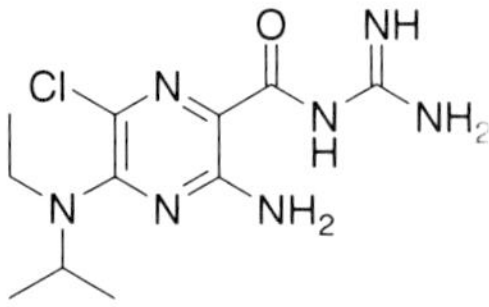 EIPA

5 Target Product Profile for an SMA Therapeutic

Drug discovery teams in industry typically work toward a desired target product profile (TPP). The TPP states in simple terms the desired mechanism of action, therapeutic benefit, and how it will be used. A sample TPP for an SMA therapeutic might be the following:

Mechanism of action: Increase SMN2 gene expression and SMN protein.
Clinical indication: For treatment of SMA in children with autosomal recessive mutation of SMN1 and variable SMN2 gene copy number.
Dosing form: Liquid and syrup dosage forms for oral dosing of neonates. Tablet dosage form for older children able to swallow and/or adults.
Clinical evaluation: The compound will extend survival in early-onset, severe Type I SMA patients whose life expectancy is normally 1–2 years of age.

The TPP describes to the drug discovery team that the testing funnel will focus on cellular and animal assays that measure *SMN2* gene expression and production of Smn protein, and that animal efficacy studies would need to demonstrate improvement in lifespan for a compound that would be advanced into clinical trials. The pharmaceutic properties of the drug would need to allow oral dosing and provide adequate solubility and stability to allow formulation for liquid or syrup dosage forms. Finally, clinical evaluation of efficacy would necessitate dosing in neonates for 6–12 months with enrollment genetically stratified by *SMN2* gene copy number.

6 Conclusions

Genetic understanding of the pathogenesis of SMA has led to several encouraging therapeutic approaches. These are designed to increase *SMN2* promotor function, suppress missplicing of the *SMN2* gene transcript, or stabilize correctly spliced

SMN2 mRNA such that there will be a higher yield of Smn protein produced. Therapeutic targets such as the HDACs and DcpS have emerged with some compounds showing promising benefit in cellular assays and mouse models. It is hoped that one or more compounds entering human clinical trials will show therapeutic advantages in patients.

References

1. Pearn J (1978) Genetic studies of acute infantile spinal muscular atrophy (SMA type I): an analysis of sex ratios, segregation ratios, and sex influence. J Med Genet 15:414–417
2. Roberts DF, Chavez J, Court SD (1970) The genetic component in child mortality. Arch Dis Child 45:33–38
3. Lefebvre S, Burglen L, Reboullet S et al (1995) Identification and characterization of a spinal muscular atrophy-determining gene. Cell 80:155–165
4. Bussaglia E, Clermont O, Tizzano E et al (1995) A frame-shift deletion in the survival motor neuron gene in Spanish spinal muscular atrophy patients. Nat Genet 11:335–337
5. Parsons DW, McAndrew PE, Monani UR et al (1996) An 11 base pair duplication in exon 6 of the SMN gene produces a type I spinal muscular atrophy (SMA) phenotype: further evidence for SMN as the primary SMA-determining gene. Hum Mol Genet 5:1727–1732
6. Hahnen E, Schonling J, Rudnik-Schoneborn S et al (1997) Missense mutations in exon 6 of the survival motor neuron gene in patients with spinal muscular atrophy (SMA). Hum Mol Genet 6:821–825
7. McAndrew PE, Parsons DW, Simard LR et al (1997) Identification of proximal spinal muscular atrophy carriers and patients by analysis of SMNT and SMNC gene copy number. Am J Hum Genet 60:1411–1422
8. Parsons DW, McAndrew PE, Iannaccone ST et al (1998) Intragenic telSMN mutations: frequency, distribution, evidence of a founder effect, and modification of the spinal muscular atrophy phenotype by cenSMN copy number. Am J Hum Genet 63:1712–1723
9. Ogino S, Wilson RB (2002) Genetic testing and risk assessment for spinal muscular atrophy (SMA). Hum Genet 111:477–500
10. Monani UR, Lorson CL, Parsons DW et al (1999) A single nucleotide difference that alters splicing patterns distinguishes the SMA gene SMN1 from the copy gene SMN2. Hum Mol Genet 8:1177–1183
11. Feldkotter M, Schwarzer V, Wirth R et al (2002) Quantitative analyses of SMN1 and SMN2 based on real-time Light-Cycler PCR: Fast and highly reliable carrier testing and prediction of severity of spinal muscular atrophy. Am J Hum Genet 70:358–368
12. Wirth B, Brichta L, Hahnen E (2006) Spinal muscular atrophy and therapeutic prospects. Prog Mol Subcell Biol 44:109–132
13. Anhuf D, Eggermann T, Rudnik-Schoneborn S, Zerres K (2003) Determination of SMN1 and SMN2 copy number using TaqMan technology. Hum Mutat 22:74–78
14. MacLeod MJ, Taylor JE, Lunt PW, Mathew CG, Robb SA (1999) Prenatal onset spinal muscular atrophy. Europ J Paediatr Neurol 3:65–72
15. Thomas NH, Dubowitz V (1994) The natural history of type I (severe) spinal muscular atrophy. Neuromuscul Disord 4:497–502
16. Russman BS, Buncher CR, White M et al (1996) Function changes in spinal muscular atrophy II and III. The DCN/SMA Group. Neurology 47:973–976
17. Brahe C, Servidei S, Zappata S et al (1995) Genetic homogeneity between childhood-onset and adult-onset autosomal recessive spinal muscular atrophy. Lancet 346:741–742

18. Jarecki J, Chen X, Bernardino A et al (2005) Diverse small-molecule modulators of SMN expression found by high-throughput compound screening: early leads towards a therapeutic for spinal muscular atrophy. Hum Mol Genet 14:2003–2018
19. Young PJ, Le TT, Man N, Burghes AH, Morris GE (2000) The relationship between SMN, the spinal muscular atrophy protein, and nuclear coiled bodies in differentiated tissues and cultured cells. Exp Cell Res 256:365–374
20. Monani UR, Pastore MT, Gavrilina TO et al (2003) A transgene carrying an A2G missense mutations in the SMN gene modulates phenotypic severity in mice with severe (type I) spinal muscular atrophy. J Cell Biol 160:41–52
21. Schmid A, DiDonato CJ (2007) Animal models of spinal muscular atrophy. J Child Neurol 22:1004–1012
22. Monani UR, Sendtner M, Coovert DD et al (2000) The human centromeric survival motor neuron gene (SMN2) rescues embryonic lethality in Smn(-/-) mice and results in a mouse with spinal muscular atrophy. Hum Mol Genet 9:333–339
23. Hsieh-Li HM, Chang JG, Jong YJ et al (2000) A mouse model for spinal muscular atrophy. Nat Genet 24:66–70
24. Bach JR, Baird JS, Plosky D et al (2002) Spinal muscular atrophy type 1: management and outcomes. Pediatr Pulmonol 34:16–22
25. Hua Y, Vickers TA, Baker BF et al (2007) Enchancement of SMN2 Exon7 Inclusion by antisense oligonucleotides targeting the exon. PLoS Biology 5:729–744
26. Baughan T, Shabibi M, Coady TH et al (2006) Stimulating full-length SMN2 expression by delivering bifunctional RNAs via a viral vector. Mol Ther 14:54–62
27. Dickson A, Osman E, Lorson CL (2008) A negatively-acting bifunctional RNA increases survival motor neuron *in vitro* and *in vivo*. Hum Gen Ther 19:1307–1316
28. Coady TH, Baughan TD, Shabibi M et al (2008) Development of a single vector system that enhances trans-splicing of SMN2 transcripts. PLoS ONE 3:1–11
29. Madocsai C, Lim SR, Geib T et al (2005) Correction of SMN2 pre-mRNA splicing by antisense U7 small nuclear RNAs. Mol Ther 12:1013–1022
30. Hofmann Y, Wirth B (2002) hnRNP-G promotes exon 7 inclusion of survival motor neuron (SMN) via direct interaction with Htra2-b1. Hum Mol Genet 11:2037–2049
31. Tzeng AC, Cheng JF, Fryczynski H et al (2000) Randomized double-blind prospective study of thyrotropin-releasing hormone for the treatment of spinal muscular atrophy: a preliminary report. Am J Phys Med Rehabil 79:435–440
32. Escolar DM, Henricson EK, Mayhew J et al (2001) Clinical evaluator reliability for quantitative and manual muscle testing measures of strength in children. Muscle Nerve 24:787–793
33. Iannaccone ST, Hynan LS (2003) Reliability of 4 outcome measures in pediatric spinal muscular atrophy. Arch Neurol 60:1130–1136
34. Swoboda KJ, Prior TW, Scott CB et al (2005) Natural history of denervation in SMA: relation to age, SMN2 copy number, and function. Ann Neurol 57:704–712
35. Kernochan LE, Russo ML, Woodling NS et al (2005) The role of histone acetylation in SMN gene expression. Hum Mol Genet 14:1171–1182
36. Brichta L, Hofmann Y, Hahnen E et al (2003) Valproic acid increases the SMN2 protein level: a well-known drug as a potential therapy for spinal muscular atrophy. Hum Mol Genet 12:2481–2489
37. Sumner CJ, Huynh TN, Markowitz JA et al (2003) Valproic acid increases SMN levels in spinal muscular atrophy patient cells. Ann Neurol 54:647–654
38. Andreassi C, Angelozzi C, Tiziano FD et al (2004) Phenylbutyrate increases SMN expression *in vitro*: relevance for treatment of spinal muscular atrophy. Eur J Hum Genet 12:59–65
39. Brahe C, Vitali T, Tiziano FD et al (2005) Phenylbutyrate increases SMN gene expression in spinal muscular atrophy patients. Eur J Hum Genet 13:256–259
40. Mercuri E, Messina S, Kinali M et al (2004) Congenital form of spinal muscular atrophy predominantly affecting the lower limbs: a clinical and muscle MRI study. Neuromuscul Disord 14:125–129

41. Stevens MR (1999) Hydroxyurea: an overview. J Biol Regul Homeost Agents 13:172–175
42. Grzeschik SM, Ganta M, Prior TW et al (2005) Hydroxyurea enhances SMN2 gene expression in spinal muscular atrophy cells. Ann Neurol 58:194–202
43. Lunn MR, Root DE, Martino AM et al (2004) Indoprofen upregulates the survival motor neuron protein through a cyclooxygenase-independent mechanism. Chem Biol 11:1489–1493
44. Russman BS, Iannaccone ST, Samaha FJ (2003) A phase 1 trial of riluzole in spinal muscular atrophy. Arch Neurol 60:1601–1603
45. Miller RG, Moore DH, Dronsky V et al (2001) A placebo-controlled trial of gabapentin in spinal muscular atrophy. J Neurol Sci 191:127–131
46. Merlini L, Solari A, Vita G et al (2003) Role of gabapentin in spinal muscular atrophy: results of a multicenter, randomized Italian study. J Child Neurol 18:537–541
47. Chang JG, Hsieh-Li HM, Jong YJ et al (2001) Treatment of spinal muscular atrophy by sodium butyrate. Proc Natl Acad Sci 98:9808–9813
48. Mercuri E, Bertini E, Messina S et al (2007) Randomized, double-blind, placebo-controlled trial of phenylbutyrate in spinal muscular atrophy. Neurology 68:51–55
49. Sumner CJ (2007) Molecular Mechanisms of Spinal Muscular Atrophy. J Child Neurol 22:979–989
50. Brichta L, Hofmann Y, Hhnen E et al (2003) Valproic acid increases the SMN2 protein level: a well-known drug as a potential therapy for spinal muscular atrophy. Hum Mol Genet 12:2481–2489
51. Darras BT, Kang PB (2007) Clinical Trials in Spinal Muscular Atrophy, HDACs. In: Pomeroy S (ed) Current Opinion in Pediatrics, vol 19. Wiley, New York, pp 675–679
52. Hahnen E, Eyupoglu IY, Brichta L et al (2006) In Vitro and ex vivo evaluation of second-generation histone deacetylase inhibitors for the treatment of spinal muscular atrophy. J Neurochem 98:193–202
53. Reissland M, Brichta L, Hahnen E, Wirth B (2006) The benzamide M344, a novel histone deacetylase inhibitor, significantly increases SMN2 RNA/protein levels in spinal muscular atrophy cells. Hum Genet 120:101–110
54. Avila AA, Burnett BG, Taye AA et al (2007) Trichostatin A increases SMN expression and survival in a mouse model of spinal muscular atrophy. J Clin Invest 117:659–671
55. Trophos and TRO19622: FAQ, in Trophos Drug Discovery and Neuroscience, http://www.csma.org.ua/news/trophos/trophos_faqs_v5.pdf
56. Andreassi C, Jarecki J, Zhou J et al (2001) Aclarubicin treatment restores SMN levels to cells derived from type I spinal muscular atrophy patients. Hum Mol Genet 10:2841–2849
57. Berniac JA, Hastings ML, Liu YH et al (2008) A tetracycline that corrects SMN2 splicing for potential treatment of spinal musculasr atrophy (SMA). AAN Poster http://www.paratekpharm.com/docs/AAN/AAN%202008%20Poster.pdf
58. Thurmond J, Butchbach MER, Palomo M et al (2008) Synthesis and biological evaluation of novel 2, 4-diaminoquinazoline derivatives as SMN2 promoter activators for the potential treatment of spinal muscular atrophy. J Med Chem 51:449–469
59. Singh J, Salcius M, Liu SW et al (2008) DcpS as a therapeutic target for spinal muscular atrophy. ACS Chem Biol 3:711–722
60. Dayangaç-Erden D, Bora G, Ayhan P et al (2009) Histone deacetylase inhibition activity and molecular docking of (E)-Resveratrol: its therapeutic potential in spinal muscular atrophy. Chem Biol Drug Design 73:355–364
61. Sakla MS, Lorson CL (2008) Induction of full-length survival motor neuron by polyphenol botanical compounds. Hum Genet 122:635–643
62. Angelozzi C, Borgo F, Tiziano FD et al (2008) Salbutamol increases SMN mRNA and protein levels in spinal muscular atrophy cells. J Med Genet 45:29–31
63. Yuo CY, Lin HH, Chang YS et al (2008) 5-(N-ethyl-N-isopropyl)-amiloride enhances SMN2 exon 7 inclusion and protein expression in spinal muscular atrophy cells. Ann Neurol 63:26–34

Index